D1206521

Volume

31

in the Wiley Series in

Advances in Environmental Science and Technology

JEROME O. NRIAGU, Series Editor

VANADIUM IN THE ENVIRONMENT

VANADIUM IN THE ENVIRONMENT

Part 2: Health Effects

Edited by

Jerome O. Nriagu

Department of Environmental and Industrial Health
School of Public Health, The University of Michigan
Ann Arbor, Michigan

A WILEY-INTERSCIENCE PUBLICATION

JOHN WILEY & SONS

New York • Chichester • Weinheim • Brisbane • Singapore • Toronto

This book is printed on acid-free paper. ∞

Copyright © 1998 by John Wiley & Sons, Inc. All rights reserved.

Published simultaneously in Canada.

No part of this publication may be reproduced, stored in a retrieval system or transmitted in any form or by any means, electronic, mechanical, photocopying, recording, scanning or otherwise, except as permitted under Sections 107 or 108 of the 1976 United States Copyright Act, without either the prior written permission of the Publisher, or authorization through payment of the appropriate per-copy fee to the Copyright Clearance Center, 222 Rosewood Drive, Danvers, MA 01923, (508) 750-8400, fax (508) 750-4744. Requests to the Publisher for permission should be addressed to the Permissions Department, John Wiley & Sons, Inc., 605 Third Avenue, New York, NY 10158-0012, (212) 850-6011, fax (212) 850-6008, E-Mail: PERMREQ @ WILEY.COM.

Library of Congress Cataloging-in-Publication Data:

Vanadium in the environment / edited by Jerome O. Nriagu.
    p.   cm.—(Advances in environmental science and technology ;
v. 30)
    Includes index.
    1. Vanadium—Environmental aspects.   I. Nriagu, Jerome O.
II. Series.
TD180.A38   vol. 30
[TD196.V35]
628 s—dc21
[363.738]   97-14872
    ISBN 0-471-17778-4 (cloth : alk. paper) (Part 1)
    ISBN 0-471-17776-8 (cloth : alk. paper) (Part 2)

Printed in the United States of America

10 9 8 7 6 5 4 3 2 1

CONTRIBUTORS

MARIO A. ALTAMIRANO-LOZANO, Laboratorio de Citogenética, Mutagénesis y Toxicología Reproductiva, UIBR, Facultad de Estudios Superiores-Zaragoza, UNAM, A.P., 9-020, México 15000, D.F. México

ENRIQUE J. BARAN, Centro de Química Inorgánica (CEQUINOR), Facultad de Ciencias Exactas, Universidad Nacional de la Plata, C. Correo 962, 1900-La Plata, Argentina

KÁTIA PADILHA BARRETO, Instituto de Biociências, Universidade Federal do Rio Grande do Sul, Porto Alegre, Brasil

MARY BATTELL, Division of Pharmacology and Toxicology, Faculty of Pharmaceutical Sciences, The University of British Columbia, Vancouver, B.C., Canada V6T 1Z3

EVA BAYEROVÁ, NIKOM spc., Medical Center, CZ-262 02, Mníšek pod Brdy, Czech Republic

ANUM BISHAYEE,* Division of Biochemistry, Department of Pharmaceutical Technology, Jadavpur University, Calcutta 700 032, India

PAOLO BOSCOLO, Centre of Occupational Medicine and Ergophthalmology, University of Chieti, 66100 Chieti, Italy

JANUSZ Z. BYCZKOWSKI, ManTech Environmental Technology, Inc., Dayton, OH 45437-0009

MARCO CARMIGNANI, Section of Pharmacology, Department of Basic and Applied Biology, University of L'Aqulia, 67010 Coppito (AQ), Italy

MALAY CHATTERJEE, Division of Biochemistry, Department of Pharmaceutical Technology, Jadavpur University, Calcutta 700 032, India

SOMCHAI EIAM-ONG, Nephrology Unit, Department of Medicine, Faculty of Medicine, Chulalongkorn University Hospital, Bangkok, Thailand 10330

GERARD ELBERG,† Department of Biochemistry, The Weizmann Institute of Science, Rehovot 76100, Israel

---

* *Present address:* MSB F-451, Division of Radiation Research, Department of Radiology, New Jersey Medical School, University of Medicine and Dentistry of New Jersey, Newark, NJ 07103-2714.

† *Present address:* Department of Cell Biology, Baylor College of Medicine, Houston, TX 77030.

v

MARIA DE GRAÇA FAUTH, Instituto de Biociências, Pontifíca Universidade Católica do Rio Grande do Sul, Porto Alegre, Brasil

MARIO FELACO, Institute of Biology and Genetics, University of Chieti, 66100 Chieti, Italy

G. B. GERBER, Teratogenicity and Mutagenicity Unit, Catholic University of Louvain, Avenue E. Mounier 72, B-1200 Brussels, Belgium

MARCELO DE LACERDA GRILLO, Instituto de Biociências, Universidade Federal do Rio Grande do Sul, Porto Alegre, Brasil

FREDERICK G. HAMEL, Research Service, Veterans Affairs Medical Center, and Departments of Internal Medicine and Pharmacology, University of Nebraska Medical Center, Omaha, NE

M. KODL, Regional Hygiene Institute, Prague

JAN KUČERA, Nuclear Physics Institute, Academy of Sciences of the Czech Republic, CZ-250 68 Řež near Prague, Czech Republic

ARUN P. KULKARNI, Toxicology Program, College of Public Health MDC-56, University of South Florida, Tampa, FL 33612

JAROSLAV LENER, National Institute of Public Health, Šrobárova 48 CZ-100, 42 Prague 10, Czech Republic

A. LÉONARD, Teratogenicity and Mutagenicity Unit, Catholic University of Louvain, Avenue E. Mounier 72, B-1200 Brussels, Belgium

JINPING LI, Department of Biochemistry, The Weizmann Institute of Science, Rehovot 76100, Israel

JOHN H. McNEILL, Division of Pharmacology and Toxicology, Faculty of Pharmaceutical Sciences, The University of British Columbia, Vancouver, B.C., Canada V6T 1Z3

JANA MŇUKOVÁ, National Institute of Public Health, Šrobárova 48, CZ-100 42 Prague 10, Czech Republic

E. ROLDÁN-REYES, Laoratorio de Citogenética, Mutagénesis y Toxicología Reproductiva, UIBR, Facultad de Estudios Superiores-Zaragoza, UNAM, A.P. 9-020, México 15000, D.F. México

E. ROJAS, Laboratorio de Genética, Toxicológica Molecular, Departamento de Toxicología Ambiental, Instituto de Investigaciones Biomédicas, UNAM, A.P. 70228, Ciudad Universitaria, México 04510, D.F. México

ENRICO SABBIONI, Commission of the European Communities, Joint Research Center Ispra, Environment Institute, I-21020 Ispra (Varese), Italy

HIROMU SAKURAI, Department of Analytical and Bioinorganic Chemistry, Kyoto Pharmaceutical University, Nakauchi-cho 5, Misasagi, Yamashina-ku, Kyoto 607, Japan

AGNIESZKA ŚCIBIOR, Department of Cell Biology, Institute of Biology, Maria-Curie Skłodowska University, Akademicka 19, 20-033 Lublin, Poland

YORAM SHECHTER, Department of Biochemistry, The Weizmann Institute of Science, Rehovot 76100, Israel

VISITH SITPRIJA, Nephrology Unit, Department of Medicine, Faculty of Medicine, Chulalongkorn University Hospital, Bangkok, Thailand 10330

V. SKOKANOVÁ, Regional Hygiene Institute, Prague

K.H. THOMPSON, Division of Pharmacology and Toxicology, Faculty of Pharmaceutical Sciences, The University of British Columbia, Vancouver, B.C., Canada V6T 1Z3

AKIHIRO TSUJI, Department of Analytical and Bioinorganic Chemistry, Kyoto Pharmaceutical University, Nakauchi-cho 5, Misasagi, Yamashina-ku, Kyoto 607, Japan

ANNA RITA VOLPE, Section of Pharmacology, Department of Basic and Applied Biology, University of L'Aqulia, 67010 Coppito (AQ); Receptor Chemistry Centre, CNR, Catholic University School of Medicine, 00168 Rome, Italy

GUILLERMO FEDERICO WASSERMAN, Instituto de Biociências, Universidade Federal do Rio Grande do Sul, Porto Alegre, Brasil

HALINA ZAPOROWSKA, Department of Cell Biology, Institute of Biology, Maria-Curie Skłodowska University, Akademicka 19, 20-033 Lublin, Poland

CONTENTS

INTRODUCTION
TO THE SERIES

The deterioration of environmental quality, which began when mankind first congregated into villages, has existed as a serious problem since the industrial revolution. In the second half of the twentieth century, under the ever-increasing impacts of exponentially growing population and of industrializing society, environmental contamination of the air, water, soil, and food has become a threat to the continued existence of many plant and animal communities of various ecosystems and may ultimately threaten the very survival of the human race. Understandably, many scientific, industrial, and governmental communities have recently committed large resources of money and human power to the problems of environmental pollution and pollution abatement by effective control measures.

Advances in Environmental Science and Technology deals with creative reviews and critical assessments of all studies pertaining to the quality of the environment and to the technology of its conservation. The volumes published in the series are expected to service several objectives: (1) stimulate interdisciplinary cooperation and understanding among the environmental scientists; (2) provide the scientists with a periodic overview of environmental developments that are of general concern or that are of relevance to their own work or interests; (3) provide the graduate student with a critical assessment of past accomplishment which may help stimulate him or her toward the career opportunities in this vital area; and (4) provide the research manager and the legislative or administrative official with an assured awareness of newly developing research work on the critical pollutants and with the background information important to their responsibility.

As the skills and techniques of many scientific disciplines are brought to bear on the fundamental and applied aspects of the environmental issues, there is a heightened need to draw together the numerous threads and to present a coherent picture of the various research endeavors. This need and the recent tremendous growth in the field of environmental studies have clearly made some editorial adjustments necessary. Apart from the changes in style and format, each future volume in the series will focus on one particular theme or timely topic, starting with Volume 12. The author(s) of each pertinent section will be expected to critically review the literature and the most important recent developments in the particular field; to critically evaluate new concepts, methods, and data; and to focus attention on important unresolved or controversial questions and on probable future trends. Monographs embodying the results of unusually extensive and well-rounded investigations will also be published in the series. The net result of the new editorial policy should be more integrative and comprehensive volumes on key environmental issues and pollutants. Indeed, the development of realistic standards of environmental quality for many pollutants often entails such a holistic treatment.

JEROME O. NRIAGU, Series Editor

PREFACE

Vanadium is widely distributed and the 21st most abundant element in the earth's crust. Although vanadium is more abundant than many of the more common metals such as iron, zinc, and copper, there is no convincing evidence that it is an essential element for man. It is, however, believed to be essential for a number of species such as chicken and rats; deficiency symptoms in such species include retarded growth, impairment of reproduction, disturbance of lipid metabolism and inhibition of Na^+/K^+-ATPase activity in the kidney, brain, and heart. In spite of being a nutritional element, vanadium is not accumulated by the biota; the only organisms known to bioaccumulate it to any significant degree being some mushrooms, tunicates, and sea squirts.

Vanadium has two different catalytic functions in biological systems. It can exist in many oxidation states and the oxyanions and oxycations are known to be powerful oxidants under physiological conditions. Secondly, vanadium forms sulfur-containing anionic and cationic centers and iron-vanadium-sulfur clusters associated in biology with nitrogen fixation. The multiple oxidation states, ready hydrolysis and polymerization confer a level of complexity to the chemistry of vanadium well above that of many elements. Although its current biological role seems to be diminished, the high accumulation of vanadium in some oils suggests that vanadium was involved in early photosynthetic processes.

Vanadium is used widely in industrial processes including the production of special vanadium–iron steels, in the production of hard metals and temperature-resistant alloys, in iron and steel refining and tempering, in glass industry, in the manufacture of pigments, paints, and printing inks, for lining arc welding electrodes, and as catalysts in the pharmaceutical industry. Its use with nonferrous metals is of particular importance in the atomic energy industry, aircraft construction, and space technology. The high temperature industrial processes as well as the combustion of fossil fuels, especially oils, now release large quantities of vanadium into the environment. The behavior and effects of vanadium pollution in most ecosystems remain poorly understood. A number of studies have even associated the ambient levels of vanadium in the air with cardiovascular diseases, bronchitis, and lung carcinoma in the general population. Vanadium also interferes with a multitude of biochemical processes, can penetrate the blood–brain and placental barriers and is present in milk. The reproductive toxicology of vanadium must be of concern and a fertile field for study.

Vanadium is an element of considerable environmental and scientific interest because of its wide industrial applications and large releases into the environment, complex chemistry, and narrow thresholds between essential and toxic doses. While there are now a number of review papers and reports on vanadium toxicity in animal species including human beings, few attempts have been made to relate the toxicological aspects to the physical–chemical features of the element in the ecosystem. An objective of this volume has been to provide a comprehensive picture of current biological, chemical, and clinical research on vanadium in various environmental media. Individual chapters cover the sources, distribution, transformations, mechanisms of transport, fate as well as human and ecosystem effects. The chapters are written by acknowledged experts from a wide variety of scientific disciplines; indeed, the literature on vanadium is so disparate that no single scientist can provide a detailed account of all the recent developments. The authors have been asked to focus on general principles of vanadium behavior rather than on systematic compilation of published data. The volume is intended to be of interest to graduate students and practicing scientists in the fields of environmental science and engineering, ecology, nutrition, toxicology, public health, and environmental control. More importantly, it is addressed to everyone who is concerned about the impact of metallic pollutants on our health and our life support system.

JEROME O. NRIAGU

Ann Arbor, Michigan

ADVANCES IN ENVIRONMENTAL SCIENCE AND TECHNOLOGY
Jerome O. Nriagu, Series Editor

VANADIUM IN THE ENVIRONMENT

1

HEALTH EFFECTS OF ENVIRONMENTAL EXPOSURE TO VANADIUM

Jaroslav Lener

National Institute of Public Health, Šrobárova 48, CZ-100 42, Prague 10, Czech Republic

Jan Kučera

Institute of Nuclear Physics, Academy of Sciences of the Czech Republic, CZ-250, Rež near Prague

M. Kodl
V. Skokanová

Regional Hygiene Institute, Prague

Vanadium in the Environment. Part 2: Health Effects, Edited by Jerome O. Nriagu.
ISBN 0-471-17776-8.   © 1998 John Wiley & Sons, Inc.

1. INTRODUCTION

Contamination of the environment with toxic heavy metals and the related health hazard to the population is an important problem in most industrially developed countries. Lately, besides traditionally studied elements such as lead, cadmium, zinc, and mercury attention is also being devoted to other toxic elements, including vanadium. This element is being studied primarily from the point of view of its physiological and biochemical activities within the context of its potential essentiality to the human organism. Besides that, it is necessary to also study its negative effects on the population under circumstances of growing pollution of the environment with this metal due to combustion of fossil mineral oils as well as to industrial production, especially certain metallurgical processes (Chapter 3 Vol. 30)

The effects of vanadium have been summarized in a number of monographs (Friberg et al., 1986; Chasteen, 1990; CBEAP, 1974; Hopkins and Mohr, 1971; WHO, 1988). The consequences to health of exposure to vanadium have up to now been studied mostly from the point of view of exposure of professionals. Little attention has been given to the study of the effect of this element on the general population. An WHO monograph on vanadium (WHO, 1988) mentions only three epidemiological studies, while it stresses that "an area of great importance is the exposure of the general population" and "epidemiological studies on populations living in areas with high vanadium exposure should be carried out, relating possible adverse effects to exposure levels."

In the present study we assessed the potential effect of the long-term influence of vanadium emissions in the atmosphere originating from the production of V_2O_5 by the hydrometallurgical processing of vanadium-rich slag. The production cycle, namely crushing and milling of vanadium slag, is the source of great dustiness that, depending on the direction of the wind, affects an area over a radius of three kilometers, which has a population of 4,850 permanent inhabitants.

2. MATERIALS AND METHODS

Over a period of 3 years we have followed up three groups of 10- to 12-year-old schoolchildren that include 15 children (11 boys, 4 girls) from the Czech localities of Cisovice and Lisnice (Group A), the area potentially most affected by the emission of vanadium; 28 children (14 boys, 14 girls) from the locality of Mnisek (Group B), an area of medium exposure; and 32 children (17 boys, 15 girls) from the locality of Stechovice (Group C), a control area not affected by any emission from vanadium production. The geographical situation of the vanadium plant surroundings is shown in Figure 1.

The size of the groups was limited by the number of children 10–12 years of age attending school in the given localities and by the consent of their parents to inclusion in the study. At the beginning of the study anamnestic

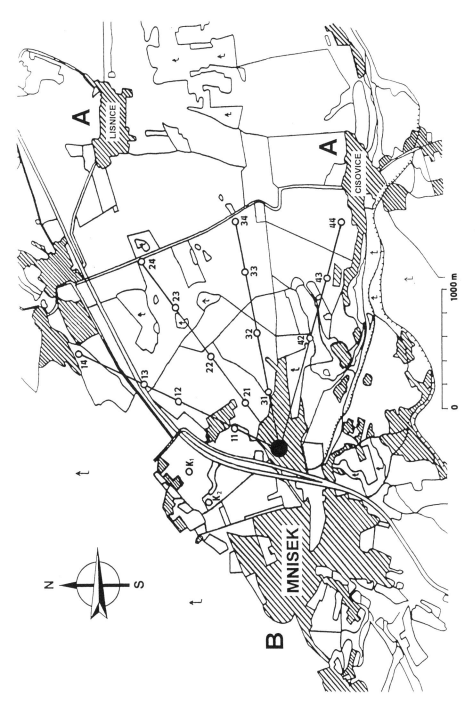

Figure 1. Localities exposed to vanadium from vanadium production plant.

and socioeconomic data on each child were obtained through personal questionnaires.

2.1. Timetable

Samples were taken at the following intervals:

  1st sampling—June 15, 1992
  2nd sampling—November 19, 1992
  3rd sampling—June 14, 1993
  4th sampling—December 2, 1993
  5th sampling—May 27, 1994

2.2. Samples

Samples of venous blood, saliva, hair, and fingernails were taken in the children under study and processed by a method described previously (Kučera et al., 1992).

For assessing the vanadium concentration in dust aerosol, samples of ambient air were collected with low-volume sampling devices.

Soil samples were taken from the superficial layer of top soil, and the vanadium content was determined from an acid extract of dry matter.

The vanadium concentration in drinking water was determined in the drinking water taken from wells of the families of the children under study; in addition, water from the municipal water supply in the locality of Mnisek was investigated.

2.3. Analytical Determinations

Vanadium in the biological samples and in the dust aerosol was determined by neutron activation analysis, either instrumentally or with radiochemical separation (Kučera et al., 1996).

The vanadium concentration in soil and drinking water was determined by atomic absorption spectrophotometry.

Cystine in the hair and fingernails was determined with the aid of an automatic amino acid analyzer upon oxidation to cysteic acid (Kučera et al., 1994).

2.4. Hematological and Biochemical Investigations

To assess the children's health status the following indicators were investigated.

- Hematological examination: erythrocyte, leukocyte, and platelet counts; concentration of hemoglobin; hematocrit value; mean corpuscular volume, mean corpuscular hemoglobin (by Automatic Coulter Counter System, Coulter Electronics).
- Determination of specific immunity: immunoglobulins (IgA, IgE, IgG, secretory IgA, IgM), transferrin, α-1-antitrypsin, β-2-microglobulin (by radial diffusion and ELISA).
- Determination of cellular immunity: phagocytosis of peripheral leukocytes; stimulation of T-lymphocyte mitogenic activity with phytohemagglutinin, concanavalin A, and pokeweed mitogen (Luster et al., 1988).
- Cytogenetic analysis: frequency of chromosomal aberrations in peripheral lymphocytes; sister chromatid exchanges.
- Examination of serum lipids: total cholesterol, triglycerides, high-density cholesterol (HDL), low-density cholesterol (LDL) (by Reflotron, Boehringer).

2.5. Statistical Evaluation

All findings were evaluated by analysis of variance and by Student's t test.

3. RESULTS AND DISCUSSION

The results of investigations of the health indicator values in children from localities exposed to vanadium show some significant differences from those found in the control group.

The results of hematological examinations are summarized in Tables 1–7, from which it is apparent that the differences between children from localities exposed to vanadium and children from the unpolluted control area have mainly manifested themselves in lower RBC counts, hemoglobin levels, and hematocrit values in the former.

Table 1   Erythrocyte Count $(10^6/\text{mm}^3)^a$

Sampling	Group A	Group B	Group C	P
1	4.8 ± 0.33	4.9 ± 0.27	4.9 ± 0.32	
2	4.8 ± 0.33	4.6 ± 0.25	4.8 ± 0.26	[b]
3	4.8 ± 0.37	4.8 ± 0.31	5.0 ± 0.32	[c]
4	4.7 ± 0.30	4.8 ± 0.26	4.9 ± 0.31	[d]
5	4.7 ± 0.34	4.8 ± 0.32	4.9 ± 0.33	

[a] Data are $M \pm SD$.
[b] $P < 0.05$; C:B.
[c] $P < 0.05$; C:A.
[d] $P < 0.05$; C:A, B.

Table 2   Hemoglobin (g/L)[a]

Sampling	Group A	Group B	Group C	P
1	13.5 ± 0.88	13.9 ± 0.72	13.7 ± 0.85	
2	13.5 ± 1.0	13.2 ± 0.78	13.7 ± 0.80	
3	13.5 ± 1.1	13.8 ± 0.71	14.4 ± 0.83	[b]
4	12.9 ± 1.2	13.6 ± 0.49	14.2 ± 0.70	[c]
5	13.9 ± 1.0	14.1 ± 0.71	14.5 ± 0.96	

[a] Data are $M \pm SD$.
[b] $P < 0.05$; C : A, B.
[c] $P < 0.001$; C : A, B.

Prior, experience in the study of effects of metals in the environment on the organism shows that a suitable method for facilitating, in mass-scale investigations, the assaying of the magnitude of the risk to health is to follow up the state of immunity of the studied organisms by examining immunoglobulin levels and other indicators of immunity. The results of our examinations of the immunity indicators are summarized in Tables 8–12.

A statistically significant decrease in the values of serum and secretory immunoglobulin A was found in children most heavily exposed to vanadium (Group A) as well as children from a moderately exposed locality (Group B). A basic factor of specific immunity that reacts very sensitively to certain noxae in the polluted environment is immunoglobulin G. As is apparent from Table 10, lower levels of serum IgG have been found in the children of group B, but also in those of group A. This phenomenon can be related to the generally known immunosuppressive action of heavy metals.

As regards the values of immunoglobulin M, no statistically significant differences have been found between children from the exposed and the control areas, although higher concentrations of IgM have been observed in the sera of children from the exposed localities.

The investigation into immunoglobulin E (Table 11) did not give unambiguous data. Although values found in the first three samplings suggested, along

Table 3   Hematocrit (%)[a]

Sampling	Group A	Group B	Group C	P
1	40.4 ± 2.5	41.2 ± 1.9	40.7 ± 2.4	
2	39.9 ± 2.8	38.6 ± 1.7	40.2 ± 2.3	[b]
3	40.7 ± 3.7	40.7 ± 1.8	42.4 ± 3.1	[b]
4	39.0 ± 3.3	40.7 ± 1.5	42.0 ± 2.1	[c]
5	39.7 ± 2.7	40.3 ± 1.8	41.9 ± 2.8	[c]

[a] Data are $M \pm SD$.
[b] $P < 0.05$; C : B.
[c] $P < 0.05$; C : A, B.

Table 4 Platelet Count (1,000/mm^3)^a

Sampling	Group A	Group B	Group C	P
1	264 667 ± 47 429	264 536 ± 43 885	278 563 ± 47 656	
2	259 538 ± 35 213	247 958 ± 45 670	289 929 ± 48 195	b
3	261 417 ± 38 547	249 480 ± 46 440	272 464 ± 34 015	
4	278 375 ± 62 493	253 273 ± 46 828	284 625 ± 46 828	
5	253 800 ± 47 562	253 190 ± 59 249	246 815 ± 41 775	

^{a} Data are $M \pm SD$.
^{b} $P < 0.05$; C:B.

Table 5 Mean Corpuscular Volume (fL)^a

Sampling	Group A	Group B	GroupC
1	83.9 ± 2.7	84.0 ± 3.2	83.7 ± 3.3
2	84.4 ± 4.7	82.9 ± 3.2	83.0 ± 2.9
3	83.8 ± 2.8	83.4 ± 3.9	83.9 ± 2.9
4	83.0 ± 3.6	84.9 ± 2.9	84.5 ± 2.9
5	84.2 ± 3.3	84.8 ± 3.8	85.4 ± 3.2

^{a} Data are $M \pm SD$.

Table 6 Mean Corpuscular Hemoglobin (MCH) (pg)^a

Sampling	Group A	Group B	Group C
1	28.1 ± 1.0	28.3 ± 1.2	28.1 ± 1.3
2	28.3 ± 1.3	28.1 ± 2.1	28.5 ± 1.9
3	28.4 ± 1.3	28.3 ± 1.3	28.6 ± 1.2
4	27.7 ± 1.6	28.4 ± 1.2	28.5 ± 1.2
5	29.6 ± 1.4	29.6 ± 1.6	29.6 ± 1.3

^{a} Data are $M \pm SD$.

Table 7 Leukocyte Count (10^3/mm^3)^a

Sampling	Group A	Group B	Group C	P
1	6.2 ± 1.2	6.6 ± 2.3	6.5 ± 1.6	
2	6.6 ± 1.4	6.1 ± 1.5	7.2 ± 1.4	b
3	5.8 ± 0.6	6.3 ± 1.2	6.4 ± 1.4	
4	5.5 ± 1.0	6.2 ± 1.2	6.2 ± 1.1	
5	6.0 ± 0.96	6.4 ± 0.99	6.4 ± 1.3	

^{a} Data are $M \pm SD$.
^{b} $P < 0.05$; C:B.

Table 8   Immunoglobulin A (IU/ml)[a]

Sampling	Group A	Group B	Group C	P
1	105 ± 48	81 ± 41	112 ± 40	[b]
2	142 ± 76	89 ± 50	117 ± 54	[c]
3	103 ± 37	73 ± 37	126 ± 36	[d]
4	122 ± 38	99 ± 53	136 ± 41	[b]
5	111 ± 50	100 ± 55	139 ± 46	[b]

[a] Data are $M \pm SD$. 1 IU = 0.017 g/L.
[b] $P < 0.05$; C:B.
[c] $P < 0.05$; A:B.
[d] $P < 0.001$; C:A, B.

with seasonal variations, an increased concentration of IgE in Group A and Group B children, this increase was not confirmed in the two samplings that followed.

Among the nonspecific defense mechanisms of organisms against various noxae are the so-called acute reactants. Usually classified as inhibitors of proteases, they are substances released from phagocytes in the course of inflammatory processes. Their levels rise very early at the beginning of an infection and may also accompany injury from a toxic metal that could cause a decline in immune capacity against infectious agents.

Of this group of indicators in our study we followed up the levels of α-1-antitrypsin, transferrin, and β-2-microglobulin (Tables 13–15). The values found did not differ among the separate groups and were in the range of normal values. A statistically nonsignificant decrease in the values of α-1-antitrypsin and transferrin could be observed in the exposed children. No differences in the levels of β-2-microglobulin were found.

In our follow-up of the effect of vanadium exposure on the state of the children's immunity we also focused our attention on indicators of cell-mediated immunity. We assessed the effect of vanadium on the activity of T-lymphocytes in the peripheral blood, a factor of natural cellular immunity.

Table 9   Secretory IgA in Saliva (g/L)[a]

Sampling	Group A	Group B	Group C	P
1	0.81 ± 0.25	0.98 ± 0.21	0.68 ± 0.22	[b]
2	0.77 ± 0.14	0.74 ± 0.14	0.96 ± 0.23	
3	0.73 ± 0.14	0.74 ± 0.14	0.96 ± 0.23	[c]
4	0.89 ± 0.19	0.96 ± 0.22	1.01 ± 0.22	
5	0.68 ± 0.23	0.70 ± 0.22	0.83 ± 0.27	

[a] Data are $M \pm SD$.
[b] $P < 0.001$; C:B.
[c] $P < 0.001$; C:A.

Table 10   Immunoglobulin G (IU/ml)[a]

Sampling	Group A	Group B	Group C	P
1	166 ± 23	149 ± 31	171 ± 34	[b]
2	203 ± 53	173 ± 43	188 ± 37	
3	166 ± 34	152 ± 37	170 ± 27	
4	167 ± 52	163 ± 43	192 ± 29	[b]
5	177 ± 27	147 ± 39	153 ± 28	

[a] Data are $M ± SD$. 1 IU = 0.087 g/L.
[b] $P < 0.05$; C:B.

Table 11   Immunoglobulin E (IU/ml)[a]

Sampling	Group A	Group B	Group C
1	209 ± 489	230 ± 477	144 ± 346
2	163 ± 177	77 ± 107	99 ± 134
3	134 ± 303	122 ± 266	117 ± 248
4	38 ± 24	132 ± 253	165 ± 231
5	99 ± 174	95 ± 117	194 ± 401

[a] Data are $M ± SD$.

Table 12   Immunoglobulin M (IU/ml)[a]

Sampling	Group A	Group B	Group C
1	271 ± 95	296 ± 107	260 ± 96
2	271 ± 118	240 ± 98	244 ± 139
3	258 ± 73	235 ± 71	231 ± 71
4	263 ± 85	258 ± 102	254 ± 97
5	280 ± 117	229 ± 77	219 ± 88

[a] Data are $M ± SD$. 1 IU = 0.009 g/L.

Table 13   Transferrin (g/L)[a]

Sampling	Group A	Group B	Group C
1	2.3 ± 0.72	2.5 ± 1.1	2.7 ± 0.82
2	2.0 ± 0.25	2.5 ± 0.76	2.3 ± 0.68
3	2.1 ± 0.49	2.3 ± 0.73	2.5 ± 0.57
4	2.9 ± 0.73	3.1 ± 1.3	3.0 ± 0.68
5	2.0 ± 0.75	2.1 ± 0.67	2.3 ± 0.78

[a] Data are $M ± SD$.

Table 14   α-1-Antitrypsin (g/L)[a]

Sampling	Group A	Group B	Group C
1	2.1 ± 0.64	2.4 ± 0.76	2.4 ± 0.94
2	2.4 ± 0.69	2.1 ± 0.44	2.3 ± 0.82
3	2.5 ± 0.48	2.7 ± 1.1	2.8 ± 0.73
4	2.9 ± 0.75	2.7 ± 0.83	2.4 ± 1.0
5	2.1 ± 0.75	2.2 ± 0.80	2.7 ± 1.2

[a] Data are $M \pm SD$.

As to the phagocytic activity of leukocytes, no significant differences were found among the proportions of phagocytosing leukocytes in the blood, although this proportion was larger in the children of Group A (Table 16).

Significant differences were observed in T-lymphocytes upon stimulation of their mitotic activity with specific mitogens: phytohemagglutinin, concanavalin A, and pokeweed mitogen. The level of stimulation of T-lymphocyte mitotic activity expressed as a stimulation index (Luster et al., 1988) is significantly higher in children from both exposed areas and provides some evidence of the activation of cellular immunity. These differences in the values of the stimulation index were found upon stimulation of the mitotic activity of T-lymphocytes with phytohemagglutinin (Table 17), concanavalin A (Table 18), as well as pokeweed mitogen (Table 19).

On the basis of our investigations it can be assumed that in children living in an exposed area there takes place, in many cases, an alteration of humoral as well as cellular factors of natural immunity in the sense of stimulation. Accordingly, there may be an initial stage of processes in both spheres of immunity that in connection with exposure to vanadium may at first appear as stimulation and then, upon long-term immunomodulatory action of the element, may result in the decline or even collapse of natural immunity.

We have attempted to evaluate the state of natural immunity in the children by an assessment of the incidence of viral and bacterial infections of the respiratory tract. From an analysis of the records kept by local pediatricians

Table 15   β-2-Microglobulin (mg/L)[a]

Sampling	Group A	Group B	Group C
1	—	—	—
2	1.2 ± 0.29	1.4 ± 0.49	1.5 ± 0.59
3	2.3 ± 0.58	2.3 ± 1.2	2.6 ± 0.66
4	2.5 ± 1.8	2.6 ± 2.5	2.0 ± 0.84
5	2.6 ± 1.2	3.5 ± 1.7	2.6 ± 0.92

[a] Data are $M \pm SD$.

Table 16   Phagocytes (%)[a]

Group	n	M	± SD
A	15	9.1	4.4
B	28	8.3	5.4
C	32	8.3	3.7

[a] Data are from first sampling.

Table 17   Lymphocyte Stimulation with Phytohemagglutinin[a]

Group	n	M	± SD	P
A	15	51	47	[b]
B	28	51	53	[b]
C	32	5.0	7.3	

[a] Data are from first sampling.
[b] $P < 0.01$; C:A, C:B.

Table 18   Lymphocyte Stimulation with Concanavalin A[a]

Group	n	M	± SD	P
A	15	31.3	31.0	[b]
B	28	27.5	28.2	[b]
C	32	1.7	1.7	

[a] Data from first sampling.
[b] $P < 0.001$; C:A, C:B.

Table 19   Lymphocyte Stimulation with Pokeweed Mitogen[a]

Group	n	M	± SD	P
A	15	10.5	7.5	[b]
B	28	9.0	8.8	[b]
C	32	1.7	1.4	

[a] Data from first sampling.
[b] $P < 0.01$; C:A, C:B.

Table 20   Incidence of Viral and Bacterial Respiratory Infections

	1991		1992		
Group	Viral	Bacterial	Viral	Bacterial	Total
A + B (n = 46)	40	8	39	15	102
C (n = 29)	10	13	20	11	54

we found that there was a greater number of cases of respiratory infections in children from the exposed areas (Table 20).

Data in the literature on the influence of vanadium on the biosynthesis of cholesterol led us also to follow up serum lipid levels in children from the exposed and control areas. As far as total cholesterol and triglycerides are concerned, we found lower levels in exposed children in one sampling only (Tables 21–22). On the other hand, at the first, third, and fifth sampling we observed lower HDL and LDL cholesterol levels in the blood of these children (Tables 23–24).

For the assessment of any eventual genotoxic action of vanadium emission a one-off cytogenetic analysis of peripheral lymphocytes, was performed no difference was found between the groups under study on evaluation of the proportion of aberrant cells in the Ames test, and of the mean frequency of sister chromatid exchanges (Table 25).

To verify the possibility of utilizing determination of cystine concentrations in the hair and fingernails as an exposure test for vanadium, we conducted an investigation of the cystine concentration in samples taken from the groups under study (Kučera et al., 1994).

From Table 26 it is apparent that there were no significant differences between the groups from the point of view of cystine concentration in the subjects hair and fingernails.

Higher levels of vanadium in the blood and hair of the children living in the vicinity of the production plant in comparison with those from unpolluted

Table 21   Total Cholesterol (mmol/L)[a]

Sampling	Group A	Group B	Group C	P
1	4.8 ± 0.74	4.7 ± 0.64	4.4 ± 0.65	
2	4.1 ± 0.79	4.1 ± 0.94	4.0 ± 0.66	
3	4.8 ± 0.65	4.2 ± 0.88	5.6 ± 1.19	[b]
4	3.9 ± 0.79	4.1 ± 0.75	3.6 ± 0.64	
5	4.0 ± 0.56	4.1 ± 0.75	3.6 ± 0.64	

[a] Data are $M \pm SD$.
[b] $P < 0.001$; C: A, B.

Table 22   Triglycerides (mmol/L)[a]

Sampling	Group A	Group B	Group C	P
1	—	—	—	
2	1.1 ± 0.39	0.96 ± 0.28	0.96 ± 0.20	
3	1.1 ± 0.26	0.95 ± 0.17	1.3 ± 0.38	[b]
4	0.95 ± 0.28	0.95 ± 0.25	1.1 ± 0.55	
5	1.1 ± 0.09	1.2 ± 0.30	1.2 ± 0.34	

[a] Data are $M \pm SD$.
[b] $P < 0.05$; C:B.

locality serving as controls testified to the extent of the exposure of the population to vanadium (Table 27). The difference in the vanadium concentration in the hair was marked between children whose parents were employed at the vanadium plant and those from other families (Table 28). In this case this difference is probably amplified by secondary contamination in those households in which at least one of the parents works in the vanadium plant.

For the assessment and evaluation of the overall load of contamination experienced the population in connection with the production of vanadium, this element was followed up in the ambient air, in the soil, and in drinking water.

Continuous measurements of the vanadium concentration in the dust aerosol were conducted in the locality of Mnisek at a distance of about 1 km west of the plant from March 1993 through September 1994 (Fig. 2). In the course of the period under follow-up the mean concentration of vanadium in the ambient air reached 15.0 ng/m^3 (1.16–68.9 ng/m^3) and was markedly higher than in other regions of the Czech Republic, where the mean level was 2.73 ng/m^3 (1.09–6.59 ng/m^3) (Kučera et al., 1996).

The course of the concentration curve constructed on the basis of average values of duplicate samples demonstrated a seasonal decline in the summer

Table 23   Low-Density Cholesterol (LDL) (mmol/L)[a]

Sampling	Group A	Group B	Group C	P
1	—	—	—	
2	2.4 ± 0.86	2.2 ± 0.89	2.3 ± 0.57	
3	3.1 ± 0.60	2.7 ± 0.79	3.6 ± 1.0	[c]
4	2.4 ± 0.89	2.3 ± 0.65	2.4 ± 0.69	
5	2.7 ± 0.44	2.6 ± 0.88	2.1 ± 0.62	[b]

[a] Data are $M \pm SD$.
[b] $P < 0.05$; C:A, B.
[c] $P < 0.001$; C:B.

Table 24   High-Density Cholesterol (HDL) (mmol/L)[a]

Sampling	Group A	Group B	Group C	P
1	1.2 ± 0.18	1.2 ± 0.14	1.5 ± 0.28	[b]
2	1.2 ± 0.20	1.1 ± 0.23	1.2 ± 0.25	
3	1.2 ± 0.25	1.1 ± 0.05	1.5 ± 0.29	[c]
4	1.1 ± 0.26	0.89 ± 0.32	0.88 ± 0.19	
5	0.90 ± 0.17	0.78 ± 0.23	0.97 ± 0.23	[d]

[a] Data are $M \pm SD$.
[b] $P < 0.001$; C:A, B.
[c] $P < 0.001$; C:A.
[d] $P < 0.05$; C:B.

months, apparently in connection with a limitation in vanadium production; on the other hand the amount of vanadium found in dust aerosol increased in the spring and autumn months. A slump in vanadium emission levels in the ambient air was recorded at the turn of the year 1993, when production by hydrometallurgical methods ceased and the ecologically much more acceptable production of ferrovanadium commenced. Increased levels of vanadium in the air in the summer months of 1994 were apparently a consequence of secondary dust coming from the long-standing contamination of the top layer of the soil, which underwent erosion due to the extremely high air temperatures at that time.

Examination of the vanadium concentration in the soil 600–2,400 m downwind from the plant (see Fig. 1) showed considerable contamination depending on distance from the source of emissions. Detailed values are presented in Table 29.

Assessment of the overall vanadium load in the environment of the population studied was supplemented by a follow-up of levels of this element in individual and municipal sources of drinking water in localities in the neighborhood of the production plant. As is apparent from the values presented in Table 30, vanadium contamination of the soil affects only its topmost layer and did not penetrate into the substratum, so that the sources of drinking

Table 25   Cytogenetic Analysis of Peripheral Lymphocytes[a]

Group	n	Aberrant Cells (%)	Mean Frequency of Sister Chromatid Exchanges per Cell
A	13	1.2 ± 1.2	4.6 ± 1.0
B	26	1.3 ± 1.1	4.6 ± 0.87
C	19	0.95 ± 0.97	—

[a] Data from first sampling.

Table 26 Cystine Concentration in Hair and Fingernails (mg/g)[a]

Sampling	Sample	Group A	Group B	Group C
1	Hair	187 ± 121	150 ± 88	192 ± 21
	Fingernails	100 ± 19	100 ± 15	103 ± 19
2	Hair	175 ± 26	168 ± 11	165 ± 14
	Fingernails	—	—	—
3	Hair	173 ± 12	165 ± 14	164 ± 10
	Fingernails	103 ± 13	107 ± 14	99 ± 8.5
4	Hair	137 ± 10	135 ± 9.0	139 ± 8.5
	Fingernails	106 ± 17	96 ± 8.3	96 ± 7.4
5	Hair	160 ± 6.7	160 ± 6.5	159 ± 6.6
	Fingernails	105 ± 20	110 ± 7.4	110 ± 8.9

[a] Data are $M \pm SD$.

water in the environs of the plant presents no significant hazard to the population from the point of view of the overall load.

4. CONCLUSIONS

The major result of a 3-year follow-up of selected indicators of the health status in schoolchildren living in the vicinity of a metallurgical plant producing vanadium is the finding that in our conditions long-term exposure to vanadium emissions had no negative impact on their health. Differences between children from the exposed region and those from an ecologically unpolluted control locality varied within the range of normal values in all cases.

Children from groups exposed to vanadium had lower values of RBC counts, hemoglobin content, and hematocrit. Investigation of their specific immunity revealed a decrease in levels of serum and secretory IgA and a seasonal decrease in IgG. Besides that, however, on the basis of the evaluation

Table 27 Vanadium Concentration in Blood, Hair, and Fingernails ($\mu g/L$, $\mu g/kg$)[a]

Sample	Group A	Group B	Group C	P
Blood	—	0.10 ± 0.07	0.05 ± 0.05	[b]
Hair	96 ± 42	181 ± 114	69 ± 50	[d]
Fingernails	189 ± 41	186 ± 38	109 ± 68	[c]

[a] Data are $M \pm SD$.
[b] $P < 0.05$; C : B.
[c] $P < 0.05$; C : A.
[d] $P < 0.001$; C : A, B.

Table 28   Vanadium Concentration in Hair of Children of Vanadium Plant Employees and Other Families (Group B)

Group of Children	n	Vanadium (μg/kg)	P
Children of employees	11	226 ± 145	[a]
Other children	13	127 ± 74	

[a] $P < 0.05$.

of so-called acute reactants no differences were found at the level of nonspecific immunity of the organism.

Marked differences between the groups under study were observed, however, upon evaluation of their levels of natural cell-mediated immunity. On the basis of significantly higher values of mitotic activity of T-lymphocytes in children from the immediate vicinity of the production plant, it can be stated that they have an elevated activity of cellular immunity. The levels of humoral and cellular factors of natural immunity point to an alteration in the sense of stimulation. On the other hand, in children from the exposed locality greater numbers of viral and bacterial infections were registered, pointing rather to a decrease in natural immunity.

Cytogenetic analysis of peripheral lymphocytes in the children exposed to vanadium revealed no genotoxic influence from its emission.

Evidence about the level of vanadium exposure in children from the polluted localities was produced by an examination of the concentration of this element in the blood, hair, and fingernails. The vanadium concentration was significantly higher in these cases than in children from the control group; it was marked expecially in those children of Group B where in whom the influence of employment of one or both parents in the plant was manifested. As yet, we are unable to explain unequivocally why children from the Mnisek locality (Group B) whose parents were not employed at the vanadium plant had a higher vanadium content in the hair than children from Group A, although is situated windward, in relation to the plant, against the prevailing winds. In this case it is apparently the distance from the source of vanadium emissions that is decisive, it being 1.5 or 2.0 km for the residence of Group A children, whereas, although windward, Mnisek (Group B) is situated only 500 m away.

The high vanadium concentration in the soil in the vicinity of the plant was not manifested in its concentration in groundwater; however, it may be a long-term source of secondary contamination of the ambient air with vanadium even after termination of the hazardous hydrometallurgical technology of vanadium production.

In view of the retention of vanadium in the organism and its long excretion half-life, we deem it necessary to stress that epidemiological studies of the

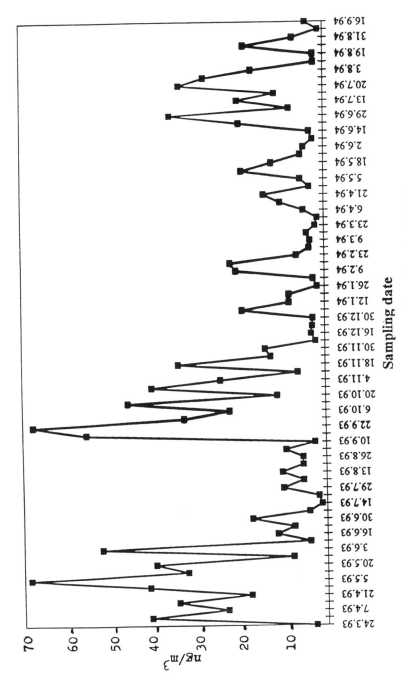

Figure 2. Vanadium concentration in ambient air of Mnisek.

17

Table 29   Soil Vanadium Concentration in the Vicinity of Vanadium Plant

Sampling Location	Distance from Plant (m)	V (mg/kg Dry Weight)
11	600	69
12	1,200	19
13	1,800	19.5
14	2,400	18
21	600	92
22	1,200	45.5
23	1,800	31.5
24	2,400	47
31	600	136
32	1,200	45
33	1,800	30.5
34	2,400	38
42	1,200	29
43	1,800	22
44	2,400	20

effect of vanadium in the population should continue to be conducted. It is important from the point of view of health, especially in regions of high-level exposure, and in this connection studies broadened by the addition of specific functional examinations should be contemplated.

Table 30   Vanadium Concentration in Drinking Water[a,b]

Locality		V (μg/L)
Cisovice	32	0.02
	41	0.04
	53	0.01
	121	0.23
	142	0.06
	160	0.03
	163	0.01
	213	0.44
Lisnice	98	0.04
	133	0.04
Mnisek		0.01

[a] MAC of vanadium in drinking water in Czech Republic: 10 μg/L.
[b] Drinking sources were individual wells except for Mnisek, where source was the municipal water supply.

REFERENCES

CBEAP (Committee on Biological Effects of Atmospheric Pollutants). (1974). *Vanadium: Medical and Biologic Effects of Environmental Pollutants.* National Academy of Sciences, Washington, D.C.

Chasteen, N. D. (Ed.). (1990). *Vanadium in Biological Systems, Physiology and Biochemistry.* Kluwer Academic Publishers, Dordrecht, Boston, London.

Friberg L., Nordberg G. F., and Vouk, V. (Eds.). (1986), *Handbook on the Toxicology of Metals.* Elsevier, Amsterdam, New York, Oxford.

Hopkins, L. L., and Mohr, H. E. (1971). The biological essentiality of vanadium. In W. Mertz and W. Cornatzer (Eds.), *Newer Trace Elements in Nutrition.* Marcel Dekker, New York.

Kučera J., Byrne A. R., Mravcová A., and Lener, J. (1992). Vanadium levels in hair and blood of normal and exposed persons. *Sci. Total Environ.* **15**, 191–205.

Kučera, J., Lener, J., and Mňuková, J. (1994). Vanadium levels in urine and cystine levels in fingernails and hair of exposed and normal persons. *Biol. Trace Element Res.* **43/45**, 327–334.

Kučera, J., Lener, J., Soukal, L., and Horáková, J. (1996). Air pollution and biological monitoring of environmental exposure to vanadium using short-time neutron activation analysis. *J. Trace Microprobe Techniques* **14**(1), 191–201.

Luster, M. I., Munson, A. E., Thomas, P. T., et al. (1988). Methods evaluation. Development of a testing battery to assess chemical-induced immunotoxicity. *Fundam. Appl. Toxicol.* **10**, 2–19.

WHO. (1988). *Environmental Health Criteria* Vol. 81, *Vanadium.* WHO, Geneva.

2

TOXICOLOGY OF VANADIUM IN MAMMALS

K. H. Thompson, Mary Battell, and John H. McNeill

Division of Pharmacology and Toxicology, Faculty of Pharmaceutical Sciences, The University of British Columbia, Vancouver, B.C., Canada V6T 1Z3

Vanadium in the Environment. Part 2: Health Effects, Edited by Jerome O. Nriagu.
ISBN 0-471-17776-8.   © 1998 John Wiley & Sons, Inc.

1. INTRODUCTION

Vanadium compounds are generally known to have insulin-mimetic activity, both in vitro and in vivo (Tolman et al., 1979; Heyliger et al., 1985; Meyerovitch et al., 1987) and may become useful in the future treatment of diabetes (Shechter, 1990; Brichard et al., 1991; Cohen et al., 1995). Questions regarding the possible toxicity of vanadium compounds, in particular following prolonged administration for the treatment of diabetes, are thus a serious concern.

Toxicity studies to date have included both inorganic and organic vanadium compounds (Waters, 1977; Jhandhyala and Hom, 1983; Domingo et al., 1990; Dai et al, 1994a,b). An extensive literature exists concerning the effects of vanadium inhalation (Athanassiadis, 1969; Waters, 1977; Venugopal and Luckey, 1978; Haider and Kashyap, 1989). Vanadium is known to be more toxic when inhaled, and relatively less so when ingested (Boyd and Kustin, 1984). This review will concentrate on the most recent results of toxicity investigations, particularly in the range of doses relevant to long-term therapeutic use of orally administered vanadium compounds, and principally in experimentally diabetic animals. The dose range is quite broad in animals, 10–100 mg V/kg-d (Henquin et al., 1994; McNeill et al., 1992; Meyerovitch et al., 1987; Thompson et al., 1993); in the few human studies, doses of 0.3–0.7 mg V/kg-d improved glucose utilization (Cohen et al., 1995; Goldfine et al., 1995).

Vanadium is present in most cells of all plants and animals. Physiologically relevant oxidation states include $+3$, $+4$, and $+5$, mainly as the oxyions vanadyl (VO^{2+}) and vanadate ($H_2VO_4^-$) (Nechay, 1984). Intracellularly, vanadium tends to be present as vanadyl, bound to glutathione, catecholamines or other small peptides (Rehder, 1991; Sakurai, 1994). In plasma, vanadium is usually found bound to transferrin and, in red cells, to hemoglobin (Breuch et al., 1984). The solution chemistry of vanadium is exceedingly complex (Butler, 1990), and the actual species present depends on pH, concentration, and the presence of oxidants and antioxidants as well as chelating agents (Willsky, 1990).

Most studies of vanadium's pharmacological potential have utilized inorganic vanadium salts such as sodium metavanadate, sodium orthovanadate, ammonium tartarovanadate, and vanadyl sulfate (for review, see Orvig et al., 1995). More recently, researchers have investigated the potential of various organic vanadium compounds to enhance the glucose-lowering effects previously seen with inorganic vanadium salts (McNeill et al., 1992; Shechter et al., 1992; Cam et al., 1993a; Orvig et al., 1995; Sakurai et al., 1995). The advantage of all of these compounds over insulin therapy is in the ease of administration: most can be given orally in solution and do not require parenteral administration.

1.1. Kidney Function

Chronic vanadium treatment results in significant accumulation of vanadium in the kidneys (Thompson and McNeill, 1993; Mongold et al., 1990; Domingo et al., 1991); however, most of the accumulated vanadium is likely bound to small peptides or macromolecules in the form of vanadyl and thus is not available as vanadate, a more potent inhibitor of ATPases (Rehder, 1991). Indicators of kidney function include blood or plasma urea, bilirubin, and plasma creatinine.

Plasma urea is often elevated in experimentally diabetic rats compared with controls (Jensen et al., 1981; Cam et al., 1993b; Domingo et al., 1991a,b). Elevated urea levels were partially returned to normal by vanadium ($VOSO_4$) 1.0–1.5 mg/ml (4.6–6.5 mM vanadium) in the drinking water (Cam et al., 1993a,b; Dai et al., 1994b), but not at lower concentrations (1.2 mM vanadium; Domingo et al., 1991b). Changes may be time-dependent, as urea concentrations were significantly elevated above baseline in all vanadyl-treated groups at 3 months and were still significantly above normal in vanadium-treated diabetic animals at 6 and 9 months, but they became normalized by 12 and 16 months, in a year-long toxicity study (Dai et al., 1994b). Creatinine concentrations were generally not affected (or were normalized from diabetes-induced increases; Dai et al., 1993) by vanadium in these studies, with the exception of one study (Domingo et al., 1991b), in which rats consuming 1.2 mM vanadium as vanadyl sulfate pentahydrate for 4 weeks showed elevated plasma creatinine compared with untreated rats. In the latter study, the same concentration (1.2 mM) of vanadium given as sodium orthovanadate or metavanadate did not affect creatinine concentrations.

Renal effects of vanadium are dependent on the mode of administration. Intraperitoneal (i.p.) administration of sodium orthovanadate resulted in inhibition of tubular reabsorption of sodium (Bräunlich et al., 1989) and hypokalemic distal renal tubular acidosis with increased urinary pH (Dafnis et al., 1992). In the latter study, glomerular filtration rate and serum creatinine concentrations were not affected by vanadium treatment. Since vanadate is a potent Na^+,K^+-ATPase inhibitor, these results may indicate a much higher concentration of vanadate, rather than vanadyl, when administration is parenteral rather than enteral. Another significant factor is the severely limited absorption of vanadium from an oral dose, usually less than 1% (Underwood, 1977). Conversion of vanadate to vanadyl in vivo is not instantaneous (Breuch et al., 1984), but takes place on the order of 15–30 min under physiological conditions. Recent human studies showed no effect of vanadium treatment on renal function (2 weeks, 125 mg sodium metavanadate per day; 13 μmol V/kg-d, Goldfine et al., 1995; 3 weeks, 100 mg vanadyl sulfate per day, 5.9 μmol V/kg-d, Cohen et al., 1995). Vanadium given by the subcutaneous route was toxic to the kidney in doses of 12 or 18 μmol V/kg-d for 16 days in the form of ammonium metavanadate. Significant histological changes including necrosis, cellular proliferation, and fibrosis were noted at these doses but not at 6 μmol V/kg-d. Forty days after treatment was stopped recovery was noted. Vanadium is obviously more toxic to the kidneys when given by a parenteral route (Al-Bayati et al, 1989).

The impact of vanadium on kidney function is also species-dependent. Whereas in the rat, vanadate produced a profound diuresis (Kumar and Corder, 1980; Nechay, 1984), it did not do so in the dog or cat (Lopez-Novoa et al., 1982; Larsen and Thomsen, 1980a,b). In the dog and cat, vanadate treatment resulted in decreased glomerular filtration rate. Chickens fed 25, 50, or 100 ppm vanadium (0.5–2.0 mM) in the feed for 15 months demonstrated decreased Na^+,K^+-ATPase activity in kidney homogenates in a dose-dependent fashion (Phillips et al., 1982). Rats given 10 and 40 μg V/ml (as sodium metavanadate in the drinking water, 0.2–0.8 mM) for 180 days had decreased Na^+,K^+-ATPase activity in the distal tubules of the nephrons, increased urinary excretion of potassium, and increased plasma renin activities (Boscolo et al., 1994).

1.2. Liver Function

Liver has one of the highest accumulations of vanadium in chronically treated animals after bone and kidney (Parker and Sharma, 1978; Ramanadham et al., 1991; Thompson and McNeill, 1993). Enzymatic measures of liver function include plasma aspartyl aminotransferase (AST, previously known as glutamic-oxaloacetic transaminase, or GOT) and alanine aminotransferase (ALT, previously known as glutamic-pyruvic transaminase, or GPT). These parameters were generally seen to be elevated in diabetes induced with streptozotocin (STZ) (Bell and Hye, 1983), but at least partially normalized or unaffected with vanadium treatment (Domingo et al., 1991a,b; Dai et al., 1994b). Increased activities of AST and ALT with STZ-diabetes were completely prevented by administration of bis(maltolato)oxovanadium (iv), an organic insulin-mimetic agent, during a 6-month toxicity study (Dai et al., 1993).

1.3. Hematological Indices

Disruption of hematopoiesis is particularly sensitive to exposure to metals and is of major concern toxicologically (Maines, 1994). In a 12-week study of the hematological effects of ammonium metavanadate, 0.14 mg/ml, vanadyl sulfate, 0.26 mg/ml, and bis(maltolato)oxovanadium (iv), 0.38 mg/ml (all 1.2 mM, in drinking water), in Wistar rats, no differences were found between treated and untreated rats in hematocrit, hemoglobin level, erythrocyte count, reticulocyte percentage, leukocyte count and composition, platelet count, or osmotic fragility of the erythrocyte (Dai and McNeill, 1995). These results contrast with earlier studies of vanadium pentoxide (3 mg/kg) in the diet and ammonium metavanadate, 0.15–0.3 mg/ml (1.3–2.6 mM) in the drinking water, which showed decreased erythrocyte count and hemoglobin level in treated compared with untreated rats, but no change in hematocrit (Chakraborty et al., 1977; Zaparowska and Wasilewski, 1989, 1992). Increased percentages of reticulocytes and polychromatophilic erythrocytes in the peripheral blood were also noted (Zaparowska and Wasilewski, 1991). Similarly, in rats treated with ammonium metavanadate, 0.03–0.12 mmol V/kg-d, for 4 weeks decreased erythrocyte count, hemoglobin level, and hematocrit index, along with increased reticulocyte count and polychromatophilic erythrocytes in the peripheral blood, were observed, but these changes were not dose-dependent, and they were interpreted from statistically invalid analyses (Student's t tests; Zaparowska et al., 1993). In accord with the study of Dai et al. (1995), ammonium metavandate, 0.15 mg/ml, in drinking water for 12 weeks administered to male Wistar rats (Russanov et al., 1994) resulted in no changes in erythrocyte count or hemoglobin level; the hematocrit index was slightly increased. In a separate study using a range of vanadyl sulfate doses (0.16–0.71 mmol/kg-d in drinking water, tailored to blood glucose response) in STZ-diabetic male Wistar rats, no changes were seen in hematological indices over a treatment period of 1 year (Dai and McNeill, 1994).

Hematological effects of vanadium treatment may also be species-dependent. BALB/c mice injected subcutaneously with ammonium metavanadate, 20 mg/kg, showed no consistent changes in erythrocyte count, hemoglobin level, or hematocrit value (Wei et al., 1982). By contrast, in broiler chickens fed 10, 20, or 40 ppm vanadium (0.2, 0.4, or 0.8 mM vanadium), increased hemoglobin and hematocrit levels were noted (Blalock and Hill, 1987). In humans treated with 100 mg vanadyl sulfate hydrate daily (0.46 mmol V/d, 5.9 μmol V/kg-d) for 3 weeks, hemoglobin and hematocrit levels were significantly lowered (from 14.3 $\pm$ 0.6 g/dl to 13.3 $\pm$ 1.0 g/dl, $P < 0.05$, a decrease that may not be physiologically significant), but platelet count and white blood cells were unchanged (Cohen et al., 1995).

Comparison of erythrocyte counts and in vitro hemolytic indices for female ICR mice following a single ip injection equivalent to 9.2 mg V/kg body weight (0.18 mmol/kg) for three different valances of vanadium, namely, vanadium(III) chloride, vanadyl(IV) sulfate, and sodium orthovanadate(V), indicated that vanadium(IV) was most damaging in promoting hemolytic rupture, apparently due to an increased vulnerability to rupture after removal from the animal (Hogan, 1990). Vanadyl sulfate, 4 mM, and sodium vanadate, 10 mM (concentrations far above what has been seen in vivo), were shown to have a comparable hemolytic action on human erythrocytes in vitro (Hansen et al., 1985).

1.4. Cardiovascular Function

Vanadium compounds cause positive or negative inotropic effects depending on the concentration of the injected or fed compound and the animal species (Larsen and Thomsen, 1980a,b; Hom et al., 1982; Inciarte et al., 1980; Erdmann, 1980). In cats and dogs, iv sodium metavanadate, 2.5–5.0 mg/kg (20–40 μmol V/kg), decreased the force of ventricular contraction, presumably owing to coronary constriction (Jackson, 1912, as cited in Venugopal and Luckey, 1978). In male Sprague-Dawley rats given 10, 40, or 100 ppm vanadium (0.2, 0.8, or 2.0 mM, as sodium metavanadate) for 7 months, arterial hypertension was related to central sympathetic hypertonia, increased urinary potassium excretion, and preferential accumulation of vanadium (as vanadyl) in the aorta (Carmignani et al., 1993), perhaps because of changes in Ca^{2+} concentrations in the aorta of vanadate-treated rats (Sandirasegarane and Gopalakrishnan, 1995). In the latter study, acute additions of 10–200 μM vanadate to cultured aortic smooth muscle cells resulted in a rapid and concentration-dependent cytosolic free-calcium increase.

At high concentrations ($>10^{-4}$ M), VO_3^- was associated with positive inotropic effect in cultured heart cells (Werden et al., 1982). Prolonged dietary administration of vanadate has been shown to cause a dose-dependent increase in blood pressure in uninephrectomized rats (Steffan et al., 1981). Increased peripheral vascular resistance, without increased blood pressure, was observed in vanadate-treated (sodium metavanadate, 1 μg/ml in the drinking water for 12 months) male rabbits. However, in male Sprague-Dawley rats treated with the same dosage for 6 months, increased peripheral vascular resistance was accompanied by increased blood pressure (Carmignani et al., 1996). By contrast, in spontaneously hypertensive rats (SHR) rats treated with vanadyl sulfate, 0.25–1 mg/ml in the drinking water for 8 weeks, a dose-dependent decrease in the blood pressure accompanied by decreased plasma insulin was observed (Bhanot and McNeill, 1994). Discrepant results here may relate to the linkage between hyperinsulinemia and hypertension (Reaven, 1993).

1.5.   Reproductive and Developmental Toxicity

Information concerning the reproductive toxicity of vanadium is accumulating. In rats, inhaled vanadium had gonadotoxic effects characterized by disturbances of spermatogenesis and of the motility and osmotic resistance of spermatozoa, and by a decrease in fertility (Roshchin et al., 1980). Sodium metavanadate (60–80 mg/kg-d, or 0.51–0.68 mmol V/kg-d) in drinking water resulted in decreased sperm count and fertility in male mice (Llobet et al., 1993). Neither the sperm motility nor the morphology of the testes was affected. Accumulation of vanadium in testes (without morphological changes) accompanied prolonged oral administration of vanadyl sulfate (40–100 mg V/kg-d, or 0.82–2.0 mmol V/kg-d; Dai et al., 1994a). Functional changes were not measured in the latter study.

Vanadium crosses the placenta and can accumulate in the fetus (Edel and Sabbioni, 1989; Paternain et al., 1993). Investigations of the possible developmental toxicity of vanadium compounds have yielded inconsistent results. Intraperitoneal administration of ammonium metavanadate, 0.47, 1.88, or 3.75 mg/kg-d (4–32 μmol V/kg-d), to Syrian golden hamsters resulted in an increased incidence of micrognathia, supernumerary ribs, altered sternebral ossification, nd meningocele in the live fetuses. A significant, but not dose-dependent, increase in the number of dead or resorbed fetuses was observed in the middle dose group (Carlton et al., 1982). The authors pointed out that the number of abnormal fetuses was too small to assess the teratogenicity of ammonium vanadate.

In mice, a single iv injection of 1 mM vanadium pentoxide, 0.15 ml, on day 8 of pregnancy slowed fetal skeletal ossification but did not affect implantation of the fetuses or have any teratogenic effect (Wide, 1984). Vanadyl sulfate pentahydrate, 0.15–0.6 mmol/kg-d by gavage, on days 6–15 of pregnancy caused maternal toxicity, lower fetal length and weight, increased incidence of cleft palate, micrognathia, hydrocephaly, and decreased ossification of the supraoccipital bone, carpus, and tarsus in a dose-dependent fashion (Paternain et al., 1993). By contrast, sodium orthovanadate, 0.04–0.32 mmol/kg-d by gavage, on days 6–15 of pregnancy, did not result in significant increases in these same parameters, except for delayed ossification at an intermediate dosage level (Sanchez et al., 1991). Sodium metavanadate, 16–64 μmol/kg-d by ip injection, at the same time interval in mice caused maternal toxicity, fetotoxicity, and an increased incidence of cleft palate, but no changes in skeletal growth rate (Gomez et al., 1992).

Vanadium treatment (0.25 and 0.50 mg sodium orthovanadate per milliliter in the drinking water) of pregnant diabetic rats (Ganguli et al., 1994a, 1994b) produced toxic effects at doses previously used for insulin-mimetic effect in male rats with little or no toxicity. Vanadate had a dose-dependent negative effect on the ability of the dames to carry pregnancy to term that was magnified in diabetic animals. These results suggest that pregnant animals may be more sensitive to the potential toxicity of vanadium. One problem with these studies was that there was no true diabetic group to compare with the vanadium-treated groups, since diabetic animals not receiving vanadium were treated with insulin and thus the effect of diabetes on the fetus was not truly examined.

1.6.   Effects on Food and Fluid Intake

One of the most contentious issues with regard to vanadium treatment of diabetes regards feeding and fluid intake suppression, or lack of it, by pharmacological administration of vanadium compounds. Since these effects are highly dependent on diet (Berg et al., 1963; Blakely et al., 1995; Pugazhenthi et al., 1993; Nielsen, 1987) and on the severity of diabetes induced (Thompson et al., 1993), it is a difficult issue to resolve. Studies to date have shown both that it does (Zaparowska and Wasilewski, 1991; Domingo et al., 1991; Meyerovitch et al., 1989), and does not (Pugazhenthi and Khandelwal, 1990; McNeill et al., 1992; Cam et al., 1993b), suppress food and fluid intake at doses sufficient to lower blood glucose levels in diabetic rats. In patients with non-insulin-dependent diabetes mellitus, 100 mg vanadyl sulfate per day did not suppress food intake, and resulted in only a 1% reduction in weight over a 3-week treatment period (Cohen et al., 1995). No difference in food intake or body weight was detected between treated and untreated insulin-dependent and non-insulin-dependent subjects (Goldfine et al., 1995).

A major confounding factor in determining the effect of vanadium administration on food and fluid intake in STZ-diabetic rats is that induction of diabetes results in marked increases in food and fluid intake (Bell and Hye, 1983). Amelioration of the diabetic state, by whatever means, results in a lower (normalized) intake compared with that of untreated diabetic animals (Wohaieb and Godin, 1987). Those investigations in which there was a gradual introduction of vanadium salts in food or fluid (Pugazhenthi and Khandelwal, 1990; Henquin et al., 1994) have reported much less dramatic effects on overall intake, whether in control or diabetic animals. Pair-feeding studies (Rossetti and Laughlin, 1989; Brichard et al., 1989; Brichard et al., 1992; Yuen et al., 1997) have shown clearly that any anorectic effect of vanadium treatment does not fully account for the hypoglycemic effects seen. In insulin-resistant animals, vanadium administration does not result in lower body weight gain (Henquin et al., 1994).

1.7.   Lethal Dose

The lethal dose (LD_{50}) of vanadium is highly species-dependent. In mice, LD_{50} levels of 0.2–0.3 mmol V/kg administered ip were measured (Venugopal and Luckey, 1978). Vanadyl sulfate has been shown to be relatively less toxic than sodium orthovanadate (0.48 mmol V/kg) in mice but, curiously, not in rabbits or guinea pigs (Hudsen, 1964). In rats, LD_{50} levels of 0.15 mmol/kg for ip sodium metavanadate; 0.8 mmol/kg for the same compound administered by gavage; 0.3 mmol/kg for ip vanadyl sulfate pentahydrate; and 1.8 mmol/kg for the latter by gavage were determined (Llobet and Domingo, 1984). In chicks, 200 μg V/g practical diet caused high mortality and levels of 30 μg V/g resulted in toxicity, as evidenced by depression of growth (Nielsen, 1987). Sodium metavanadate administered at doses of 0, 2.5, 5.0, 10, and 20 mg V/kg-d by gavage (equivalent to 0, 20, 40, 80, and 160 μmol V/kg-d, or roughly 160, 320, 640, and 1320 μg V/g feed) to severely diabetic rats resulted in death for roughly half of the rats at all doses (Ortega et al., 1991). Nonetheless, ammonium metavanadate, 6 mg V/kg-d (120 μmol V/kg-d), administered to Wistar (nondiabetic) rats in the drinking water did not induce toxicity (Dai et al., 1995). Differences in diet and method of vanadium administration may have contributed to discrepancies here. It has been noted that some animals refuse to drink vanadium solutions and subsequently die of dehydration. Temporary removal from treatment of vanadium-treated diabetic rats at the first signs of dehydration results in improved survival (M. Battell, personal communication; Brichard et al., 1989), and vanadium solutions can later be introduced at lower concentrations without toxic effect.

Determination of the lethal dose by parenteral administration may be expected to be fairly consistent in a particular animal species, owing to lack of gastrointestinal absorption (Rehder, 1991). However, oral doses may vary considerably because of the extremely low bioavailability of most vanadium compounds. The absorption of vanadium salts has been reported to be as low as 0.1% and as high as 25% (Underwood, 1977). Vanadium absorption is known to be strongly affected by such dietary components as type of carbohydrate, fiber, protein concentration, other trace elements, chelating agents, and electrolytes (Nielsen, 1987). Associated pathology or physiological state may also affect vanadium absorption and hence may render a consistent determination of an oral LD_{50} very difficult.

Species besides rats, mice, and chickens may be less susceptible to vanadium toxicity. Vanadium supplementation up to 300 $\mu g/g$ feed in the ration of Coturnix (quail) did not affect egg production, egg weight, or egg cholesterol content, nor did it cause any significant growth depression or mortality (Hafez and Kratzer, 1976). Ammonium vanadate supplements of 10, 100, and 200 $\mu g/g$ diet in lambs produced no signs of clinical toxicity after 84 days. At 400 and 800 $\mu g/g$ diet, some signs of toxicity (namely, depressed growth) were seen (Hansard et al., 1978). In humans, treatment for 6–10 weeks with 4.5–18 mg V/day as ammonium vanadyl tartrate produced only slight toxicity (cramps, diarrhea) at the higher doses (Dimond et al., 1963). Food intake recovered after an initial accommodation period. Oxytartarovanadate was administered at a level of 4.5 mg V/day for 16 months to patients, who showed no signs of toxicity throughout the study (Schroeder et al., 1963). Other human vanadium trials have also demonstrated a high tolerance for oral vanadium at pharmacological doses greater than 100 times the usual dietary intake (Cohen et al., 1995; Goldfine et al., 1995; Somerville and Davies, 1962; Curran et al., 1959).

The lethal dose of vanadium is not only species-dependent, but also age- and diet-dependent. Mature rats fed up to 500 $\mu g V/g$ diet were able to tolerate this amount if it was introduced gradually into the diet (Strasia, 1971). Similarly, old hens fed 300 $\mu g V/g$ diet had no mortality (but depressed egg production), whereas growing chicks had high mortality at that level (Nielsen, 1987).

1.8. Vanadium Accumulation and Homeostasis

Tissue vanadium concentrations observed as a result of representative chronic vanadium treatment investigations are summarized in Table 1. All were determined by atomic absorption spectrophotometry with deuterium background correction. It is worth noting that these levels are not strongly time-dependent, and in fact tissue accumulation appears to level off after a short time of treatment (Thompson and McNeill, 1993; Dai et al., 1994a). It should also be noted that higher doses are required at the beginning of treatment than are necessary several weeks into treatment. In some cases, vanadium effects persist even after vanadium has been withdrawn (Cam et al., 1993b; McNeill et al., 1992; Cohen et al., 1995).

In general, tissue levels of trace elements reflect intake, bioavailability and excretion rate until levels are high enough to override homeostatic regulation, when possible toxic effects may ensue (Underwood, 1977). This may account for a response to oral supplementation with vanadium that is dependent on

Table 1 Vanadium Content of Vanadium-Treated Rat Tissues (μg/g Tissue or μg/g Plasma), as Measured by Various Investigators[a]

Treatment (conc.)[b]	Kidney	Liver	Spleen	Lung	Muscle	Bone	Plasma	Ref
DT (1 mg/ml)[c]	10.5	2.5	3.6	1.1	—	—	0.70	1
CT (1 mg/ml)[c]	8.0	1.0	2.3	0.4	—	—	0.56	1
DT (0.75 mg/ml)[c]	6.6	1.0	2.1	0.3	0.1	10.0	0.18	2
DT (1 mg/ml)[c]	4.3	1.2	2.0	0.5	0.2	14.1	0.16	2
DT (1 mg/ml)[c]	4.7	1.2	1.9	—	0.2	6.4	0.29##	3
DT (0.2 mg/ml)[d]	7.0	1.6	2.3	—	0.3	5.8	0.17##	3
DT (0.5 mg/ml**)	7.3	1.5	2.8	—	0.9	10.0	0.24##	3
DT (0.2 mg/ml**)	—	—	—	—	—	—	0.84	4
DT (0.8 mg/ml**)	7.2	—	—	—	—	—	1.20	4
CT (0.60 mg/ml**)	6.7	—	—	—	—	—	1.88	5
CT (0.60 mg/ml**)	9.1	1.3	—	—	0.3	4.4	1.65	5
CT (0.20 mg/ml*)	11.4	3.9	—	—	0.6	26.4	0.54##	6
DT (0.75 mg/ml#)	13.8	3.2	—	—	0.4	18.3	0.76	7
CT (0.75 mg/ml#)							0.84	7

To obtain data in nmol/g or /ml, multiply above $\times$ 20.

[a] Treatment periods varied from 3 to 12 weeks. Data are presented as means only.

[b] DT, diabetic treated; CT, control treated; concentration of vanadium compound in drinking water.

[c] Vanadyl.

[d] Vanadate.

[e] BMOV.

[f] Blood levels.

Sources: (1) Ramanadham et al. (1991); (2) Mongold et al. (1990); (3) Domingo et al. (1991a); (4) Meyerovitch et al. (1987); (5) Saxena et al. (1992); (6) Parker and Sharma (1978); (7) Yuen et al. (1993).

30

the severity of the diabetes (Bendayan and Gingras, 1989; Thompson et al., 1993), as low doses of vanadium result in glucose-lowering without toxicity in mildly diabetic animals (Henquin et al., 1994), whereas the higher doses seemingly required by more severely diabetic animals may incur toxicity that compromises the glucose-lowering potential.

1.9. Dietary Considerations

Diet may affect responsiveness to vanadate treatment (Blakely et al., 1995; Pugazhenthi et al., 1993). The type of carbohydrate in the diet, and the level of protein, have been shown to affect bioavailability of trace elements, including vanadium (Oster et al., 1994). The presence of chelating agents may also ameliorate toxicity (McNeill et al., 1992; Domingo et al., 1994; Watanabe et al., 1994; Sakurai et al., 1990). Two different strategies have been tried: one to lessen accumulation of orally administered vanadium (Gomez et al., 1991); the other to increase absorption, thereby lessening the oral dosage of vanadium required for therapeutic effect (McNeill et al., 1992; Cam et al., 1993). As the most commonly noted toxic side effect in human studies to date has been gastrointestinal disturbance, reducing the required oral dose could be expected to lessen this problem. In consideration of the probable necessity of long-term administration of vanadium compounds for alleviation of diabetic symptoms, a number of chelating agents have been investigated for their ability to prevent vanadium accumulation in tissues (Domingo et al., 1994). At present, these also appear to have reduced pharmacological effectiveness. In a new chick development assay, deferoxamine mesylate was most effective in preventing tissue accumulation (Hamada, 1994). Coadministration with other metals, especially zinc (Yamaguchi et al., 1989) and lithium (Srivastava et al., 1993); or with antioxidants, such as vitamin E (Haider and El-Fahkri, 1991), vitamin C (Jones and Basinger, 1983), or newer synthetic compounds (Ugazio et al., 1994), may alleviate or prevent associated toxicities. High-protein diets, supplemental ascorbic acid, and presence of chromate and Mn^{3+} ions have also been shown to decrease the oral toxicity of vanadium (Venugopal and Luckey, 1978).

2. CONCLUSION

Pharmacological use of orally administered vanadium compounds requires consideration of dietary factors, including concentrations of other trace elements, added antioxidants, and chelating agents, as well as the possible modulating factors of age, sex, pregnancy, and other pharmaceutical agents. Given what is known of trace metal effects on animal physiology, individual responses to vanadium treatment might be influenced by severity of disease (including gastrointestinal and renal function), plasma and cellular binding proteins, including glutathione and catecholamines, exercise, stress, and genetic predisposition (Maines, 1994). These factors would have to be taken into account when determining the appropriate dose of vanadium compounds in the treatment of diabetes. Nonetheless, the potential therapeutic advantage of orally administered vanadium, and the available options for mitigating toxicities, suggest that further development should yield safe and effective pharmacological formulations.

REFERENCES

Al-Bayati, M. A., Giri, S. N., Raabe, O. G., Rosenblatt, L. S., and Shifrine, M. (1989). Time and dose-response study of the effects of vanadate on rats: Morphological and biochemical changes in organs. *J. Environ. Pathol. Toxicol. Oncol.* **9,** 435–455.

Athanassiadis, Y. C. 1969. Air Pollution Aspects of Vanadium and Its Compounds. Report PB-188-093. Bethesda, MD. Litton Systems, Inc.

Bell Jr., R. H., and Hye, R. J. (1983). Animal models of diabetes mellitus: Physiology and pathology. *J. Surg. Res.* **35,** 433–460.

Bendayan, M., and Gingras, D. (1989). Effect of vanadate on blood glucose and insulin levels as well as on the exocrine pancreatic function in streptozotocin-diabetic rats. *Diabetologia* **32,** 561–567.

Berg, L. R., Bearse, G. E., and Merrill, L. H. (1963). Vanadium toxicity in laying hens. *Poult. Sci.* **42,** 1407–1411.

Bhanot, S., and McNeill, J. H. (1994). Vanadyl sulfate lowers plasma insulin and blood pressure in spontaneously hypertensive rats. *Hypertension* **24,** 625–632.

Blakely, S. R., Mislo, B. L., Basi, N. S., and Pointer, R. H. (1995). Dietary fructose alters the insulin-like effects of dietary vanadate in adipocytes from rats. *Nutr. Res.* **15,** 25–35.

Blalock, T. L., and Hill, C. H. (1987). Studies on the role of iron in the reversal of vanadium toxicity in chicks. *Biol. Trace Elem. Res.* **14,** 225–235.

Boscolo, P., Carmignani, M., Volpe, A. R., Felaco, M., Delrosso, G., Porcelli, G., and Giuliano, G. (1994). Renal toxicity and arterial hypertension in rats chronically exposed to vanadate. *Occup. Environ. Med.* **51,** 500–503.

Boyd, D. W., and Kustin, K. (1984). Vanadium: A versatile biochemical effector with an elusive biological function. In C. L. Eichhorn and L. G. Marzilli (Eds.), *Advances in Inorganic Biochemistry*, Vol. 6. Elsevier, Amsterdam, pp. 311–363.

Bräunlich, H., Pfeiffer, R., Grau, P., and Reznik, L. (1989). Renal effects of vanadate in rats of different ages. *Biomed. Biochim. Acta* **48,** 569–575.

Breuch, M., Quintanilla, M. E., Legrum, W., Koch, J., Netter, K. J., and Fuhrmann, G. F. (1984). Effects of vanadate on intracellular reduction equivalents in mouse liver and the fate of vanadium in plasma, erythrocytes and liver. *Toxicology* **31,** 283–295.

Brichard, S. M., Assimacopoulos-Jeannet, F., and Jeanrenaud, B. (1992). Vanadate treatment markedly increases glucose utilization in muscle of insulin-resistant *fa/fa* rats without modifying glucose transporter expression. *Endocrinology* **131,** 311–317.

Brichard, S. M., Lederer, J., and Henquin, J. C. (1991). The insulin-like properties of vanadium: A curiosity or a perspective for the treatment of diabetes? Diabete Metab. **17,** 495–440.

Brichard, S. M., Pottier, H. M., and Henquin, J. C. (1989). Long-term improvement of glucose homeostasis by vanadate in obese hyperinsulinemic fa/fa rats. *Endocrinology* **125,** 2510–2516.

Butler, A. (1990). Coordination chemistry of vanadium in aqueous solution. In N. D. Chasteen (Ed.), *Vanadium in Biological Systems. Physiology and Biochemistry*, Chap. 2. Kluwer Academic Publishers, Boston, pp. 25–49.

Cam, M. C., Cros, G. H., Serrano, J. J., Lazaro, R., and McNeill, J. H. (1993a). In vivo antidiabetic actions of naglivan, an organic vanadyl compound in streptozotocin-diabetic rats. *Diab. Res. Clin. Pract.* **20,** 111–121.

Cam, M. C., Pederson, R. A., Brownsey, R. W., and McNeill, J. H. (1993b). Long-term effectiveness of oral vanadyl sulphate in streptozotocin-diabetic rats. *Diabetologia* **36,** 218–224.

Carlton, B. D., Beneke, M. B., and Fisher, G. L. (1982). Assessment of the teratogenicity of ammonium vanadate using Syrian golden hamsters. *Environ. Res.* **29,** 256–262.

Carmignani, M., Boscolo, P., Ripanti, G., Porcelli, G., and Volpe, A. R. (1993). Mechanisms of the vanadate-induced arterial hypertension only in part depend on the levels of exposure. In M. Anke, D. Meissner, and C. F. Mills (Eds.), *Trace Elements in Man and Animals* (TEMA 8). Proceedings of the Eighth International Symposium on Trace Elements in Man and Animals, Dresden, Germany, July, 1993. Verlag Media Touristik, Gersdorf.

Carmignani, M., Volpe, A. R., Masci, O., Boscolo, P., Di Giacomo, F., Grilli, A., Del Rosso, G., and Felaco, M. (1996). Vanadate as factor of cardiovascular regulation by interactions with the catecholamine and nitric oxide systems. *Biol. Trace Elem. Res.* **51,** 1–12.

Chakraborty, D., Battacharyya, A., Majumdar, K., and Chatterjee, G. C. (1977). Effects of chronic vanadium pentoxide administration on L-ascorbic acid metabolism in rats. Influence of L-ascorbic acid supplementation. *Int. J. Vit. Nutr. Res.* **47,** 81–87.

Cohen, N., Halberstam, M., Shlimovich, P., Chang, C. J., Shamoon, H., and Rossetti, L. (1995). Oral vanadyl sulfate improves hepatic and peripheral insulin sensitivity in patients with non-insulin-dependent diabetes mellitus. *J. Clin. Invest.* **95,** 2501–2509.

Curran, G. L., Azarnoff, D. L., and Bolinger, R. E. (1959). Effect of cholesterol synthesis inhibition in normocholesterolemic young men. *J. Clin. Invest.* **38,** 1251–1261.

Dafnis, E., Spohn, M., Lonis, B., Kurtzman, N. A., and Sabatini, S. (1992). Vanadate causes hypokalemic distal renal tubular acidosis. *Am. J. Physiol.* **262** (*Renal Fluid Electrolyte Physiol.* **31**), F449–F453.

Dai, S., and McNeill, J. H. (1994). One-year treatment of non-diabetic and STZ-diabetic rats with vanadyl sulphate did not alter blood pressure or haematological indices. *Pharmacol. Toxicol.* **74,** 110–115.

Dai, S., Thompson, K. H., and McNeill, J. H. (1994a). One-year treatment of streptozotocin-induced diabetic rats with vanadyl sulphate. *Pharmacol. Toxicol.* **74,** 101–109.

Dai, S., Thompson, K. H., Vera, E., and McNeill, J. H. (1994b). Toxicity studies on one-year treatment of non-diabetic and streptozotocin-diabetic rats with vanadyl sulphate. *Pharmacol. Toxicol.* **75,** 265–273.

Dai, S., Vera, E., and McNeill, J. H. (1995). Lack of haematological effect of oral vanadium treatment in rats. *Pharmacol. Toxicol.* **76,** 263–268.

Dai, S., Yuen, V. G., Orvig, C., and McNeill, J. H. (1993). Prevention of diabetes-induced pathology in STZ-diabetic rats by bis(maltolato)oxovanadium(IV). *Pharmacol. Commun.* **3,** 311–321.

Dimond, E. G., Caravaca, J., and Benchimol, A. (1963). Vanadium. Excretion, toxicity, lipid effect in man. *Am. J. Clin. Nutr.* **12,** 49–53.

Domingo, J. L., Gomez, M., Llobet, J. M., Corbella, J., and Keen, C. L. (1991a). Improvement of glucose homeostasis by oral vanadyl or vanadate treatment in diabetic rats is accompanied by negative side effects. *Pharmacol. Toxicol.* **68,** 249–253.

Domingo, J. L., Gomez, M., Llobet, J. M., Corbella, J., and Keen, C. L. (1991b). Oral vanadium administration to streptozotocin-diabetic rats has marked negative side-effects which are independent of the form of vanadium used. *Toxicology* **66,** 279–287.

Domingo, J. L., Gomez, M., Sanchez, D. J., Llobet, J. M., Corbella, J., and Keen, C. L. (1994). Normalization of hyperglycemia by vanadate or vanadyl treatment in diabetic rats: Pharmacological and toxicological aspects. *Trace Elem. Electr.* **11,** 16–22.

Domingo, J. L., Llobet, J. M., Gomez, M., Corbella, J., and Keen, C. L. (1990). Effects of oral vanadium administration in streptozotocin-diabetic rats. In P. Collery, L. A. Poinier, M. Manbait, and J. C. Etienne, (Eds.), *Metal Ions in Biology and Medicine.* John Libbey Eurotext, Paris, pp. 312–314.

Edel, J., and Sabbioni, E. (1989). Vanadium transport across placenta and milk of rats to the fetus and newborn in rats. *Biol. Trace Elem. Res.* **22,** 265–275.

Erdmann, E. (1980). Cardiac effects of vanadate. *Basic Res. Cardiol.* **75,** 411–412.

Ganguli, S., Reuland, S. P., Franklin, L. A., Deakins, D. D., Johnston, W. J., and Pasha, A. (1994a). Effects of maternal vanadate treatment on fetal development. *Life Sci.* **55,** 1267–1276.

Ganguli, S., Reuland, S. P., Franklin, L. A., and Tucker, M. (1994b). Effect of vanadate on reproductive efficiency in normal and streptozotocin-treated diabetic rats. *Metabolism* **43,** 1384–1388.

Goldfine, A. B., Simonson, D. C., Folli, F., Patti, M.-E., and Kahn, C. R. (1995). Metabolic effects of sodium metavanadate in humans with insulin-dependent and noninsulin-dependent diabetes mellitus in vivo and in vitro studies. *J. Clin. Endocrinol. Metab.* **80,** 3311–3320.

Gomez, M., Domingo, J. L., Llobet, J. M., and Corbella, J. (1991). Evaluation of the efficacy of various chelating agents on urinary excretion and tissue distribution of vanadium in rats. *Toxicol. Lett.* **57,** 227–234.

Gomez, M., Sanchez, D., Domingo, J. L., and Corbella, J. (1992). Embryotoxic and teratogenic effects of intraperitoneally administered metavanadate in mice. *J. Toxicol. Environ. Health* **37,** 47–56.

Hafez, Y., and Kratzer, F. H. (1976). The effect of dietary vanadium on fatty acid and cholesterol synthesis and turnover in the chick. *J. Nutr.* **106,** 249–257.

Haider, S. S., and El-Fakhri, M. (1991). Action of alpha-tocopherol on vanadium-stimulated lipid peroxidation in rat brain. *Neurotoxicology* **12,** 79–86.

Haider, S. S., and Kashyap, S. K. (1989). Vanadium intoxication inhibits sulfhydryl groups and glutathione in the rat brain. *Indust. Health* **27,** 23–25.

Hamada, T. (1994). A new experimental system of using fertile chick eggs to evaluate vanadium absorption and antidotal effectiveness to prevent vanadium uptake. *J. Nutr. Biochem.* **5,** 382–388.

Hansard, S. L., Ammerman, G. B., Fick, K. R., and Millar, S. M. (1978). Performance and vanadium content of tissues in sheep as influenced by dietary vanadium. *J. Anim. Sci.* **46,** 1091–1095.

Hansen, T. V., Aaseth, J., and Skaug, V. (1985). Hemolytic activity of vanadyl sulfate and sodium vanadate. *Acta Pharmacol. Toxicol.* **59,** 562–565.

Henquin, J. C., Carton, F., Ongemba, L. N., and Becker, D. J. (1994). Improvement of mild hypoinsulinaemic diabetes in the rat by low non-toxic doses of vanadate. *J. Endocrinol.* **142,** 555–561.

Heyliger, C. E., Tahiliani, A. G., and McNeill, J. H. (1985). Effect of vanadate on elevated blood glucose and depressed cardiac performance of diabetic rats. *Science* **227,** 1474–1476.

Hogan, G. R. (1990). Peripheral erythrocyte levels, hemolysis and three vanadium compounds. *Experientia* **46,** 444–446.

Hom, G. J., Chelly, J. E., and Jandhyala, B. S. (1982). Evidence for centrally mediated effects of vanadate on the blood pressure and heart rate in anesthetized dogs. *Proc. Soc. Exp. Biol. Med.* **169,** 401–405.

Hudson, T. G. F. (1964). *Vanadium: Toxicology and Biological Significance.* Elsevier Publishing Co., Amsterdam, p. 69.

Inciarte, D. J., Steffen, R. P., Dobbins, D. E., Swindall, B. T., Johnston, J, and Haddy, F. J. (1980). Cardiovascular effects of vanadate in the dog. *Am. J. Physiol.* **239,** H47–56.

Jandhyala, B. S., and Hom, G. J. (1983). Physiological and pharmacological properties of vanadium. *Life Sci.* **33,** 1325–1330.

Jensen, P. K., Christiansen, J. S., Steven, K., and Parving, H.-H. (1981). Renal function in streptozotocin-diabetic rats. *Diabetologia* **21**, 409–414.

Jones, M. M., and Basinger, M. A. (1983). Chelate antidotes for sodium vanadate and vanadyl sulfate intoxication in mice. *J. Toxicol. Environ. Health.* **12**, 749–756.

Kumar, A., and Corder, C. N. (1980). Diuretic and vasoconstrictor effects of sodium orthovanadate on the isolated perfused rat kidney. *J. Pharmacol. Exp. Ther.* **213**, 85–90.

Larsen, J. A., and Thomsen, O. O. (1980a). Vascular effects of vanadate. *Basic Res. Cardiol.* **75**, 428–432.

Larsen, J. A., and Thomsen, O. O. (1980b). Vanadate-induced oliguria and vasoconstriction in the cat. *Acta Physiol. Scand.* **110**, 367–374.

Llobet, J. M., and Domingo, J. L. (1984). Acute toxicity of vanadium compounds in rats and mice. *Toxicol. Lett.* **23**, 227–231.

Llobet, J. M., Colomina, M. T., Sirvent, J. J., Domingo, J. L., and Corbella, J. (1993). Reproductive toxicity evaluation of vanadium in male mice. *Toxicology* **80**, 199–206.

Lopez-Novoa, J. M., Mayol, V., and Martinez-Maldonado, M. (1982). Renal actions of orthovanadate in the dog. *Proc. Soc. Exp. Biol. Med.* **170**, 418–426.

Maines, M. D. (1994). Modulating factors that determine interindividual differences in response to metals. In W. Mertz, C. O. Abernathy, and S. S. Olin (Eds.), *Risk Assessment of Essential Elements.* ILSI Press, Washington, DC, pp. 21–39.

McNeill, J. H., Yuen, V. G., Hoveyda, H. R., and Orvig, C. (1992). Bis(maltolato)oxovanadium (IV) is a potent insulin mimic. *J. Med. Chem.* **35**, 1489–1491.

Meyerovitch, J., Farfel, Z., Sack, J., and Shechter, Y. (1987). Oral administration of vanadate normalizes blood glucose levels in streptozotocin-treated rats. *J. Biol. Chem.* **262**, 6658–6662.

Meyerovitch, J., Shechter, Y., and Amir, S. (1989). Vanadate stimulated in vivo glucose uptake in brain and arrests food intake and body weight gain in rats. *Physiol. Behavior* **45**, 1113–1116.

Mongold, J. J., Crose, G. H., Vian, L., Tep, A., Ramanadham, S., Siou, G., Diaz, J., McNeill, J. H., and Serrano, J. J. (1990). Toxicological aspects of vanadyl sulphate on diabetic rats: Effects on vanadium levels and pancreatic B-cell morphology. *Pharmacol Toxicol.* **67**, 192–198.

Nechay, B. R. (1984). Mechanisms of action of vanadium. *Annu. Rev. Pharmacol. Toxicol.* **24**, 501–524.

Nielsen, F. H. (1987). Vanadium. In W. Mertz (Ed.), *Trace Elements in Human and Animal Nutrition*, Vol. 1. 5th ed. Academic Press, San Diego.

Ortega, A., Llobet, J. M., Domingo, J. L., and Corbella, J. (1991). Lack of improvement of glucose homeostasis in STZ-diabetic rats after administration by gavage of metavanadate. *Trace Elem. Med. Care* **8**, 181–186.

Orvig, C., Thompson, K. H., Battell, M., and McNeill, J. H. (1995). Vanadium compounds as insulin mimics. *Metal Ions Biol. Syst.* **31**, 575–594.

Oster, M. H., Uriu-Hare, J. Y., Trapp, C. L., Stern, J. S., and Keen, C. L. (1994). Dietary macronutrient composition influences tissue trace element accumulation in diabetic Sprague-Dawley rats. *Proc. Soc. Exptl. Biol. Med.* **207**, 67–75.

Parker, R. D. R., and Sharma, R. P. (1978). Accumulation and depletion of vanadium in skeletal tissues of rats treated with vanadyl sulphate and sodium orthovanadate. *J. Environ. Pathol. Toxicol.* **2**, 235–245.

Paternain, J. L., Domingo, J. L., Gomez, M., Ortega, A., and Corbella, J. (1993). Developmental toxicity of vanadium in mice after oral administration. *J. Appl. Toxicol.* **10**, 181–186.

Phillips, T. D., Nechay, B. R., Neldon, S. L., Kubena, L. F., and Heidlbaugh, N. D. (1982). Vanadium-induced inhibition of renal $Na+,K+$-adenosinetriphosphatase in the chicken after chronic dietary exposure. *J. Toxicol. Environ. Health* **9**, 651–661.

Pugazhenthi, S., Angel, J. F., and Khandelwal, R. L. (1993). Effects of high sucrose diet on insulin-like effects of vanadate in diabetic rats. *Mol. Cell. Biochem.* **122**, 77–84.

Pugazhenthi, S., and Khandelwal, R. L. (1990). Insulinlike effects of vanadate on hepatic glycogen metabolism in nondiabetic and streptozocin-induced diabetic rats. *Diabetes* **39,** 821–827.

Ramanadham, S., Heyliger, C., Gresser, M. J., Tracey, A. S., and McNeill, J. H. (1991). The distribution and half-life for retention of vanadium in the organs of normal and diabetic rats orally fed vanadium (IV) and vanadium (V). *Biol. Trace Elem. Res.* **30,** 119–124.

Reaven, G. M. (1993). Role of insulin resistance in human disease (syndrome X): An expanded definition. *Annu. Rev. Med.* **44,** 121–131.

Rehder, D. (1991). The bioinorganic chemistry of vanadium. *Angew. Chem. Int. Ed. Engl.* **30,** 148–167.

Roshchin, A. V., Ordzhonikidze, E. K., and Shalganova, I. V. (1980). Vanadium—Toxicity, metabolism, carrier state. *J. Hyg. Epidemiol. Microbiol. Immunol.* **24,** 377–383.

Rossetti, L., and Laughlin, M. R. (1989). Correction of chronic hyperglycemia with vanadate, but not with phlorizin, normalizes in vivo glycogen repletion and in vitro glycogen synthase activity. *J. Clin Invest.* **84,** 892–899.

Russanov, E., Zaparowska, H., Ivancheva, E., Kirkova, M., and Konstantinova, S. (1994). Lipid peroxidation and antioxidant enzymes in vanadate-treated rats. *Comp. Biochem. Physiol. C—Pharmacol. Toxicol. Endocrinol.* **107,** 415–421.

Sakurai, H. 1994. Vanadium distribution in rats and DNA cleavage by vanadyl complexes: Implications for vanadium toxicity and biological effects. *Environ. Health Perspec.* **102,** 35–36.

Sakurai, H., Tsuchiya, K., Nukatsuka, M., Kawada, J., Ishikawa, S., Yoshida, H., and Komatsu, M. (1990). Insulin-mimetic action of vanadyl complexes. *J. Clin. Biochem. Nutr.* **8,** 193–200.

Sakurai, H., Fujii, K., Watanabe, H., and Tamura, H. (1995). Orally active and long-term acting insulin-mimetic vanadyl complexe: Bis(picolinato)oxovanadium(IV). *Biochem. Biophys. Res. Comm.* **214,** 1095–1101.

Sanchez, D., Ortega, A., Domingo, J. L., and Corbella, J. (1991). Developmental toxicity evaluation of orthovanadate in the mouse. *Biol. Trace Elem. Res.* **30,** 219–226.

Sandirasegarane, L., and Gopalakrishnan, V. (1995). Vanadate increases cytosolic free calcium in rat aortic smooth muscle cells. *Life Sci.* **56,** PL169–PL174.

Saxena, A. K., Srivastava, P., Kale, R. K., and Baquer, N. Z. (1992). Effect of vanadate administration on polyol pathway in diabetic rat kidney. *Biochem. Int.* **26,** 59–68.

Schroeder, H. A., Balassa, J. J., and Tipton, I. H. (1963). Abnormal trace metals in man—Vanadium. *J. Chron. Dis.* **16,** 1047–1071.

Shechter, Y. (1990). Insulin mimetic effects of vanadate: Possible implications for future treatment of diabetes. *Diabetes* **39,** 1–5.

Shechter, Y., Shisheva, A., Lazar, R., Libman, J., and Shanzer, A. (1992). Hydrophobic carriers of vanadyl ions augment the insulinomimetic actions of vanadyl ions in rat adipocytes. *Biochemistry* **31,** 2063–2068.

Somerville, J., and Davies, B. (1962). Effect of vanadium on serum cholesterol. *Am. Heart J.* **64,** 54–56.

Srivastava, P., Saxena, A. K., Kale, R. K., and Baquer, N. Z. (1993). Insulin like effects of lithium and vanadate on the altered antioxidant status of diabetic rats. *Res. Commun. Chem. Pathol. Pharmacol.* **80,** 283–293.

Steffan, R. P., Pamnani, M. B., Clough, D. L., Huot, S. J., Muldoon, S. M., and Haddy, F. J. (1981). Effect of prolonged dietary administration of vanadate on blood pressure in the rat. *Hypertension* **3,** 1173–1178.

Strasia, C. A. (1971). Vanadium: Essentiality and Toxicity in the Laboratory Rat. University of Michigan, Ann Arbor. Doctoral dissertation.

Thompson, K. H., Leichter, J., and McNeill, J. H. (1993). Studies of vanadyl sulfate as a glucose-lowering agent in STZ-diabetic rats. *Biochem. Biophys. Res. Commun.* **197,** 1549–1555.

Thompson, K. H., and McNeill, J. H. (1993). Effect of vanadyl sulfate feeding on susceptibility to peroxidative change in diabetic rats. *Res. Commun. Chem. Pathol. Pharmacol.* **80,** 187–200.

Tolman, E. L., Barris, E., Burns, M., Pansini, A., and Partridge, R. (1979). Effects of vanadium on glucose metabolism in vitro. *Life Sci.* **25,** 1159–1164.

Ugazio, G., Bosia, S., Burdino, E., and Grignolo, F. (1994). Amelioration of diabetes and cataract by Na$_3$VO$_4$ plus U-83836E in streptozotocin treated rats. *Res. Commun. Mol. Pathol. Pharmacol.* **85,** 313–328.

Underwood, E. J. (1977). *Trace Elements in Human and Animal Nutrition.* 4th ed. Academic Press, New York.

Venugopal, B., and Luckey, T. D. (1978). *Metal Toxicity in Mammals.* Vol. 2. of *Chemical Toxicity of Metals and Metalloids.* Plenum Press, New York.

Watanabe, H., Nakai, M., Komazawa, K., and Sakurai, H. 1994. A new orally active insulin-mimetic vanadyl complex: Bis(pyrrolidine-N-carbodithioato)-oxovanadium(IV). *J. Med. Chem.* **37,** 876–877.

Waters, M. D. (1977). Toxicology of vanadium. *Adv. Mod. Toxicol.* **2,** 147–189.

Wei, C-I., Al Bayati, M. A., Culbertson, M. R., Rosenblatt, L. S., and Hansen, L. D. (1982). Acute toxicity of ammonium metavanadate in mice. *J. Toxicol. Environ. Health* **10,** 673–687.

Werden, K., Bauriedel, G., Fischer, B., Krawietz, W., and Erdmann, E. (1982). Stimulatory (insulin-mimetic) and inhibitory (ouabain-like) action of vanadate on potassium uptake and cellular sodium and potassium in heart cells in culture. *Biochim. Biophys. Acta* **646,** 261–267.

Wide, M. (1984). Effect of short-term exposure to five industrial metals on the embryonic and fetal development of the mouse. *Environ. Res.* **33,** 47–53.

Willsky, G. R. (1990). Vanadium in the biosphere. In N. D. Chasteen (Ed.), *Vanadium in Biological Systems. Physiology and Biochemistry,* Chap. 1. Kluwer Academic Publishers, Boston, pp. 1–24.

Wohaieb, S. A., and Godin, D. V. (1987). Alterations in free radical tissue-defense mechanisms in streptozotocin-induced diabetes in rat. Effects of insulin treatment. *Diabetes* **36,** 1014–1018.

Yamaguchi, M., Oishi, H., and Suketa, Y. (1989). Effects of vanadium on bone metabolism in weanling rats: Zinc prevents the toxic effect of vanadium. *Res. Exp. Med.* **189,** 47–53.

Yuen, V. G., Orvig, C., and McNeill, J. H. (1997). Glucose-lowering effects of bis(maltolato)oxovanadium(IV) are distinct from food restriction in STZ-diabetic rats. *Am. J. Physiol.* **272** (*Endocrinol. Metab.* **35**), E30–E35.

Yuen, V. G., Orvig, C., Thompson, K. H., and McNeill, J. H. (1993). Improvement in cardiac dysfunction in streptozotocin-induced diabetic rats following chronic oral administration of bix(maltolato)oxovanadium(IV). *Can. J. Physiol. Pharmacol.* **71,** 270–276.

Zaparowska, H. (1994). Effect of vanadium on L-ascorbic acid concentration in rat tissues. *Gen. Pharmacol.* **25,** 467–470.

Zaparowska, H., and Wasilewski, W. (1989). Some selected peripheral blood and haemopoietic system indices in Wistar rats with chronic vanadium intoxication. *Comp. Biochem. Physiol.* **93C,** 175–180.

Zaparowska, H., and Wasilewski, W. (1991). Significance of reduced food and water consumption in rats intoxicated with vanadium. *Comp. Biochem. Physiol.* **99C,** 349–352.

Zaparowska, H., and Wasilewski, W. (1992). Haematological results of vanadium intoxication in Wistar rats. *Comp. Biochem. Physiol.* **101C,** 57–61.

Zaparowska, H., Wasilewski, W., and Slotwinska, M. (1993). Effects of chronic vanadium administration in drinking water to rats. *BioMetals* **6,** 3–10.

3

MUTAGENICITY, CARCINOGENICITY, AND TERATOGENICITY OF VANADIUM

A. Léonard and G. B. Gerber

Teratogenicity and Mutagenicity Unit, Catholic University of Louvain, Avenue E. Mounier 72, B-1200 Brussels, Belgium

Vanadium in the Environment. Part 2: Health Effects, Edited by Jerome O. Nriagu.
ISBN 0-471-17776-8. © 1998 John Wiley & Sons, Inc.

1. INTRODUCTION

Assessment of protection against possible negative effects of an agent must consider those consequences that can ensue at low doses at full severity without a threshold—that is, the "stochastic risks" of cancer and hereditary damage as well as "deterministic" risks, such as damage to the developing organism, that arise only above a threshold exposure, usually with a severity that is dependent on dose. To protect against deterministic risk, it suffices to keep exposure well below the threshold dose, whereas protection against stochastic risks must consider whether a risk from a given exposure level is to be deemed acceptable.

Only epidemiological data can furnish an irreproachable proof that a substance causes cancer, results in hereditary damage to the progeny, or is deleterious to the child in utero. However, such human data are scarce for most agents with respect to cancer and teratogenic effects and are absent with respect to hereditary damage. Moreover, people in an occupational setting are frequently exposed to more than one agent, for example, during smelting, so that the imputation of cause to a given agent or to a combination thereof usually remains in doubt.

Studies on experimental animals under controlled conditions represent the second-best choice for evaluating such risks but they are expensive and time-consuming and may not reflect the situation under which people are exposed at work or at leisure. It must also be understood that neither epidemiology nor animal experiments can definitely state that a substance would not be carcinogenic or embryotoxic-teratogenic under any conditions.

To overcome some of these difficulties, an ample spectrum of short-term assays has been developed to evaluate the mutagenic/carcinogenic potential of chemical compounds (Hollstein et al., 1973; World Health Organization, 1980; Kirkland, 1990). These tests utilize biochemical changes and mutational responses of prokaryotic, yeast, plants, or mammalian cells and are based on the generally accepted assumption that damage to DNA is, in general, responsible for the initial transformation of a normal to a malignant cell. However, such tests will fail for agents that cause cancer disease by promoting an already transformed cell or by modifying the progression of a tumor.

Some short-term assays have also been proposed for the purpose of evaluating in vitro the teratogenic potential of chemicals (Flint and Horten, 1984; Flint et al., 1984). Experience with these tests is still too limited to make them useful for prediction of embryotoxic-teratogenic risks under routine conditions. Obviously, such tests also fail to take account of the ability of a substance to pass the placental barrier.

2. MUTAGENICITY OF VANADIUM

2.1. Aneuploidy and Mitotic Abnormalities Caused by Vanadium

Cell division requires that the genetic material be perfectly distributed among the daughter cells. The formation and action of the mitotic spindle apparatus that assures this distribution is governed by a complicated interplay of enzymatic reactions and regulations. Faults in this procedure might result in diseases with abnormal chromosome numbers, such as Down's syndrome. However, aneuploidy appears usually rather late in the development of malignant tumors. Obviously, such effects on cell division would be of deterministic nature and not occur unless a threshold exposure is exceeded. Consequently, mitotic poisons cannot be considered as true mutagens because they do not act directly upon the hereditary material, although they may contribute to hereditary damage as do true mutagens. From the substantial body of information available on metal compounds, it appears that the cytologically detectable consequences of exposure of eukaryotic cells to metal salts are primarily such aneugenic effects. Aneugenic properties and carcinogenic potential appear, however, not to be obviously correlated, since many metal salts have such properties, although they are not considered to be carcinogenic (Léonard et al., 1984).

Vanadium salts have been shown (Cantley and Aisen, 1979; Chasteen, 1983; Jandhyala and Hom, 1983; Nechay, 1984; Dafnis and Sabatini, 1994; Hei et al., 1995; Stankiewicz and Tracey, 1995) to interfere with an important spectrum of enzymatic systems such as Na^+, K^+-ATPase, Ca^{2+}-ATPase, H^+-K^+-ATPase, H^+-ATPase, K^+-ATPase, actomyosin ATPase, ribonuclease, and several phosphatases and specific protein kinases. Some of these effects could explain why vanadium salts are cytotoxic and how they could modify (inhibit or activate) gene expression (Klarlund, 1985; Bosch et al., 1990; Parfett and Pilon, 1995). More specifically, vanadium salts in micromolar concentrations inhibit (Ca- and Mg-requiring) dynein-ATPases and thus affect the motility of demembraned sea urchins, cilia of sea urchin embryos, flagella from pig sperm, and bracken fern, as well as chromosome movement in lysed PtK1 cells (Cande and Wolniak, 1978; Gibbons et al., 1978; Kobayashi et al., 1978). In vitro studies on hamster trachea demonstrated that $VOSO_4$ and V_2O_5 inhibit mucociliary function, a vital clearance mechanism of the respiratory tract (Schiff and Graham, 1984).

More specifically related to mitotic disturbances and spindle formation, it has been observed that vanadyl sulfate inhibits microtubule polymerization (Hantson et al., 1996) in the tubulin in vitro assay using material isolated from pig brain (Shelanski et al., 1973). At a concentration of 5 mM of vanadyl sulfate, the polymerization of microtubules was completely inhibited; at 1 and 5 mM, inhibition amounted to 39% and 73%, respectively. Moreover, $VOSO_4$ produces polyploid cells in *Saccharomyces cerevisiae* (Sora et al., 1986), and high concentrations of pentavalent vanadium cause pyknosis, loss of chromatin, and damage to the spindle of *Allium cepa* cells (Singh, cited in Sharma and Talukder, 1987). Cytokinesis is inhibited in the meristem of *Allium cepa* root tips after a 4-h incubation with Na_3VO_3 (Navas et al., 1986). Similar results were observed in mammalian cells after an in vitro exposure to vanadium salts (Bracken and Sharma, 1985; Sharma and Talukder, 1987; Roldán and Altamiro, 1990). Vanadate (10–100 μM) inhibits the anaphase movement of chromosomes and the elongation of the spindle in lysed PtK1 cells. While studying the ability of 2.5, 5.1, 20, 40, 80, or 160 μM of $NaVO_3$, NH_4VO_3, SVO_3, and Na_3VO_4 to produce numerical chromosome aberrations in vitro in human lymphocytes, Migliore et al. (1993, 1995) observed a significant increase in micronuclei, more than 68% of which contained centrosomes as shown by an alphoid centromere-specific probe confirming the ability of vanadium salts to interfere with spindle formation. Treatment of Chinese hamster V79 cells with vanadium pentoxide increased kinetochore-positive micronucleated cells in a dose-dependent manner (Zhong et al., 1994).

In vivo application of vanadium pentoxide by intraperitoneal injection (0.17, 2.13, or 6.4 mg/kg during five consecutive days), subcutaneous injection (0.24–8 mg/kg), or inhalation (1.44–11.3 mg/kg) caused an increase of micronuclei in mouse bone marrow cells (Sun, 1987), but oral administration (1.44–11.3 mg/kg) gave negative results, probably owing to the poor intestinal absorption of this compound. Vanadium salts also produced numerical chromosome aberrations (Ciranni et al., 1995) when male CD1 mice were given a single intragastric dose of 75 mg/kg sodium orthovanadate, 100 mg/kg vanadyl sulfate, or 50 mg/kg ammonium metavanadate dissolved in sterile water. The mice were sacrificed at different time intervals after treatment, and the incidence of micronulei and of numerical chromosome aberrations detected in second metaphase were scored in bone marrow cells.

2.2. Malignant Transformation of Cells Caused by Vanadium

The abnormal growth characteristics of certain connective tissue cell lines cultured on plates in the presence of potentially carcinogenic agents have been widely used to assess the carcinogenic potential of a wide range of agents.

Vanadate appears to produce such cell transformation (Klarlund, 1985; Parfett and Pilon, 1995), but the large majority of inorganic salts evaluated causes a similar positive response regardless of their known carcinogenic potential (Léonard et al., 1984). Thus, as Brookes (1991) stated, the relationship between in vitro transformation and malignancy remains uncertain; and it remains to be demonstrated that an observed change in morphology and growth characteristics of already immortal cultured cells is equivalent to a transformed genotype.

2.3. DNA Damage Caused by Vanadium

Repairable damage to DNA caused by an agent can be tested by the rec assay, which evaluates the difference in survival after treatment with the agent between wild-repair-competent (Rec^+) and repair-incompetent (Rec^-) strains of *Bacillus subtilis*. The sensitivity of the latter strains is increased by genetic recombination. Thereby, it is assumed that any enhanced lethality is due to increased DNA damage. However, this test, which was initially proposed by Japanese scientists for the assessment of the genotoxicity of chemicals, is of doubtful significance with respect to mutagenic risks. Vanadium salts $VOCl_2$

(0.4 M), V_2O_5 (0.5 M), NH_4VO_3 (0.5 M) were slightly positive in the rec assay on H17 (Rec^+, arg^-, try^-) strains of *Bacillus subtilis* (Kada et al., 1980; Kanematsu et al., 1980).

It is evident, however, that, in vitro, V^{4+} can interact with nucleotides (Baran, 1995), modify the synthesis and repair of DNA, and under certain conditions stimulate mitogenic activity (Hori and Oka, 1980; Carpenter, 1981; Ramanadham and Kern, 1983; Smith, 1983; Marini et al., 1987). It can also produce DNA strand breaks (Birnboin, 1978). Sodium metavanadate ($NaVO_3$), vanadyl sulfate (VOS_4), and vanadium pentoxide (V_2O_5), instilled into the trachea of rats, have been shown to cause an increase in the expression of certain mRNAs (Pierce et al., 1996). Using single-cell gel electrophoresis (the Comet assay) to detect DNA damage expressed as DNA strand breaks and alkali-labile sites Rojas et al. (1996) found that vanadium pentoxide (0.3, 30, and 3,000 μM) produced a dose-dependent increase in DNA migration— corresponding to degraded DNA—in fresh whole blood lymphocytes. In cultured human lymphocytes such a significant effect could, however, be detected only at the highest concentration tested (3,000 μM). When the cells were kept for 48–90 h before electrophoresis, migration of DNA had returned again to control values. Sakurai (Sakurai, 1994; Sakurai et al., 1995) postulates that the cleavage of DNA molecules may have been produced by hydroxy radicals ($\cdot OH$) generated by hydrogen peroxydes reacting with vanadlyl ions.

2.4. Mutations Caused by Vanadium

Several tests with various bacterial strains have been designed to detect mutations caused by chemical compounds on the basis of a reversion of specific point mutations (i.e., substitution of a base pair) or of frame shifts in the DNA chain (i.e., intercalation or loss of a base). According to the ICPEMC (International Committee for Protection against Environmental Mutagens and Carcinogens) report (Purchase, 1982), these assays are the only ones that meet the strict criteria of an established predictive test for carcinogenicity. In general the method was found to be very sensitive and, in many cases, to correlate positively with the carcinogenic potential of the substances tested. It is mainly on the basis of negative results with these assays that cancer produced by exposure to certain metals has been postulated to occur via nongenotoxic mechanisms. Although vanadium ions appear to be able to damage DNA, they do not cause frame shift mutations in the *Salmonella* TA98 strain or base-change mutations in the TA 1535 and TA 100 *Salmonella* or in the WP 2 or WP2 her *E. coli* strains (Kanematsu and Kada, 1978; Kada et al., 1980).

Recently, however, it has been reported that several mutagenic substances do not cause reverse mutations in prokaryotic organisms although they produce forward mutations in eukaryotic cells. Indeed, mechanisms that cause forward mutations reflect a much broader spectrum of events, such as unrepaired or unrepairable insertions, deletions, rearrangements, or point mutations than do those that can be detected by the reversion assays. It is not surprising, therefore, that ammonium vanadate could produce gene conversion and point mutations in the D7 strain of *Saccharomyces cerevisiae,* an effect that was reduced when the cells were treated with the hepatic S9 fraction or when logarithmically growing cells with a high cytochrome P450 content were utilized (Bronzetti et al., 1990; Galli et al., 1991). In contrast, vanadium pentoxide at concentrations of 1–4 μg/ml failed to produce 6-guanine-resistant mutations in V79 Chinese hamster lung fibroblast cell lines (Zhong et al., 1994). Ammonium metavanadate slightly increased the forward mutation rate at the HPRT locus in V79 cells in serum-free medium (less in glucose salt medium) and of the bacterial GPT (xanthine guanine phosphoribosyl transferase) gene in transgenic G12 cells derived from V79 cells (Cohen et al., 1992) The latter assay is more sensitive than the former and detects not only point mutations but also deletions. All these positive effects were much less pronounced for metavanadate than for chromate ions and may result from unrepaired DNA-protein crosslinks caused by these metals. Indeed, these authors observed such crosslinks in Chinese hamster ovary cells and in human MOLT4 cells. However, it should be pointed out that, in a recent article, the same group (Klein et al., 1994), using the G12 and G10 hprt$^-$/gpt$^+$ transgenic cell lines, could not confirm their previous, slightly positive, results (Cohen et al., 1992) obtained with ammonium metavanadate although the gpt gene integrated in G19 displays, in general, a more marked mutagenic response than when it is integrated in the transgenic G12 cell.

2.5.   Clastogenic Properties of Vanadium

Typically, breakage of chromosomes with the formation of chromosome abnormalities has been studied in cultured human lymphocytes. Whereas, as discussed above, the ability of vanadium to cause aneuploidy seems well established, clastogenic properties are weak if they exist at all.

Negative results were reported by Paton and Allison (1972) after treatment of human leukocytes with unspecified concentrations of $NaVO_3$ or Na_3VO_4 and by Sun (1987) after treatment with vanadium pentoxide. Another study (Roldán and Altamiro, 1990) also failed to observe structural chromosome aberrations and sister chromatid exchanges in lymphocytes treated with 2.4 or 6 μg/ml of vanadium pentoxide, although this treatment reduced the mitotic index and caused some satellite associations. Migliore et al. (1993, 1995) found that human lymphocytes exposed in vitro to $NaVO_3$, NH_4VO_3, SVO_5, or Na_3VO_4, did not display structural chromosome aberrations but showed a small increase in the number of sister chromatid exchanges (SCEs). However, using cultures of Chinese hamster cells to which 12–18 μg/ml of V_2O_3, 6–24 μg/ml of $VOSO_4$, or 4–16 μg/ml of NH_4VO_3 had been added, Owusu-Yaw et al. (1990) observed a dose-dependent increase in structural chromosome aberrations in the presence as well as in the absence of a S9 fraction. A small increase in sister chromatid exchanges was also seen, but this seemed not to depend on dose at concentrations of 0.5–4 μg/ml of V_2O_3, 0.5–6 μg/ml of $VOSO_4$, and 0.1–1 μg/ml of NH_4VO_3.

A few in vivo experiments also yielded conflicting results. Rats chronically force-fed with 4 mg/kg of vanadium pentoxide for 21 days showed no structural chromosome aberrations in their bone marrow cells, but displayed a small decrease in mitotic index (Giri et al., 1979). Male mice injected daily subcutaneously with vanadium pentoxide (0.2, 1, or 4 μg/kg) for 3 months and subsequently mated with normal females did not transmit dominant lethality before birth (Sun, 1987); such an effect would have reflected major chromosomal damage. On the other hand, Ciranni et al. (1995) observed an increase in structural chromosome aberrations after a single intragastric application to CD-1 mice of the rather large dose of 100 mg/kg of SVO_5 but not after 75 mg/kg of Na_3VO_4 or 50 mg/kg NH_4VO_3.

3. CARCINOGENICITY OF VANADIUM COMPOUNDS

Only a few metals have been clearly demonstrated to cause cancer in humans; Be, As, Cr, and Ni cause pulmonary neoplasms after inhalation. Arsenic also produces cancer of the skin. Carcinogenic properties have also been attributed to cadmium and cobalt, etc. but with insufficient evidence. Some other metals (e.g., lead) can elicit cancer in animals after high-dose local administration, probably as a consequence of stimulated cell replacement after extensive necrosis from the cytotoxic action of these materials.

We are not aware of any studies dealing with cancer in humans following vanadium exposure. Consequently, one has to rely on the results of mutagenic tests and some scanty and unsatisfactory animal experiments. In this context, the guidelines for evaluating carcinogenic properties of an agent suggested by different authorities, among them the OECD (OECD, 1981–1985), should be considered:

- Each dose group and concurrent control groups should consist of at least 50 animals of each sex.
- The design and conduct of the exposure should allow determination of dose relationships. Therefore, at least three dose levels should be used in addition to the control group: The uppermost dose should be sufficiently high to elicit signs of minimal toxicity without substantially altering the normal life span because of effects due to other causes than tumors; the lowest dose should not interfere with normal growth, development, and longevity of the animals.
- The experiment should cover the majority of the expected life span of the strain studied; in general, the study should not be terminated before 18 months for mice and 24 months for rats.

- The exposure should be initiated as soon as possible after weaning and before the animals are 6 weeks old.
- The type of tumor should be ascertained by histological analysis.

Only one of the studies on cancer induction by vanadium compounds reported in the literature and discussed below fulfills these criteria.

No studies appear to specifically address cancer after oral exposure to vanadium compounds. The extensive studies in Schroeder's laboratory (Schroeder et al., 1963, 1970; Schroeder and Balassa, 1967) were designed primarily to test the general toxicity of metal salts and appear inadequate for evaluating the carcinogenicity of vanadium, because the number of animals was small, no histological tests were performed, and it was not ascertained whether the maximum tolerated dose had been given. It should be mentioned, nevertheless, that these studies did not indicate any increase in tumor frequency in rats or mice given vanadium sulfate ($VOSO_4$) at 5 μg/ml drinking water.

No information seems to be available on exposure to vanadium via the skin.

Intraperitoneal injection of vanadium 2,4-pentanedione at doses of 24, 60, or 120 mg/kg did not significantly increase the incidence of lung adenomas in mice (Stoner et al., 1976). Pellets of solid vanadium solids from a leaching plant (also containing a small amount of chromate) implanted into the lower left bronchus of rats of 8–10 weeks of age produced inflammatory changes in about half of the animals but did not increase the incidence of bronchial carcinoma in a statistically significant manner (Levy et al., 1986). This careful study, consisting of about 100 animals in each of the 23 groups studied, showed a substantial increase in lung cancer when given certain samples containing strontium chromate or zinc chromate.

Some experiments suggest that some vanadium compounds may have an antitumor activity (Djordjevic, 1995). Thus, Thompson et al. (1984) injected 50 female rats at an age of 60 days with the potent carcinogen 1-methyl-1-nitrosourea (MNU, 50 mg/kg in acidified saline). Ten other rats received only saline. Seven days later, 25 of the MNU-treated rats and 5 of the control rats were given a diet supplemented with $VOSO_4 \cdot 3H_2O$ (initially 15 ppm; after 28 days raised to 25 ppm). Both cancer incidence and average number of cancers per rate were claimed to be reduced in the vanadium-treated rats. However, the experiment was terminated within 180 days after the MNU treatment, and the number of animals per group used was small. In our opinion, the most that might be concluded is that vanadium treatment retarded the appearance of the tumors.

In another study (Kingsnorth et al., 1986) 1,2 dimethylhydrazine (DMH, 20 mg/kg) was injected subcutaneously to CD-1 mice weekly for a period of 20 weeks to induce large-bowel neoplasms (carcinomas as well as adenomas). Some animals received ammonium vanadate (10 or 20 mg/L) in the drinking water. Vanadate increased thymidine incorporation and reduced RNA content but did not affect incidence or type of tumor induced by DMH. Again the number of animals was small: 16–23 per group.

A recent study (Bishayee and Chatterjee, 1995) also reported an increase in hepatoma incidence when 0.5 ppm of ammonium vanadate was added to the drinking water of rats in which hepatocarcinogenesis had been initiated by a single intraperitoneal injection of diethylnitrosamine (200 mg/kg) followed by promotion with phenobarbital (0.05%) in the diet. These results remain doubtful, however, since the study was terminated after a period of observation of 20 weeks, and the number of animals per group was extremely small (7–12!).

These in vivo studies on an antitumor effect of vanadium compounds in laboratory animals have been complemented with some in vitro and in vivo experiments with transplantable tumors. Vanadocene dichloride ($V(C_5H_5)_2Cl_2$), like many other neutral metallocene complexes and several vanadate compounds, has been reported to inhibit the growth of Ehrlich ascites cells and sarcoma 180 cells in fluid or solid culture as well as that of B16 melanoma, Lewis lung carcinoma, transplantable murine lymphoma and so forth (Köpf-Maier and Köpf, 1979, 1994; Hanauske et al., 1987; Sardar et al., 1993)

4. TERATOGENICITY OF VANADIUM COMPOUNDS

The organism developing in utero can be harmed either as a consequence of toxic effects to the mother or a teratogenic action directed to the embryo/fetus. In addition, embryonic death and malformations may result from genetic damage to parental germ cells (see Section 2.4). Toxicity to the mother can cause the death of the embryo/fetus as well as retardation in its development, including delayed ossification. Such effects may occur even when the agent does not cross the placental barrier. In contrast, an agent producing true teratogenesis must directly reach the embryo/fetus. Such malformations require damage to more than one cell at critical stages of development; like maternal toxicity, they will not appear below a threshold dose (Beckman and Brent, 1984; Brent, 1986). Thus, the teratogenic potential of a chemical is related to cytotoxic actions and not to mutagenesis and cannot be evaluated from mutagenicity assays.

Vanadium can cross the placental barrier (Edel and Sabbioni, 1989; Paternain et al., 1990) but appears to accumulate in the fetal membranes rather than in the fetus itself (Roshchin et al., 1980; Hackett and Kelman, 1983; World Health Organization, 1988).

Probably, most consequences of administration of vanadium to pregnant females (increased rates of resorbed and dead fetuses, reduction of fetal weight) result from toxicity to the mother; some of the teratogenic effects seen at high doses might be related to the ability of vanadate to interfere with chromosome distribution.

Paternain et al. (1987) observed a significant increase in the number of resorbed and dead fetuses after oral administration of 20 mg/kg/d of sodium metavanadate ($NaVO_3$) to pregnant rats during organogenesis. Similar results were obtained by Gomez et al. (1992) and Bosque et al. (1993) with Swiss mice injected intraperitoneally with doses of 2, 4, or 8 mg/kg/d of $NaVO_3$ on days 6–15 of gestation. All these doses were also toxic to the mothers, as shown by the fact that they gained less weight during pregnancy than the controls. Pre- and postimplantation loss increased and fetal body weight was reduced after 4 and 8 mg/kg/d, and significantly more animals with cleft palates were detected after 8 mg/kg/d. The lowest-observed adverse level for maternal toxicity and that for developmental damage were both 2 mg/kg/d, whereas teratogenic effects appeared only after the highest dose, 8 mg/kg/d. A reduction of the number of live fetuses was also seen (Ganguli et al., 1994) when sodium orthovanadate (0.25 mg/ml and 0.50 mg/ml in 0.5 N saline) was added to the drinking water of rats during days 10–20 of pregnancy. In mice given sodium orthovanadate by gavage on gestational days 6–15, no adverse effect was observed as to maternal toxicity at 7.5 mg/kg; the level at which developmental toxicity was observed was 15 mg/kg (Sanchez et al., 1991).

Intraperitoneal injection of Syrian golden hamster with 0.47, 1.88, and 3.75 mg/kg body weight of ammonium metavanadate (NH_4VO_3) on days 5–10 of gestation (Carlton et al., 1982) resulted in a significant increase in the number of dead and resorbed embryos in the group that had received 1.88 mg/kg. Minor skeletal abnormalities, such as micrognathia, supernumerary ribs, and altered ossification of the sternum, were significantly increased in all groups, whereas a small increase seen in external abnormalities was not statistically significant. As the authors state, the small number of malformed offspring and the lack of a clear-cut dose-effect relationship did not allow a definite assessment of the teratogenicity of ammonium vanadate.

Very high doses of vanadyl sulfate pentahydrate ($VOSO_4 \cdot 5H_2O$) (150 mg/kg) given by gavage to pregnant mice reduced the weight gain of the mother during pregnancy but did not alter the number of implants and fetuses, although it diminished fetal weight and caused some early resorptions of embryos as well as a few malformations. No effect was seen after such an administration of 37.5 mg/kg/d (Paternain et al., 1990).

Injection of 0.15 ml of a 1 mM solution (27 μg, corresponding to about 0.8 mg/kg1) of V_2O_5 into the tail vein of NMRI mice at day 3 of pregnancy allowed implantation and subsequent development to proceed normally (Wide, 1984). The same treatment given on day 8 of pregnancy caused a delayed ossification of the supraoccipital bone, sternum, metatarsals, and caudal vertebrae in 71% and a broken spinal cord in 9% of the fetuses examined on day 17 of pregnancy.

In an article summarized by WHO in 1988, Sun (1987) reported the results of a study performed in China on pregnant rats given V_2O_5 via different routes. After daily subcutaneous injection of 0.5, 1, and 4 mg/kg of vanadium pentoxide for 10 days from day 7 to day 16, the incidence of resorbed and dead fetuses increased from 3.5% in the controls to 17% for 1 mg/kg and 27% for 4 mg/kg. In the latter group, wavy ribs were observed in more than 50% of the fetuses. When 0.3, 1, and 3 mg/kg of vanadium pentoxide were injected intraperitoneally, the yield of resorbed and dead fetuses was substantially greater than after oral administration. The type and incidence of skeletal abnormalities were similar after 0.3 mg/kg given intraperitoneally and after 9 mg/kg given orally. It should be mentioned in this context that the embryotoxic and teratogenic effects of vanadium compounds can be significantly reduced by an administration of the chelator Tiron (sodium 4,5-dihydroxybenzene-1,3-disulfonate) (Domingo et al., 1993; Domingo, 1995).

5. CONCLUSIONS

A review of the literature indicates that vanadium salts possess only a weak mutagenic activity. The most striking effect of vanadium salts is their interference with chromosome distribution by way of an inhibition of the polymerization of the microtubules; similar comparable effects have been observed, however, for most metal salts and apparently do not reflect a carcinogenic potential. The small number of in vivo studies also do not support the claim that vanadium salts are carcinogenic, although this cannot be fully excluded because of the mitogenic activity of vanadium (Ames and Gold, 1990). Vanadium at high concentrations can certainly harm the organism developing in utero, but it appears to do so mainly by way of damage to the mother. Because of the poor transfer of vanadium salts into the fetus, malformations appear only at very high doses.

REFERENCES

Ames, B. N., and Gold, L. S. (1990). Too many rodent carcinogens: Mitogenesis increases mutagenesis. *Science* **249,** 970–971.

Baran, E. J. (1995) Vanadyl(IV) complexes of nucleotides. In H. Sigel and A. Sigel (Eds.), *Metal Ions in Biological Systems: Vanadium and Its Role in Life,* Vol. 31. Marcel Dekker, New York, Basel, Hong Kong, pp. 129–146.

Beckman, D. A., and Brent, R. L. (1984). Mechanisms of teratogenesis. *Annu. Rev. Pharmacol. Toxicol.* **24,** 483–500.

Birnboin, H. C. (1978). A superoxide anion induced DNA strand-breaks metabolic pathway in human leukocytes. Effect of vanadate. *J. Nutr.* **112,** 2279–2285.

Bishayee, A., and Chatterjee, M. (1995). Inhibitory effect of vanadium on rat carcinogenesis initiated by diethylnitrosamine and promoted by phenobarbital. *Br. J. Cancer* **71,** 1214–1220.

Bosch, F., Hatzoglou, M., Park, E. A., and Hanson, R. W. (1990). Vanadate inhibits expression of the gene for phosphoenolpyruvate carboxykinase (GTP) in rat hepatoma cells. *J. Biol. Chem.* **265,** 13677–13682.

Bosque, M. A., Domingo, J. L., Llobet, J. M., and Corbella, J. (1993). Variability in the embryotoxicity and fetotoxicity of vanadate with the day of exposure. *Vet. Hum. Toxicol.* **335,** 1–3.

Bracken, W. M., and Sharma, R. P. (1985). Cytotoxicity-related alterations of selected cellular functions after in vivo vanadate exposure. *Biochem. Pharmacol.* **34,** 2465–2470.

Brent, R. L. (1986). Definition of teratogen and the relationship of teratogenicity to carcinogenicity. *Teratology* **34,** 359–360.

Bronzetti, G., Morichetti, E., Della Croce, C., Del Carratore, R., Giromini, L., and Galli, A. (1990). Vanadium: Genetical and biochemical investigations. *Mutagenesis* **5,** 293–295.

Brookes, P. (1991). Critical assessment of the value of in vitro transformation for predicting in vivo carcinogenicity of chemicals. *Mutat. Res.* **86,** 233–242.

Cande, W. Z., and Wolniak, S. M. (1978). Chromosome movement in lysed mitotic cells inhibited by vanadate. *J. Cell Biol.* **79,** 573–580.

Cantley, L. C., and Aisen, P. (1979). The fate of cytoplasmic vanadium. *J. Biol. Chem.* **254,** 1781–1784.

Carlton, B. D., Beneke, M., and Fisher, G. L. (1982). Assessment of the teratogenicity of ammonium vanadate using Syrian golden hamsters. *Environm. Res.* **29,** 256–262.

Carpenter, G. (1981). Vanadate, epidermal growth factor and the stimulation of DNA synthesis. *Biochem. Biophys. Res. Commun.* **102,** 1115–1121.

Chasteen, N. D. (1983). The biochemistry of vanadium. *Struct. Bonding* **53,** 105–138.

Ciranni, R., Antonetti, M., and Migliore, L. (1995). Vanadium salts induce cytogenetic effects in vivo treated mice. *Mutat. Res.* **343,** 53–60.

Cohen, M. D., Klein, C. B., and Costa, M. (1992). Forward mutations and DNA-protein crosslinks induced by ammonium metavanadate in cultured mammalian cells. *Mutat. Res.* **269,** 141–148.

Dafnis, E., and Sabatini, S. (1994). Biochemistry and pathophysiology of vanadium. *Nephron* **67,** 133–143.

Djordjevic, C. (1995). Antitumor activity of vanadium compounds. In H. Sigel and A. Sigel (Eds.), *Metal Ions in Biological Systems.* Vol. 31, *Vanadium and Its Role in Life,* Marcel Dekker, New York, Basel, Hong Kong, pp. 595–616.

Domingo, J. L. (1995). Prevention by chelating agents of metal-induced developmental toxicity. *Reprod. Toxicol.* **9,** 105–113.

Domingo, J. L., Bosque M. A., Luna, M., and Corbella, J. (1993). Prevention by tiron (sodium 4,5-dihydroxybenzene-1,3-disulfonate) of vanadate-induced developmental toxicity in mice. *Teratology* **48,** 133–138.

Edel, J., and Sabbioni, E. (1989). Vanadium transport across placenta and milk of rats to the fetus and newborn. *Biol. Trace Elem. Res.* **22,** 265–275.

Flint, O. P., and Horten, T. C. (1984). An in vitro assay for teratogens with cultures of rat embryo midbrain and limb bud cells. *Toxicol. Appl. Pharmacol.* **76,** 385–395.

Flint, O. P., Horten, T. C., and Ferguson, R. A. (1984). Differentiation of rat embryo cells in culture: Response following acute maternal exposure to teratogens and non-teratogens. *J. Appl. Toxicol.* **4,** 109–116.

Galli, A., Vellosi, R., Fiorio, R., Della Croce, C., Del Carratore, R., Morichetti, E., Giromini, L., Rosellini, D., and Bronzetti, G. (1991). Genotoxicity of vanadium compounds in yeast and cultured mammalian cells. *Teratogen. Carcinogen. Mutagen.* **11,** 175–183.

Ganguli, S., Reuland, D. J., Franklin, L. A., Deakins, D. D., Johnston, W. J., and Pasha, A. (1994). Effects of maternal vanadate treatment on fetal development. *Life Sci.* **55,** 1267–1276.

Gibbons, I. R., Cosson, M. P., Evans, J. A., Gibbons, B. H., Houck, B., Martinson, K. H., Sale, W. S., and Tang, W. J. Y. (1978). Potent inhibition of dynein adenosintriphosphatase and of the motility of cilia and sperm flagella by vanadate. *Proc. Natl. Acad. Sci. USA* **75,** 2220–2224.

Giri, A. K., Sanyai, R., Sharma, A., and Talukder, G. (1979). Cytological and cytochemical changes induced through certain heavy metals in mammalian systems. *Natl. Acad. Sci. Lett.,* **2,** 391–394.

Gomez, M., Sanchez, D. J., Domingo, J. L., and Corbella, J. (1992). Embryotoxic and teratogenic effects of intraperitoneally administered metavanadate in mice. *J. Toxicol. Environ. Health* **37,** 47–56.

Hackett, P. L., and Kelman, B. J. (1983). Availability of toxic trace metals to the conceptus. *Sci. Total Environ.* **28,** 433–442.

Hanauske, V., Hanauske, A. R., and Marshall, M. H. (1987). Biphasic effect of vanadium salts on in vitro tumor colony growth. *Int. J. Cell Cloning* **5,** 170–178.

Hantson, P., de Saint Georges, L., Mahieu, P., Léonard, E. D., Crutzen-Fayt, M. C., and Léonard, A. (1996). Evaluation of the ability of paracetamol to produce chromosome aberrations in man. *Mutat. Res.* **368,** 293–300.

Hei, Y. J., Chen, X., Pelech, S. L., Diamond, J., and McNeill, J. H. (1995). Skeletal muscle mitogen-activated protein kinases and ribosomal S6 kinases. Suppression in chronic diabetic rats and reversal by vanadium. *Diabetes* **44,** 1147–1155.

Hollstein, M., McCann, J., Angelosanto, F. A., and Nichols, W. W. (1973). Short-term tests for carcinogens and mutagens. *Mutat. Res.* **65,** 133–226.

Hori, C., and Oka, T. (1980). Vanadate enhances the stimulatory action of insulin on DNA synthesis in cultured mouse mammary glands. *Biochim. Biophys. Acta* **610,** 235–240.

Jandhyala, B. S., and Hom, G. J. (1983). Physiological and pharmacological properties of vanadium. *Life Sci.* **33,** 1325–1340.

Kada, T., Hirano, K., and Shirasu, Y. (1980). Screening of environmental chemical mutagens by the rec-assay system with *Bacillus subtilis*. In F. J. de Serres and A. Hollaender (Eds.), *Chemical Mutagens: Principles and Method for Their Detection*, Vol. 5. Plenum Press, New York, pp. 149–173.

Kanematsu, N., Hara, M., and Kada, T. (1980). Rec assay and mutagenicity studies on metal compounds. *Mutat. Res.* **77,** 109–116.

Kanematsu, K., and Kada, T. (1978). Mutagenicity of metal compounds. *Mutat. Res.* **53,** 207–208.

Kingsnorth, A. N., LaMuraglia, J. S., Ross, J. S., and Malt, R. A. (1986). Vanadate supplements and 1,2-dimethylhydrazine-induced colon cancer in mice: Increased thymidine incorporation without enhanced carcinogenesis. *Br. J. Cancer* **53,** 683–686.

Kirkland, D. J. (Ed.). (1990). *Basic Mutagenicity Tests: UKEMS Recommended Procedures*. Cambridge University Press, Cambridge, 144 pp.

Klarlund, J. K. (1985). Transformation of cells by an inhibitor of phosphatase acting on phosphotyrosine in proteins. *Cell* **41,** 707–717.

Klein, C. B., Kargacin, B., Su, L., Cosentino, S., Snow, E. T., and Costa, M. (1994). Metal mutagenesis in transgenic Chinese hamster cell lines. *Environ. Health Perspect.* **102**(Suppl. 3), 63–67.

Kobayashi, T., Martensen, T., Nath, J., and Flavin, M. (1978). Inhibition of dynein ATPase by vanadate, and its possible use as a probe for the role of dynein in cytoplasmic motility. *Biochem. Biophys. Res. Commun.* **81,** 1313–1318.

Köpf-Maier, P., and Köpf, H. (1979). Vanadocene dichloride: Another anti-tumor agent from the metallocene series. *Z. Naturforsch.* **34,** 805–807.

Köpf-Maier, P., and Köpf, H. (1994). Organometallic titanium, vanadium, niobium, molybdenum and rhenium complexes—Early transition metals anti-tumor drugs. In S.P. Fricker (Ed.), *Metal Compounds in Cancer Therapy*. Chapman & Hall, London.

Léonard, A., Gerber, G. B., Jacquet, P., and Lauwerys, R. R. (1984). Carcinogenicity, mutagenicity and teratogenicity of industrially used metals. In M. Kirsch-Volders (Ed.), *Mutagenicity, Carcinogenicity and Teratogenicity of Industrial Pollutants*. Plenum Press, New York, pp. 59–126.

Levy, L. S., Martin, P. A., and Bidstrup, P. L. (1986). Investigations of potential carcinogenicity of a range of chromium containing materials on rat lung. *Br. J. Indust. Med.* **43,** 243–256.

Marini, M., Zunica, G., Bagnara, G. B., and Franceschi, C. (1987). Effect of vanadate on DNA synthesis induced by mitogens in T and B lymphocytes. *Mol. Cell Biochem.* **51,** 67–71.

Migliore, L., Bocciardi, R., Macri, C., and Lo Jacono, F. (1993). Cytogenetic damage induced in human lymphocytes by four vanadium compounds and micronuclei: Analysis by fluorescence in situ hybridization with a centromic probe. *Mutat. Res.* **319,** 205–213.

Migliore, L., Scarpato, R., and Falco, P. (1995). The use of fluoroscence in situ hybridisation with a β-satellite DNA probe for the detection of acrocentric chromosomes in vanadium-induced micronuclei. *Cytogenet. Cell Genet.* **69,** 215–219.

Navas, P., Hidalgo, A., and Garcia-Herdago, G. (1986). Cytokinesis in onion roots: Inhibition by vanadate and coffeine. *Experientia* **42,** 437–439.

Nechay, B. R. (1984). Mechanisms of action of vanadium. *Annu. Rev. Pharmacol. Toxicol.* **24,** 501–524.

OECD. (1981–1985). *Guidelines for Testing of Chemicals. Section 4, Health Effects.*

Owusu-Yaw, J., Cohen, M. D., Fernando, S. Y., and Wei, C. I. (1990). An assessment of the genotoxicity of vanadium. *Toxicol. Lett.* **50,** 327–336.

Parfett, C. L. J., and Pilon, R. (1995). Oxidative stress-regulated gene expression and promotion of morphological transformation induced in C3H/10T1/2 cells by ammonium metavanadate. *Fed. Chem. Toxicol.* **33,** 301–308.

Paternain, J. L., Domingo, J. L., Gomez, A. M., Ortega, A., and Corbella, J. (1990). Developmental toxicity of vanadium in mice after oral administration. *J. Appl. Toxicol.* **10,** 181–186.

Paternain, J. L., Domingo, J. L., Llobet, J. M., and Corbella, J. (1987). Embryotoxic effects of sodium metavanadate administered to rats during organogenesis. *Rev. Esp. Fisiol.* **43,** 223–228.

Paton, G. R., and Allison, A. C. (1972). Chromosome damage in human cell cultures induced by metal salts. *Mutat. Res.* **16,** 332–336.

Pierce, L. K., Alessandrini, F., Godleski, J. J., and Paulauskis, J. D. (1996). Vanadium-induced chemokine mRNA expression and pulmonary inflammation. *Toxicol. Appl. Pharmacol.* **138,** 1–11.

Purchase, I. P. H. (1982). An appraisal of predictive tests for carcinogenicity. *Mutat. Res.* **99,** 53–71.

Ramanadham, M., and Kern, M. (1983). Differential effects of vanadate on DNA synthesis induced by mitogens in T and B lymphocytes. *Mol. Cell Biochem.* **51,** 67–71.

Rojas, E., Valverde, M., Herrera, L. A., Altamirano-Lozano, P., and Ostrosky-Wegman (1996). Genotoxicity of vanadium pentoxide evaluated by the single cell gel electrophoresis assay in human lymphocytes. *Mutat. Res.* **359,** 77–84.

Roldán, R. E., and Altamiro, L. M. A. (1990). Chromosomal alterations, sister chromatid exchanges, cell cycle kinetics and satellite associations in human lymphocyte cultures exposed to vanadium pentoxide. *Mutat. Res.* **245,** 61–65.

Roshchin, A. V., Ordshonikiodze, E. K., and Shalganova, I. V. (1980). Vanadium toxicity, metabolism, carrier state. *J. Hyg. Epidemiol. Microbiol. Immunol.* **24,** 377–383.

Sakurai, H. (1994). Vanadium distribution in rats and DNA cleavage by vanadyl complex: Implication for vanadium toxicity and biological effects. *Environ. Health Perspect.,* **102**(Suppl. 3), 35–36.

Sakurai, H., Tamura, H., and Okatani, K. (1995). Mechanism for a new antitumor vanadium complex: Hydroxyl radical-dependent DNA cleavage by 1,10-phenanthroline-vanadyl complex in the presence of hydrogen peroxide. *Biochem. Biophys. Res. Commun.* **206,** 133–137.

Sanchez, D. J., Ortega, A., Domingo, J. L., and Corbella, J. (1991). Developmental toxicity evaluation of orthovanadate in the mouse. *Biol. Trace Elem. Res.* **30,** 219–226.

Sardar, S., Ghosh, R., Mondal, A., and Chatterjee, M. (1993). Protective role of vanadium in the survival of hosts during the growth of a transplantable murine lymphoma and its profound effects on the rates and patterns of biotransformation. *Neoplasma* **40,** 27–30.

Schiff, L. J., and Graham, A. J. (1984). Cytotoxic effects of vanadium and oil-fired fly ash on hamster tracheal epithelium. *Environm. Res.* **34**, 390–402.

Schroeder, H. A., and Balassa, J. J. (1967). Arsenic, germanium, tin and vanadium in mice: Effects on growth, survival and tissue levels. *J. Nutr.* **92**, 245–252.

Schroeder, H. A., Balassa, J. J., and Tipton, I. H. (1963). Abnormal trace metals in man; Vanadium. *J. Chronic. Dis.* **16**, 1047–1071.

Schroeder, H. A., Mitchener, M., and Nason, A. P. (1970). Zirconium, niobium, antimony, vanadium and lead in rats: Life-term studies. *J. Nutr.* **100**, 59–62.

Sharma, R. P., and Talukder, G. (1987). Effects of metals on chromosomes of higher organisms. *Environ. Mutagen.* **9**, 191–226.

Shelanski, M. L., Gaskin, F., and Kantor, C. R. (1973). Microtubule assembly in the absence of added nucleotides. *Proc. Natl. Acad. Sci. USA* **70**, 765–768.

Smith, J. B. (1983). Vanadium ions stimulate DNA synthesis in Swiss mouse 3T3 and 3T6 cells. *Proc. Natl. Acad. Sci. USA* **80**, 6162–6166.

Sora, S., Carbone, M. L. A., Pacciarini, M., et al. (1986). Disomic and diploid meiotic products induced in *Saccharomyces cerevisiae* by the salts of 27 elements. *Mutagenesis* **1**, 21–28.

Stankiewicz, P. J., and Tracey, A. S. (1995). Stimulation of enzyme activity by oxovanadium complexes. In H. Sigel and A. Sigel (Eds.), *Metal Ions in Biological Systems.* Vol. 31, *Vanadium and Its Role in Life.* Marcel Dekker, New York, Basel, Hong Kong, pp. 249–285.

Stoner, G. D., Shimin, M. B., Troxell, M. C., Thompson, T. L., and Terry, L. S. (1976). Test for carcinogenicity of metallic compounds by the pulmonary tumor response in strain a mice. *Cancer Res.* **36**, 1744–1747.

Sun, M. (Ed.). (1987). *Toxicity of Vanadium and Its Environmental Health Standards.* Chengdu West China University of Medical Sciences Report, Chengdu, 20 pp.

Thompson, H. J., Chasteen, N. D., and Meeker, L. D. (1984). Dietary vanadyl (IV) inhibits chemically-induced mammary carcinogenesis. *Carcinogenesis* **5**, 849–851.

Wide, M. (1984). Effect of short-term exposure to five industrial metals on the embryonic and fetal development of the mouse. *Environ. Res.* **33**, 47–53.

World Health Organisation (Ed.). (1980). *Long-Term and Short-Term Screening Assays for Carcinogens: A Critical Appraisal.* IARC, Lyon, 426 pp.

World Health Organisation (Ed.). (1988). *Environmental Health Criteria: Task Group on Environmental Health Criteria for Vanadium,* Vol. 81. UNEP, ILO, and WHO, Geneva, 170 pp.

Zhong, Z., Gu, Z. W., Wallace, W. E., Whong, W. Z., and Ong, T. (1994). Genotoxicity of vanadium pentoxide in Chinese hamster V79 cells. *Mutat. Res.* **321**, 35–42.

4

VANADIUM EXPOSURE TESTS IN HUMANS: HAIR, NAILS, BLOOD, AND URINE

Jan Kučera

Nuclear Physics Institute, Academy of Sciences of the Czech Republic, CZ-250 68 Řež near Prague, Czech Republic

Jaroslav Lener and Jana Mňuková

National Institute of Public Health, Šrobárova 48, CZ-100 42 Prague 10, Czech Republic

Eva Bayerová

NIKOM spc., Medical Center, CZ-262 02 Mníšek pod Brdy, Czech Republic

Vanadium in the Environment. Part 2: Health Effects, Edited by Jerome O. Nriagu.
ISBN 0-471-17776-8. © 1998 John Wiley & Sons, Inc.

1. INTRODUCTION

The essentiality, distribution, and toxicology of vanadium are still areas of ongoing research at the present time. Several adverse effects of vanadium on human health, mostly following occupational exposure, were reported (WHO, 1988). The health effects of environmental exposure to vanadium of the general population were also studied (Lener et al., this volume, Ch. 1). Evaluation of the toxic effects of any harmful substance is facilitated if reliable biological indicators of exposure, such as changes of a concentration of the substance of interest, its metabolites, and other specific species in indicator or target tissues, are well established. The results of direct monitoring (i.e., the assessment of exposure from a concentration of the substance in air, water, food, workplace, etc.) may be misleading, if various mechanisms of intake and/or absorption lead to a different burden of the organism, if multiple mechanisms of intake are to be considered, and especially if short- and long-term effects are to be distinguished. Therefore, in the assessment of health risk arising from environmental, occupational, and accidental exposure to toxic metals and other harmful substances, the use of biological monitoring is steadily increasing (Clarkson et al., 1988; Dillon and Ho, 1991; Bencko, 1995).

However, our knowledge about suitable vanadium exposure tests and their sensitivities is rather limited or even contradictory. For instance, in 1953 Mountain et al. found decreased levels of cystine in the hair of rats exposed to vanadium pentoxide and also a reduction of coenzyme A in the liver, which could explain the mechanism behind the reduction of cysteine. Subsequently, decreased cystine levels in fingernails of vanadium-exposed workers were also reported by Mountain et al. (1955), while Kiviluoto et al. (1980) did not find any significant differences between the fingernail cystine levels of workers occupationally exposed to vanadium and normal persons. Use of vanadium levels in urine or blood was suggested by several authors as suitable indicators of occupational vanadium exposure (Gylseth et al., 1979; Kiviluoto et al., 1979; Pyy et al., 1981; Kiviluoto et al., 1981; Kawai et al., 1989; Todaro et al., 1991). However, highly disparate vanadium levels in those tissues for nonexposed persons, spanning more than several orders of magnitude (Sabbioni et al., 1996; Kučera and Sabbioni, this volume, Ch 5), prevented reliable evaluation of the differences found between exposed and control groups and, consequently, evaluation of the sensitivity of the individual tests.

In the present work, the results are reported of the determination,—by neutron activation analysis (NAA) with proven accuracy even at the ultratrace level—of vanadium levels in selected tissues of workers occupationally exposed in a vanadium pentoxide production plant and in tissues of control persons. The tissues included hair, nails, blood, and urine. Cystine in hair and nails was also measured to elucidate the above-mentioned discrepancies.

2. EXPERIMENTAL DESIGN

2.1. Subjects

2.1.1. Exposed Group

Exposure tests were studied in workers of a plant located 20 km southwest of Prague that was producing vanadium pentoxide from a vanadium-rich slag by a hydrometallurgical process that is schematically depicted in Figure 1. The production involves several processes that are associated with a release of vanadium-rich dust that presents a risk of occupational exposure, especially if workers do not permanently use their protection respirators. The exposed group consisted of 69 men of the average age of 41 years (range 23–58 years) with the average exposure time of 9.2 years (range 0.5–33.1 years).

2.1.2. Control Groups

One control group (further denoted as C-1), formed by administrative workers of the plants, consisted of 7 men and 10 women of the average age of 46 years. Other control groups (further denoted as C-2) were formed by 195 men and women from Prague and its surroundings (for hair analysis), by 87 men and women from the same locality (for nail analysis), by 21 men and women employed at a research institute located 25 km north of Prague (for urine analysis), and by 11 and 10 employees of another research institute in Prague (for blood and cystine analysis, respectively).

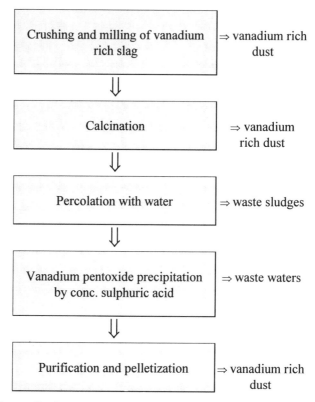

Figure 1. Hydrometallurgical process of vanadium pentoxide production from vanadium-rich slag.

2.2. Samples and Sampling Techniques

2.2.1. Air Particulate Matter (APM)

APM samples were collected at several supervision workplaces of the vanadium production plant that were considered to be the most significant pollution sources and at workplaces that could provide information on the "background" vanadium levels in the plant. Sampling was usually done in duplicate in two places located about 1 m from each other or in several sampling campaigns spanning more than a 3-month period. One, and two to five, sampling campaigns were performed at the "background" workplace and the most polluted workplace, respectively. A low-volume sampling device equipped with Synpor No. 4 membrane ultrafilters (Barvy a laky, Czech Republic) with a pore diameter of 0.8 μm and a filter diameter of 35 mm was employed. The sampling device was held with its open side down. A sampling time of 30–50 min was used and the air flow amounted to 0.9–1.3 m^3/h.

2.2.2. Hair

Hair samples weighing 300–500 mg were cut with scissors from the nape of the neck and washed according to the IAEA recommended procedure using, successively, 25 mL of acetone, 3 times 25 mL of deionized water, and 25 mL of acetone, each wash lasting 10 min (Rjabuchin, 1978). For the analysis, 100–150 ing of the air-dried hair was transferred to clean polyethylene (PE) capsules with the aid of plastic tools and was heat-sealed. The PE capsules and all plastic tools used in this work were cleaned before use by leaching in high-purity hydrochloric acid (semiconductor grade) diluted with deionized water (1 : 1 v/v) and washing several times with deionized water.

2.2.3. *Nails*

Fingernail samples weighing 20–100 mg were cut with scissors and washed by a similar procedure to that for hair cleaning. Instead of the first wash with deionized water, a 1% solution of an ionic detergent, Triton X-100, in deionized water was used; washing times were shortened from 10 to 5 min, volumes of washing agents were reduced from 25 to 5 mL, and the whole washing procedure was performed in an ultrasonic bath. After air-drying, the nails were heat-sealed into the precleaned PE capsules.

2.2.4. *Blood*

About 2–3 mL of blood was obtained by venous puncture with a disposable steel needle (after flushing the needle with about 10 mL of blood used for other purposes) at a medical center, the blood sample was collected in a precleaned PE vial. The amount of blood was ascertained by weighing and converted to a volume by using a factor of 1.06 g/mL. Then the samples were freeze-dried. Most of the freeze-dried blood samples were dry-ashed prior to analysis in quartz vials. The vials, made of natural quartz, were precleaned by boiling in high-purity dilute hydrochloric acid and washing several times with deionized water. Ashing was carried out in a special minioven (Kučera et al., 1990) with a controlled temperature regime of 2 h at 150°C, 1 h at 250°C, 1 h at 350°C, and 16 h at 450°C. Ash aliquots (90–95% as determined by weighing) were heat-sealed into the precleaned PE tubes.

2.2.5. *Urine*

About 3–5 mL of morning urine spot samples were collected in precleaned PE vials. Then 2-mL samples for irradiation were obtained by pipetting with a clean glass pipet into precleaned irradiation PE vials and heat-sealed.

2.3. Analysis

2.3.1. *Instrumental Neutron Activation Analysis (INAA)*

The APM, hair, and nail samples, some reference materials, and vanadium standards (an evaporated aliquot of a solution with known vanadium concentration heat-sealed in PE vials) were irradiated separately with 10 μg gold fluence monitors in an LWR-15 nuclear reactor of the Nuclear Research Institute (NRI) Řež, PLC, at a thermal neutron fluence rate of 6–8 $\times$ 10^{13} cm^{-2} s^{-1} for 0.5–2 min. The 1,434-keV γ-rays of ^{52}V (half-life T$_{1/2}$ = 3.75 min) were measured after a 5-min decay time for 10 min by means of a coaxial HPGe detector with a relative efficiency of 11% and a full width at half maximum (FWHM) resolution of 1.85 keV for the 1,332.5-keV γ-rays of [60]Co. A sample-to-detector distance amounted to 2–5 cm for most samples, except for filters with the APM samples collected in the plant, which had a high loading of vanadium-rich dust. These samples had to be measured at higher distances from the detector (up to 30 cm) or after longer decay times to keep the dead time of the detection system below 30%. The detection system was suitable for maintaining high count rates with minimum errors of the sort that result from high or variable count rates (Kučera and Soukal, 1988). The vanadium detection limit (3 σ) amounted to 0.3 ng, 0.1 ng, and 0.5 ng for hair, nail, and APM samples, respectively.

2.3.2. Radiochemical Neutron Activation Analysis (RNAA)

Vanadium determination in the blood and urine samples, and in most of the reference materials analyzed, was carried out either at the Nuclear Physics Institute, Řež, or at the Jožef Stefan Institute (JSI), Ljubljana, Slovenia. In the former institute, the same irradiation conditions were used as for INAA for 1–3 min, while in the latter institute irradiation was carried out at a thermal neutron fluence rate of 4×10^{12} cm^{-12} s^{-1} for 10–12 min. Then one of two recent RNAA versions of vanadium determination at the ultratrace level (Byrne and Kučera 1991) was used. Basically, they consist of rapid wet-ashing of intact (or only freeze-dried) samples, or additional wet-ashing of samples dry-ashed prior to irradiation in a mixture of concentrated H_2SO_4 and HNO_3. Wet-ashing is completed by an addition of conc. $HClO_4$ and fuming off. Then adjustment of the solution to 5 mol/L HCl and addition of $KMnO_4$ follows. Subsequently, vanadium(V) is extracted with a 0.2% solution of N-benzoyl-N-phenyl hydroxylamine (N-BPHA) in toluene. The whole procedure takes 8–10 min and 6–8 min with freeze-dried and dry-ashed samples, respectively. Measurement of 5-mL fractions containing separated [52]V for 7 min was carried out either in a well-type HPGe detector (active volume 120 cm^3, FWHM resolution 2.2. keV for the 1332.5-keV photons of ^{60}Co) at the JSI, Ljubljana, or with a coaxial HPGe detector (relative efficiency of 36%, FWHM resolution of 1.9 keV for the 1,332.5-keV γ-rays of ^{60}Co) at the NPI, Řež.

2.3.3. Determination of a Sum of Cysteine and Cystine in Nails and Hair

About 10 mg of the nail or hair samples were finely cut and treated with 3–5 mL of peroxoformic acid in a closed flask at about 200 °C for 1–2 h to oxidize cysteine and cystine to stable cysteic acid to prevent cystine losses on hydrolysis by racemization. The oxidizing agent was evaporated in a vacuum evaporator at 40 °C, 10 mL of 6 mol/L HCl was added, and the samples were hydrolyzed by boiling under reflux for 24–26 h. The hydrolase was filtered, the filter washed twice with hot water, and the solution evaporated in the vacuum evaporator at 60 °C. The residue was taken up with 10 mL of a citrate/

HCl buffer solution of pH 2.2, and cysteic acid was determined by an automatic aminoacid analyzer (AAA-339, Mikrotechna Prague) with pure cystine used as a standard and taurine as an internal standard. Other experimental details have already been given elsewhere (Kučera et al., 1994).

3. RESULTS AND DISCUSSION

3.1. Quality Assurance of Analysis

Great care was devoted to quality assurance procedures in this work. The INAA and RNAA methods employed were tested by analyzing a variety of biological and environmental reference materials with various vanadium levels as shown in Table 1. Very good agreement with reference and/or literature values where available demonstrates excellent accuracy of the INAA and RNAA procedures employed in this work. It can also be seen from Table 1 that the two RNAA modes give results in good agreement with each other. This provides strong evidence that no losses or contamination occurred on dry-ashing prior to irradiation. The RNAA mode with dry-ashed samples provided a somewhat lower determination limit of vanadium of about 10 pg for the well-type HPGe detector. This was due to a shortening of the decomposition time of the separation procedure by 2–3 min. It is also very important to note that in the reference materials (RMs) analyzed the actual vanadium levels were matched to those occurring in all types of samples analyzed. This is especially significant for urine and blood of unexposed persons, because in these materials vanadium levels are extremely low and proving accuracy of such results was hampered until the end of the 1980s because of the unavailability of suitable RMs or agreement between results obtained by various authors (Sabbioni et al., 1996; Kučera and Sabbioni, this volume, Ch. 5). Only recently were agreeable results obtained by two independent analytical techniques (Byrne and Versieck, 1990; Kučera et al., 1992; Moens et al., 1995) for the vanadium content in Versieck's second Generation Biological RM-uncontaminated freeze-dried human serum (Versieck et al., 1988).

3.2. Contamination of Workplace Air

The total mass of APM and vanadium concentration in the APM samples collected at selected workplaces in the plant are given in Table 2. Obviously, the maximum admissible limit for vanadium in workplace air, which amounts in the Czech republic to 500 mg/m^3 and 1,500 μg/m^3 for the average and peak concentrations, respectively, were occasionally exceeded at a cooling drum and a pelletizer, while a much higher risk of undesirable exposure existed at a melting furnace, and especially at a vibratory conveyer, where the highest values were found. A high spread of the vanadium concentrations measured in this work was due to the fact that most workplaces were not separated

Table 1 Vanadium in Environmental and Biological Reference Materials (mg/kg Dry Weight)

Reference Material	Method[a]	Found $\bar{x} \pm s$ (N)	Reference and Literature Values[b]
NBS SRM 1648 (urban particulate)	INAA	124.6 ± 5.3 (26)	*140 ± 3
NBS SRM 2704 (Buffalo River sediment)	INAA	92.3 ± 6.7 (6)	121 ± 8 (Gladney et al., 1987) *95 ± 4
Bowen's kale	INAA	0.374 ± 0.026 (7)	*0.386 ± 0.058 (Bowen, 1985)
	RNAA-A	0.384 ± 0.019 (5)	
CTA-FFA-1 (fine fly ash)	INAA	248 ± 4 (6)	*260 ± 10
NBS SRM 1577a (bovine liver)	INAA	0.093 ± 0.005 (3)	*0.099 ± 0.008
	RNAA-W	0.094 ± 0.003 (3)	0.097 (Gladney et al., 1987)
NIST SRM 1515 (apple leaves)	RNAA-W	0.237 ± 0.015 (3)	*0.26 ± 0.03
NIST SRM 1573a (tomato leaves)	INAA	0.810 ± 0.019 (6)	*0.835 ± 0.010
	RNAA-W	0.783 ± 0.057	
NIST SRM 1570a (spinach leaves)	INAA	0.558 ± 0.035 (6)	*0.57 ± 0.03
IAEA H-4 (animal muscle)	RNAA-W	0.00271 (1)	0.003[c] (Parr, 1980)
	RNAA-A	0.00262 ± 0.00033 (6)	0.00278 ± 0.00021 (Byrne and Versieck, 1990)
IAEA A-13 (animal blood)	RNAA-W	0.00102 ± 0.00013 (17)	
	RNAA-A	0.00116 ± 0.00015 (5)	
IAEA 336 (lichen)	INAA	1.51 ± 0.08 (5)	1.5[c] (IAEA, 1995)
IJS bovine serum (internal RM)	RNAA-W		
	RNAA-A	0.0046 ± 0.0004 (3)	0.005 ± 0.0002 (Byrne and Versieck, 1990)
Versieck's 2nd generation (human serum RM)	RNAA-W	0.00066 ± 0.00010 (3)	0.00067 ± 0.00005 (Byrne and Versieck, 1990) 0.00060 (Moens et al., 1995) 0.00065[c] (IAEA, 1995)
	RNAA-A		

[a] RNAA-A: dry-ashing mode; RNAA-W: wet-ashing mode.
[b] Values preceded by asterisk are certified values given in certificates of issuing organizations or in IAEA (1995).
[c] Information value.

Table 2 Contamination of the Workplace Air in a Vanadium Pentoxide Production Plant

Workplace	Total APM Mass, mg/m³	Vanadium Content, μg/m³	N^a
Ball mill	0.6	16.9	1
Main ventilator	0.6	20.0	1
Cooling drum	3.4–8.7	192.4–578.9	2
Melting furnace	3.1	802.1	1
Pelletizer	6.6–271	105.1–505.2	3
Roaster	1.3–29.7	31.3–149.1	5
Vibratory conveyer	7.7–61.9	1861–4846	4

[a] Number of measurements.

from the ambient air, and thus the quality of the working environment was affected by outside weather conditions. Changing vanadium concentrations and especially wearing respirators in the most polluted workplaces thus made assessment of vanadium amounts inhaled by workers from direct monitoring almost impossible. Therefore, more useful information on vanadium exposure was expected from the results of biological monitoring.

3.3. Vanadium in Hair

Hair has been recognized as a suitable and easily accessible indicator of exposure to many metals, especially from the contaminated environment (Chatt and Katz, 1988; Bencko, 1995). However, the interpretation of hair analysis, especially in occupational health studies, is complicated by the nonexistence of a washing procedure that would completely remove an external contamination without influencing the endogenous content of elements that are supposed to reflect the exposure. Vanadium is one of more loosely bound elements in hair, being washed out by various cleaning procedures to 50–70% of the initial value (Chatt et al., 1985). Therefore, to be able to arrive at comparable results, a standardized washing procedure, such as that one proposed by the IAEA (Rjabuchin 1978), should be employed. By following this procedure, vanadium values were found in the exposed and control subjects, which are summarized in Table 3.

From a comparison of the arithmetic means, $\bar{x}_a$, the geometric means, $\bar{x}_g$, and the medians, $\tilde{x}$, of the individual groups it is obvious that the values found had a skewed, nonnormal distribution, especially in the exposed group, as was also ascertained by the Shapiro–Wilk test (Shapiro and Wilk, 1965). Thus, the median appears to be the most informative quantity for evaluating the differences between the exposed and the control groups (test sensitivity). Significantly higher values can be noted in the control group C-1 than in C-2. This seems to suggest that the C-1 group was not completely separated from

Table 3   Vanadium Concentration in Hair of Exposed and Control Persons (mg/kg, Dry Weight)

Subjects	Range	$\bar{x}_a$	sd_a	$\bar{x}_g$	sd_g	$\tilde{x}$	N^a
Exposed	0.103–203	32.2	37.0	16.5	4.09	21.0	55/55
C-1	0.243–3.03	1.01	0.92	0.736	2.33	0.731	8/8
C-2	<0.009–0.389	0.031	0.045	0.021	2.24	0.018	128/195

[a] Number of samples in which vanadium could be determined/number of samples analyzed.

the pollution sources in the plant or that the possible external contamination of hair was not completely removed by the washing procedure employed. However, since similar differences between the control groups C-1 and C-2 were also found for other tissues and/or body fluids (see Tables 5–7), the former explanation seems to be most likely.

3.4.   Vanadium in Nails

Nails, as another ectoderm derivative, have also been considered as a possible bioindicator of occupational exposure. However, the risk of external contamination and the difficulties associated with interpreting the results of nail analysis are even more severe than for hair in exposed workers, because no generally accepted procedure for nail cleaning has been recognized until now. The most frequently employed procedures, such as purely mechanical cleaning by a noncontaminating tool, brushing in deionized water, or leaching in dilute acids or solutions containing several percents of detergents, may differ widely in removing both endogenously bound measurands and the external contamination. Presuming that a similar behavior can be expected for nail as for hair vanadium, a cleaning procedure was studied in this work that was regarded as potentially capable of efficient removal of external contamination without excessive washing-out of vanadium. For this purpose, the first washing with deionized water in the IAEA procedure for hair cleaning (Rjabuchin, 1978) was replaced by washing in a 1% nonionic detergent solution (Triton X-100), and the whole cleaning procedure was carried out in an ultrasonic bath. Washing-out of vanadium, and of other elements, such as sodium and chlorine, that upon neutron irradiation form short-lived radionuclides, which predominantly influence a detection limit for vanadium in INAA, was investigated. Repeated irradiations and the counting of an identical nail sample, weighing about 20 mg, after individual steps of the original and the adapted IAEA procedures for hair cleaning were carried out by taking advantage of favorable half-lives of the respective radionuclides ^{52}V ($T_{1/2}$ = 3.75 min), ^{38}Cl ($T_{1/2}$ = 37.2 min), and ^{24}Na ($T_{1/2}$ = 15.0 h). The ^{52}V and ^{38}Cl activities decay completely within 1 h and few hours, respectively, while that of ^{24}Na originating from a

Table 4   Washing-Out of Selected Elements from Nails by the Original and Adapted IAEA Procedures for Hair Cleaning

Washing Step	Element as Percentage of the Initial Content in Unwashed Nails[a]					
	V^b	V^c	Cl^b	Cl^c	Na^b	Na^c
Acetone	89 ± 16	86 ± 15	98 ± 1	96 ± 1	99 ± 1	98 ± 1
Deionized water[b]	76 ± 17		54 ± 1		64 ± 1	
1% Triton X-100[c]		86 ± 15		45 ± 1		48 ± 1
Deionized water	74 ± 20	89 ± 12	36 ± 1	29 ± 2	34 ± 2	14 ± 2
Deionized water	69 ± 15	75 ± 13	19 ± 2	15 ± 3	15 ± 2	14 ± 2
Acetone	71 ± 15	72 ± 13	19 ± 2	15 ± 3	14 ± 2	15 ± 2

[a] Error shown is the relative statistical counting error (1 s).
[b] Original IAEA procedure for hair cleaning.
[c] Adapted IAEA procedure.

previous irradiation can be subtracted. The results in Table 4, given as percentage of the initial content in unwashed nails, show efficient removal of sodium and chlorine, whereas vanadium is washed out only moderately by both procedures, to an extent comparable to that of hair cleaning. This is advantageous for achieving a low determination limit of vanadium. For this reason, and because (unlike the original IAEA procedure) no external contamination was visible after cleaning of originally dirty nails, the adapted washing procedure was employed in the present work. Table 5 shows the vanadium levels found in nails of the exposed and control groups. About the same median vanadium levels were found in fingernails and hair in the exposed group, while most values in controls C-2 were below the vanadium detection limit by INAA owing to the limited amount of nails available for analysis. Highly significant correlation ($P > 0.0011$) was found between hair and fingernail vanadium values in the exposed group by calculating the Spearman rank-correlation coefficient (this nonparametric testing was employed because of the nonnormal distribution of the values determined).

Table 5   Vanadium Concentration in Fingernails of Exposed and Control Persons (mg/kg, Dry Weight)

Subjects	Range	$\bar{x}_a$	sd_a	$\bar{x}_g$	sd_g	$\tilde{x}$	N^a
Exposed	0.260–614	34.3	79.9	15.6	3.46	19.85	60/60
C-1	0.123–16.5	2.27	4.28	0.753	4.31	0.473	15/15
C-2	<0.017–1.24	<0.044	—	<0.038	—	<0.036	13/87

[a] Number of samples in which vanadium could be determined/number of samples analyzed.

3.5. Vanadium in Blood

For technical reasons, blood vanadium was determined only in a limited number of the exposed workers and only in the control group C-2. The results are shown in Table 6. The blood vanadium values of nonexposed persons are among the lowest ones reported for vanadium concentrations in human blood or serum (Sabbioni et al., 1996; Kučera and Sabbioni this volume, Ch 5). Their reliable determination was made possible by use of a highly selective RNAA procedure (Byrne and Kučera 1991) that has been validated over a broad range of vanadium levels, including that one encountered in blood or serum samples (see Section 3.1.). It should be also pointed out that for blood sampling disposable steel needles were used in the present work. This resulted in no apparent contamination; the values were, as stated above, among the lowest reported and also showed considerable internal consistency (narrow range). Therefore, it seems that the material of the needle is probably not the critical point for possible contamination when the needle is first flushed with several milliliters of blood not taken for analysis. However, since materials of stainless steel needles may differ in content of vanadium and other alloying components, the use of disposable steel needles without inside silicon coating cannot be generally recommended. Instead, use of plastic cannulas, which eliminate the possible risk of contamination, should be preferred.

It is known from both animal and human studies (Kiviluoto et al., 1979, 1981; WHO 1988) that after acute exposure, a considerable part of vanadium is quickly (within 24 h) excreted via the kidneys and initially high blood vanadium levels fall to trace levels after approximately 2 days. The remainder of the absorbed vanadium is retained mainly in the liver and bone. However, high blood vanadium levels reappear after a few days owing to rapid element mobilization from the liver and slow release from bone, which appears to be a major sink of vanadium in mammal organisms. Thus, blood vanadium in exposed workers seems to be the best indicator of the long-term body burden. Calculation of the Spearman rank-correlation coefficient yielded no significant association ($P > 0.05$) between blood vanadium and vanadium concentrations in hair, nails, and urine in the exposed workers.

3.6. Urinary Vanadium

Urinary vanadium values for the exposed and two control groups are given in Table 7. As for blood, the lowest urinary vanadium median value for

Table 6   Vanadium Concentration in Blood of Exposed and Control Persons (ng/mL)

Subjects	Range	$\bar{x}_a$	sd_a	$\bar{x}_g$	sd_g	$\tilde{x}$	N^a
Exposed	3.10–217	33.2	69.2	12.1	3.52	13.4	9/9
C-2	0.032–0.095	0.058	0.020	0.055	1.41	0.056	11/11

^{a} Number of samples in which vanadium could be determined/number of samples analyzed.

Table 7   Urinary Vanadium in Exposed and Control Persons (ng/mL)

Subjects	Range	$\bar{x}_a$	sd_a	$\bar{x}_g$	sd_g	$\tilde{x}$	N^a
Exposed	3.0–762	65.9	123.5	29.2	3.33	30.7	62/62
C-1	1.05–53.4	3.75	3.12	2.88	2.82	1.94	14/14
C-2	0.066–0.489	0.223	0.096	0.203	1.61	0.212	21/21

[a] Number of samples in which vanadium could be determined/number of samples analyzed.

nonexposed persons (Sabbioni et al., 1996; Kučera and Sabbioni, this volume, Ch. 5) was determined in the present work by the above-described RNAA procedure with proven accuracy. In occupationally exposed workers, urinary vanadium was reported to decrease with time elapsed from the end of exposure (Kilivuoto et al., 1979, 1981; Kawai et al., 1989; Todaro et al., 1991). To get a deeper insight into the mechanism of vanadium excretion in urine, vanadium levels were also examined in our study, in a worker during 1 week in which he worked five 8-h shifts. Urinary vanadium was measured each day at the beginning, during, and at the end of a shift. After the second shift, measurement was also done when the worker was beyond the range of exposure. Figure 2 shows that urinary vanadium levels increased during the shifts (full lines) to 2–3 times greater than the initial value, and within 16 h returned to about that at the beginning of exposure. However, this value remained chronically high, approximately by two orders of magnitude higher than the median for nonexposed persons (C-2 group) during the whole week. This

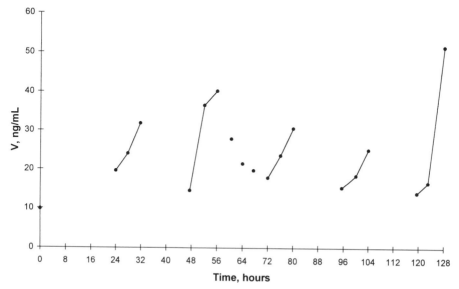

Figure 2. Changes of urinary vanadium in an exposed worker during a week (shifts indicated by solid line).

confirms the quick excretion of the inhaled vanadium via the kidneys. The results of the measurement after the second shift suggest that the vanadium excretion followed a half-life of about 12 h, whereas half-lives between 14 and 18 h could explain the drop in urinary vanadium between the end and the beginning of exposure on the other days. This demonstrates that urinary vanadium is especially suitable for biomonitoring of a very recent exposure. Examination of the urinary vanadium of this worker when he was out of the range of exposure for a longer period, for instance during his vacation, was not possible for technical reasons. Thus, it is not clear whether there is another, a longer, half-life which would influence the high "base-line" level of urinary vanadium observed at the beginning of shifts. No significant correlations ($P > 0.05$), as ascertained by the Spearman rank-correlation test, were found between urinary vanadium and vanadium concentrations in hair, nails, and blood of the exposed workers.

3.7. Cystine in Fingernails and Hair

Cystine in fingernails and hair was also measured in the exposed workers and controls to elucidate the contradictory findings on decreased cystine levels of workers exposed to vanadium (Mountain et al., 1953, 1955; Kiviluoto et al., 1980). Unlike the other tests studied, the results of cystine determination in both tissues in all groups examined were normally distributed. Therefore only the arithmetic means and their standard deviations are given in Table 8. No direct proof of accuracy of the cystine determination in fingernails and hair can be given, because no reference material for this purpose is available. However, using a method with a different principle in this work, similar values of cystine in fingernails were found compared to those determined by a colorimetric method by Kilivuoto et al. in 1980 (133 ± 16 mg/g and 138 ± 18 mg/g for exposed and controls, respectively). No significant differences were found by *t* test ($\alpha = 0.05$) between cystine levels in fingernails and hair

Table 8   Results of Cystine Determination in Fingernails and Hair (mg/g, Dry Weight)

Subjects	Range	Arithmetic Mean ± *sd*	N^a
Fingernails			
Exposed	67.7–142.5	109.5 ± 13.5	44/44
C-1	85.5–121.4	107.6 ± 10.8	11/11
C-2	104.6–143.1	118.5 ± 11.9	10/10
Hair			
Exposed	122.5–184.3	154.9 ± 14.0	50/50
C-1	128.7–169.8	154.7 ± 16.2	11/11

[a] Number of samples in which cystine could be determined/number of samples analyzed.

in exposed workers and controls (no attempt was done to determine cystine in hair of the control group C-2). Thus, the results of this study confirm the finding of Kiviluoto et al. (1980) that the cystine content in fingernails cannot be used as an exposure test for vanadium- or vanadium pentoxide-rich dust. It follows from our results that this also applies to hair cystine.

3.8. Sensitivity of Vanadium Exposure Tests

Determination of vanadium levels in hair, fingernails, blood, and urine in both exposed and nonexposed persons made it possible to evaluate the sensitivity of the biological monitoring of exposure to vanadium. This was done in two possible ways: (1) by comparing the maximum value of the exposed group to the median of the control group C-2 (criterion 1); (2) by comparing the median of the exposed group to the median of the control group C-2 (criterion 2). These comparisons are presented in Table 9.

Evaluation of the sensitivity of the tests studied suggests that vanadium determination in hair and fingernails might be considered the most sensitive bioindicator of occupational exposure to vanadium. About the same sensitivity of both tests also resulted in a significant correlation of vanadium content in both tissues. In the case of fingernails, the test sensitivity was calculated by taking the median of the controls C-2 as if vanadium were present at this level (see Table 5), while in reality a somewhat lower, but not significantly lower, value can be expected. This would result in a further increase in the sensitivity of this test. Owing to the known growth rate of both tissues, these tests should provide information on the long-term body burden with vanadium. On the other hand, the significance of these tests should not be overestimated, because of the problems with complete removal of external contamination of the tissues. Obviously, by using other cleaning procedures, different results could be obtained; therefore the results of vanadium determination in both fingernails and hair should be interpreted with a caution.

Vanadium levels in blood and urine are undoubtedly more unambiguous and straightforward indicators of occupational exposure to vanadium, al-

Table 9   Evaluation of Sensitivity of Biomonitoring Tests of Exposure to Vanadium

Tissue	Criterion 1: $\dfrac{Max._{(exp)}}{\tilde{x}_{(C-2)}}$	Criterion 2: $\dfrac{\tilde{x}_{(exp)}}{\tilde{x}_{(C-2)}}$
Hair	11,278	1,167
Fingernails	17,056	553
Blood	3,875	239
Urine	3,594	145

though the sensitivity of these tests is 3–5 times and 2.5–8 times lower for the first and second criterion, respectively, compared with the hair and finger-nail tests. Because of the quick excretion of vanadium via the kidneys observed in this work and also by other authors (Kilivuoto et al., 1979, 1981; Kawai et al., 1989; Todaro et al., 1991), vanadium determination in urine should be considered the most suitable test of a very recent exposure. However, the sensitivity of this test will decrease with the time elapsed from the end of exposure. On the other hand, vanadium determination in blood appears to be the best indicator of the long-term body burden with vanadium.

4. CONCLUSIONS

Five tests of biological monitoring of occupational exposure to vanadium were compared by determining vanadium in hair, fingernails, blood, and urine, and cystine in fingernails and hair in workers of a vanadium pentoxide production plant in which the maximum admissible limits for vanadium in the workplace air were exceeded. The same tests were also performed in control persons.

In agreement with the results of Kiviluoto et al. (1980) it was proved that exposure to vanadium has no effect on cystine levels in fingernails, nor in hair. The determination of vanadium in hair and fingernails yielded the highest sensitivity for detecting occupational exposure, because values up to four orders of magnitude higher were found in the exposed workers than in controls. However, the interpretation of these tests may be difficult because of known problems with the cleaning of these tissues prior to analysis to completely remove external contamination (vanadium-rich dust from the factory), without influencing the endogenous element content. Moreover, if a different cleaning of hair is employed than that used in this work, somewhat different vanadium levels and output of the test can be expected. This problem appears to be even more severe for fingernails.

For these reasons, blood and urinary vanadium levels should be considered the most reliable indicators of occupational exposure to vanadium. These tests exhibited up to one order of magnitude lower sensitivity than did hair and fingernail analysis. However, the interpretation of these tests appears to be straightforward, provided that no contamination occurred on sampling and analysis, because the mechanisms of excretion of the inhaled vanadium via these body fluids seems to be reasonably well established. None of these tests should be valued over the other, because they may be regarded as having a complementary role. Urinary vanadium appears to be the best indicator of very recent exposure, because this parameter increases within few hours after the onset of exposure and also drops quickly after its cessation. The latter is explained by the rapid excretion of vanadium via the kidneys, with a half-life between 12 and 18, h as was found in one exposed worker in the present study. After excretion of the major part of the inhaled vanadium, increased urinary vanadium was still found. However, no time changes of this elevated level could be studied. Our knowledge about the time response of blood vanadium in humans after the start of exposure is still lacking. However, owing to the known mechanism of slow remobilization of absorbed vanadium from the bones after several days following the end of exposure, blood vanadium levels may be regarded as the most suitable indicator of the long-term body burden.

The reliable assay of vanadium in these body fluids in occupationally nonexposed persons is a difficult analytical task. Until now, only well-elaborated procedures of RNAA, GF-AAS, preferably with preseparation, and high-resolution ICP-MS proved to be capable of accurate vanadium determination. The RNAA method should be considered superior for this purpose. From the analytical point of view, reliable vanadium determination in urine is somewhat easier to achieve than in blood, because about 3 to 4 times higher element levels in the former fluid of both normal and exposed persons can be expected. This is a reason why the determination of vanadium in urine might be preferred. If the above-mentioned analytical techniques, are best suited for vanadium determination at the ultratrace level in biological samples, are not available, it is proposed that biomonitoring of vanadium exposure by blood and urine analysis be performed in the following way. The elevated vanadium levels can usually be reliably determined by commonly available GF-AAS methods, provided that appropriate quality assurance procedures on sampling and analysis are pursued. Then, the recently published critically evaluated normal vanadium concentrations in human blood, serum, and urine (Sabbioni et al., 1996, Kučera and Sabbioni, this volume, Ch. 5) may be used for a comparison.

ACKNOWLEDGMENTS

The authors want to acknowledge financial support for this work provided by the Grant Agency of the Czech Republic (grant No. 203/93/2326) and by the Internal Grant Agency of the Ministry of Health of the Czech Republic (grant No. 1633-3). Thanks are also due to A. R. Byrne for his contribution to some analyses.

REFERENCES

Bencko, V. (1995). Use of human hair as a biomarker in the assessment of exposure to pollutants in occupational and environmental settings. *Toxicology* **101,** 29–39.

Bowen, H. J. M. (1985). Kale as a reference material. In W. R. Wolf (Ed.), *Biological Reference Materials.* Wiley Interscience, New York, pp. 3–18.

Byrne, A. R., and Kučera, J. (1991). Radiochemical neutron activation analysis of traces of vanadium in biological samples: A comparison of prior dry ashing with post-irradiation wet ashing. *Fresenius J. Anal. Chem.* **340,** 48–52.

Byrne, A. R., and Versieck, J. (1990). Vanadium determination at the ultratrace level in biological reference materials and serum by radiochemical neutron activation analysis. *Biol. Trace Elem. Res.* **26/27,** 529–540.

Chatt, A., and Katz, S. A. (1988). *Hair Analysis, Applications in the Biomedical and Environmental Sciences.* VCH Publishers, New York, Weinheim.

Chatt, A., Sajjad, M., DeSilva, K. N., and Secord, C. A. (1985). Human scalp hair as an indicator in epidemiologic monitoring of environmental exposure to elemental pollutants. In *Health-Related Monitoring of Trace Element Pollutants Using Nuclear Techniques.* IAEA-TECDOC-330. IAEA, Vienna, pp. 33–49.

Clarkson, T. W., Friberg, L., Nordberg, G. F., and Sager, P. R. (1988). *Biological Monitoring of Toxic Metals.* Plenum, New York.

Dillon, H. K., and Ho, M. H. (1991). *Biological Monitoring of Exposure to Chemicals. Metals.* Wiley, New York.

Gladney, E. S., O'Malley, B. T., Roelandts, I., and Gills, T. E. (1987). *Standard Reference Materials: Compilation of Elemental Concentration Data for NBS Clinical, Biological, Geological, and Environmental Standard Reference Materials.* NBS Spec. Publ. 260-111. Washington, D.C.

Gylseth, B., Leira, H. L., Steinnes, E., and Thomassen, Y. (1979). Vanadium in the blood and urine of workers in a ferroalloy plant. *Scand. J. Work Environ. Health* **5,** 188–194.

IAEA (1995). *Survey of Reference Materials.* Vol. 1, *Biological and Environmental Reference Materials for Trace Elements, Nuclides and Microcontaminants.* IAEA-TECDOC-854. IAEA, Vienna, December 1995.

Kawai, T., Seiji, K., Watanabe, T., Nakatsuka, H., and Ikeda, M. (1989). Urinary vanadium as a biological indicator of exposure to vanadium. *Int. Arch. Occup. Health* **61,** 283–287.

Kiviluoto, M., Pyy, L., and Pakarinen, A. (1979). Serum and urinary vanadium of vanadium-exposed workers. *Scand. J. Work Environ. Health* **5,** 362–367.

Kiviluoto, M. Pyy, L., and Pakarinen, A. (1980). Fingernail cystine of vanadium workers. *Int. Arch. Occup. Environ. Health* **46,** 179–182.

Kiviluoto, M., Pyy, L., and Pakarinen, A. (1981). Serum and urinary vanadium of workers processing vanadium pentoxide. *Int. Arch. Occup. Environ. Health* **48,** 251–256.

Kučera, J., Byrne, A. R., Mravcová, A., and Lener, J. (1992). Vanadium levels in hair and blood of normal and exposed persons. *Sci. Total Environ.* **115,** 191–205.

Kučera, J., Lener, J., Mňuková, J. (1994). Vanadium levels in urine and cystine levels in fingernails and hair of exposed and normal persons. *Biol. Trace Elem. Res.* **43–45,** 327–334.

Kučera, J., Šimková, M., Lener, J., Mravcová, A., Kinova, L., and Penev, I. (1990). Vanadium determination in rat tissues and biological reference materials by neutron activation analysis. *J. Radioanal. Nucl. Chem. Articles* **141,** 49–59.

Kučera, J., and Soukal, L. (1988). Homogeneity tests and certification analyses of coal fly ash reference materials by instrumental neutron activation analysis. *J. Radioanal. Nucl. Chem. Articles* **121,** 245–259.

Moens, L., Vanhoe, H., Riondato, J., and Dams, R. (1995). Determination of trace and ultratrace elements in human serum via high resolution inductively coupled plasma mass spectrometry. *Proceedings of the Fifth COMTOX Symposium on Toxicology and Clinical Chemistry of Metals.* Vancouver, July 10–13, p. 2.

Mountain, J. T., Delker, L. L., and Stockinger, H. E. (1953). Studies in vanadium toxicology: Reduction in the cystine content of rat hair. *AMA Arch. Ind. Hyg. Occup. Med.* **8,** 406–411.

Mountain, J. T., Stockell, F. R., and Stokinger, H. E. (1955). Fingernail cystine as an early indicator of metabolic changes in vanadium workers. *Arch. Ind. Health* **12,** 494–502.

Parr, R. M. (1980). *Intercomparison of Minor and Trace Elements in IAEA Animal Muscle H-4.* Report No. 2, IAEA/RL/69. IAEA, Vienna.

Pyy, L., Hakala, E., and Lajunen, L. J. (1981). Screening for vanadium in urine and blood serum by electrothermal atomic absorption spectrometry and d.c. plasma atomic emission spectrometry. *Anal. Chim. Acta* **158,** 297–303.

Rjabuchin J. S. (1978). *Activation Analysis of Hair as an Indicator of Contamination of Man by Environmental Trace Element Pollutants.* Report IAEA/RL/50. IAEA, Vienna.

Sabbioni, E., Kučera, J., Pietra, R., and Vesterberg, O. (1996). A critical review on normal concentrations of vanadium in human blood, serum, and urine. *Sci. Total Environ* **188,** 49–58.

Shapiro, S. S., and Wilk, M. B. (1965). An analysis of variance test for normality (complete samples). *Biometrika* **52,** 591–611.

Todaro, A., Bronzato, R., Buratti, M., and Colombi, A. (1991). *Espositione acuta a polveri conteneti vanadio: Effeti sulla salute e monitoraggio biologico in un gruppo di lavorati addetti alla manutenzione delle caldaie. Med. Lav.* **82,** 141–147.

Versieck, J., Vanballenberghe, L., De Kessel, A., Hoste, J., Wallaeys, B., Vandehaute, J., Baeck, N., Steyaert, H., Byrne, A. R., and Sunderman, Jr., J. (1988). Certification of a second generation biological reference material (freeze-dried human serum) for trace element determination. *Anal Chim. Acta* **204,** 63–75.

WHO (1988). *Environmental Health Criteria 81. Vanadium.* World Health Organization, Geneva.

5

BASELINE VANADIUM LEVELS IN HUMAN BLOOD, SERUM, AND URINE

Jan Kučera

Nuclear Physics Institute, Academy of Sciences of the Czech Republic, CZ-250 68 Řež near Prague, Czech Republic

Enrico Sabbioni

Commission of the European Communities, Joint Research Center Ispra, Environment Institute, I-21020 Ispra (Varese), Italy

Vanadium in the Environment. Part 2: Health Effects, Edited by Jerome O. Nriagu.
ISBN 0-471-17776-8.   © 1998 John Wiley & Sons, Inc.

1. INTRODUCTION

Reliable baseline values or reference levels of elements in human tissues are important indicators of the health status of the general population and a prerequisite for evaluating results of biological monitoring in occupational health studies. Several important compilations of literature values on element concentrations in numerous human tissues and body fluids or in selected clinical specimens have been published (Iyengar et al., 1978; Iyengar, 1987; Iyengar and Woittiez, 1988). Since much interest has been focused on elemental levels in plasma and serum in health and disease, specialized publications on this topic also have appeared (Versieck and Cornelis, 1980, 1989). The first of these two publications revealed that highly divergent data on vanadium in human blood, serum or plasma, and in urine have been published. For instance, published mean values of vanadium in serum and plasma have varied from 0.024 ng/mL up to 420 ng/mL—that is, differing by more than four orders of magnitude. Recently, a critical evaluation of normal concentrations of vanadium in human blood, serum, and urine was published by the present authors (Sabbioni et al., 1996) within the framework of an international project for producing reference values for concentrations of trace elements in human blood and urine: TRACY (Vesterberg et al., 1993). Criteria for the evaluation of publications for the TRACY project are rather stringent and involve 18 items on the quality of sampling, 8 items on the quality of analysis, and 7 items on the quality of statistical treatment. If a publication does not contain sufficient information on the most important factors (determinants) of sampling, analysis, and statistical treatment, it cannot be used for the TRACY project. This approach makes it possible to produce reference values or tentative reference values that are descriptive of a specified population with a particular life-style, dietary habits, and residence in a certain geographic area in a certain period (Vesterberg et al., 1993). On the other hand, publications that do not meet the TRACY criteria may still contain important information, for instance, from the point of view of analytical methodology.

Therefore, the present work also takes into account the literature data not included in the TRACY evaluation (Sabbioni et al., 1996) to show the complexity of problems associated with the accurate determination of vanadium levels in blood, serum or plasma, and urine of occupationally nonexposed persons. An attempt is made to identify the most important factors responsible for the spread of the existing values of vanadium concentrations in human blood, serum or plasma, and urine and to derive the most credible baseline values.

The expression *baseline values* is preferred in the present chapter, because reference values relate by definition to well-defined and sufficiently large population samples. Until now, however, vanadium concentrations in blood, its components, and in urine were mostly investigated in rather small groups of individuals (mostly control groups in various occupational or biomedical studies), except for one large-scale study (Minoia et al., 1990). Another expression frequently used in this context, *normal values,* may be regarded as more suitable for elements that are essential for life and are homeostatically regulated. Under ideal conditions, their levels may be expected to fluctuate within narrow limits, thereby justifying the usage of that term. Since the essentiality of vanadium for humans is still a topic of ongoing research and should be considered as an unresolved question (Sabbioni et al., 1996), the expression *baseline values* seems to be either equivalent to *normal values* or even more appropriate for the purpose of this work.

2. SOURCES OF VARIATION OF VANADIUM BASELINE LEVELS

Generally, it should be recognized that obtaining valid and representative values of trace and ultratrace element levels in biological specimens requires that a number of both biological and analytical criteria be satisfied or accounted for, such as presampling factors, external contamination, and/or elemental losses in sampling, sample handling, and storage; the use of an appropriate analytical methodology; adhering to quality assurance principles; and adequate statistical evaluation of the results obtained.

2.1. Presampling Factors

Iyengar (1982) proposed to distinguish three categories of these factors: (1) biological variations, (2) intrinsic errors, and (3) postmortem changes. Similar classification of these factors has been suggested by Heydorn (1984). The postmortem changes are irrelevant for blood and urine collection from living persons and no indication has been obtained until now that the intrinsic errors, such as medication, subclinical conditions, critically small samples, and hemolysis have any association with variations of vanadium levels in human blood and urine.

Biological variations may be divided into genetic factors, long- and short-term physiological influences, and seasonal changes. No information exists until now that genetic predispositions; short-term physiological influences, such as circadian rhythms, recent meal, posture, and stress; and seasonal physiological changes are associated with varying vanadium concentrations in human body fluids. However, some long-term physiological factors, such as geographical and environmental factors and occupation, have been identified as playing a possible role in fluctuations of baseline vanadium levels in human blood, serum or plasma, and urine (Sabbioni et al., 1996). Therefore, these sources of biological variations are mainly discussed in the present work.

2.1.1. Sex and Age

Indication of a sex difference in serum vanadium levels was reported in a Belgian study (Cornelis et al., 1980, 1981). A narrower range and somewhat lower values were found for women than for men (see Table 1). However, these differences should rather be attributed to the occupations or hobbies of the males examined. One subject with the highest serum vanadium level was a lorry driver, while the others appeared to handle vanadium-containing steel tools frequently either professionally or as a hobby. No significant sex-related differences were found for vanadium concentrations in either serum or urine in other studies (Simonoff et al., 1984; Ishida et al., 1989; Kučera et al., 1994).

2.1.2. Health Status

Some diseases or disorders have been reported to influence vanadium levels in blood, serum, and urine. Significantly lower plasma vanadium was found in manic-depressive illness (Dick et al., 1981). On the other hand, significantly higher serum vanadium was reported for patients with neurotic depression and manic-depressive illness compared with healthy controls but no differences were found between psychotics and neurotics (Simonoff et al., 1987). However, clinical findings may be questioned not only because they are contradictory, but mainly because the values for healthy persons appear to be two orders and one order of magnitude higher, respectively, than the lowest end of values published by other authors that are probably correct (see Table 1). Since the kidneys serve as the main excretory route for the absorbed element (Philips et al., 1983; WHO, 1988), renal failure may contribute to vanadium accumulation in the body. It has been suggested that this mechanism was responsible for elevated serum vanadium in patients with chronic renal failure on chronic hemodialysis therapy (Tsukamoto et al., 1990).

2.1.3. Diet

Dietary habits, among other long-term physiological factors, present the primary source of input for several elements. The possible influence of a recent meal should rather be considered among short-term physiological factors. Both these factors seem to have only minor influence on vanadium levels in blood, serum, and urine owing to the known poor vanadium absorption of the GI tract. Absorption in the range of 1–10% was reported, the lowest end of values within 1–2% being most probable (Heydorn, 1984; WHO, 1988). It has also been shown that vanadium content in diets from different parts of the world does not differ much, the mean values being in the range of 20.0–69.4 μg/kg dry weight (Byrne and Kučera 1990). Thus, considering consumption of about 500 g (dry weight) of the total diet, daily vanadium intake can be estimated as 10–20 μg per day. From the determined vanadium concentrations in dietary items (Byrne and Kosta, 1978; Minoia et al., 1994), the highest vanadium intake can be expected for some wild-growing mushrooms and some beverages, especially beer, for which the range of vanadium concentrations of 18.8–212.1 μg/L with the arithmetic mean $\pm$ SD of 84.5 $\pm$ 54.3 μg/L were found (Minoia et al., 1994). The average vanadium concentrations in drinking water in the United States and Poland were found in the range of

0.06–6 μg/L (WHO, 1988) and values of 0.62 μg/L and 0.1 μg/L were reported from Tokyo and Kyoto, respectively (Tsukamoto et al., 1990). The much lower vanadium level of 0.005 μg/L was recently found in a municipal water supply in the vicinity of Prague (Kučera, 1996, unpublished results) by means of a method with proven accuracy at this level (Byrne and Kučera, 1991, Kučera and Byrne, in press). No specific information exists on the influence of alcohol consumption, smoking habits, and medication and/or supplementation on vanadium levels in blood, serum, and urine (except for the above-mentioned beer drinking).

2.1.4. *Geographical and Environmental Factors*

The vanadium contents of soils are related to those of the parent rocks from which they are formed and range from 3 to 310 mg/kg, the highest concentrations being found in shales and clays (Waters, 1977). No specific regions with especially enhanced or depleted vanadium levels in soils have been identified that might possibly influence vanadium levels in human blood, serum, and urine by either inhaling the dust from soil weathering or by consuming locally produced foods, such as vegetable, fruit, and so forth. The levels of vanadium in fresh water in different parts of the world vary from undetectable to 220 μg/L (WHO, 1988); however, no effect was reported on the element levels in the above-given human fluids, except for a study in which elevated blood vanadium was found in children suspected of drinking water from local wells contaminated by vanadium by seepage from the nearby dumping place of a vanadium pentoxide production plant (Kučera et al., 1992). This was explained by vanadium accumulation in the bone and its subsequent slow mobilization. The vanadium concentrations in air also vary widely, the lowest (0.001–0.002 ng/m^3) being found at the South Pole (Zoller et al., 1974). The mean vanadium air concentrations of 0.1 ng/m^3 (the eastern Pacific ocean) to 0.72 ng/m^3 (rural northwest Canada) can be regarded as natural background levels for uncontaminated rural regions. Typical concentrations in urban air vary over a wide range of about 0.25 to 300 ng/m^3, with markedly higher concentrations during the winter months compared to summer months (WHO, 1988). The most important source of air pollution is combustion of fossil fuels, mainly oil but also coal, in power and heating plants, and local boilers (Sabbioni and Goetz, 1983; Sabbioni et al., 1984). A comparison of enrichment factors, EF, in ashes and stack emissions of lignite, oil, and municipal waste combustion plants showed a very high value (EF > 2,000) for vanadium released on oil burning (Obrusník et al., 1989; Kučera et al., 1993). As a consequence, for instance, burning of fuel oil may increase local rural levels of airborne vanadium to about 75 ng/m^3 (WHO, 1988). Other important vanadium pollution sources include vanadium pentoxide and metallurgical plants. In their vicinity, vanadium concentration in the air amounting to 1,000 ng/m^3 or even to higher values are often found. Assuming an average air concentration of about 50 ng/m^3, about 1 μg of vanadium may enter the respiratory tract daily and about 25% of soluble vanadium compounds may be absorbed (WHO, 1988). Such an amount may already influence the element levels in blood, serum, or urine. Therefore, owing to very high vanadium enrichment in air particulates and inhalation as one of the important routes of vanadium intake, living in contaminated areas should be considered a very important presampling factor.

2.1.5. Occupation

Well-known sources of occupational exposure to vanadium involve cleaning of oil boilers and vanadium pentoxide production. Furthermore, many metallurgical processes are associated with the production of vanadium-containing vapors, which condense to form respirable aerosols. Inhalation of the contaminated workplace air has been identified as responsible for a significant increase in vanadium levels in blood, serum, and urine in a number of studies (Gylseth et al., 1979; Kiviluoto et al., 1979; Pyy et al., 1984; Kiviluoto et al., 1981; Kawai et al., 1989; Todaro et al., 1991; Kučera et al., this volume, Ch. 4). Obviously, the exposed subjects must be excluded from studies aimed at determining vanadium baseline levels.

2.2. Sampling, Sample Handling, and Storage

Naturally occurring vanadium levels are considered to be among the lowest of all elements in mammals (Zenz, 1980). Since vanadium is contained in various percentages in disposable steel needles, collection vials, storage containers, and some chemicals and reagents, and the element may also be present in nonnegligible quantities in the ambient air, adequate conditions for sampling, sample handling, and storage of blood, its derivatives, and urine are of paramount importance in preventing external contamination.

Somewhat contradictory information exists about the risk of using stainless steel needles for blood collection. Radiorelease tests on neutron-activated stainless steel needles and syringes showed that vanadium in amounts up to 0.7 ng/mL and 0.1 ng/mL, respectively, may be released on blood collection (Minoia et al., 1992). On the other hand, the lowest reported mean value of blood vanadium was found, by employing RNAA with proven accuracy, in samples collected with disposable steel needles (after flushing with about 10 mL of blood) (Kučera et al., 1992). Thus, it seems that the material of the needle is probably not critical as regards the possibility of blood contamination. However, since the vanadium content in the steels used for manufacturing the needles may vary, the use of plastic, preferably Teflon cannulas, should be recommended, because it guarantees that the external contamination during blood sampling is excluded.

Vials made of ultrapure quartz and/or polyethylene precleaned with ultrapure acids (preferably HCl) and demineralized water were proved to introduce no detectable contamination during sample collection and storage. No loss of vanadium during freeze-drying and storage at $-20°C$ for 1 month occurs, as was shown in experiments using blood from rats or rabbits labeled with [48]V radiotracer in "metabolized form," that is, administered to the animals in doses similar to those expected in human blood, 2–7 days prior to the start of the experiments (Minoia et al., 1992).

A great risk of contamination may be involved in preconcentration procedures, especially when nitric acid is used (Blotsky et al., 1989b). For instance, unacceptably high reagent blank values for blood and urine analysis of about 2 ng of vanadium were reported for a separation procedure preceding neutron activation analysis (NAA) that included digestion of biological material in high-purity conc. HNO_3 followed by ion exchange chromatography (Blotsky et al., 1989a; Blotsky et al., 1989b). A preseparation NAA procedure employing co-precipitation of vanadium with lead or bismuth pyrolidinedithiocarbamates—$Pb(PDC)_2$ or $Bi(PDC)_3$ (Lavi and Alfassi 1988)—also yielded high vanadium values in serum (see Table 1). On the other hand, NAA with preseparation using subboiled HNO_3 and ion exchange chromatography carried out in a Class-100 clean laboratory yielded vanadium values in serum that fall within the range of the most probably correct levels (Greenberg et al., 1990).

It was also shown that dry-ashing at 450°C in quartz vials prior to analysis, but without any chemical treatment or ashing aids, introduced neither contamination nor losses of vanadium for blood analysis (Cornelis et al., 1980, 1981; Byrne and Kučera, 1991).

2.3. Analysis and Quality Assurance

A review of analytical procedures for the determination of vanadium in biological materials has recently been published (Seiler, 1995). Obviously, there are several widely available analytical techniques capable of accurate vanadium determination in various biological materials at the milligram per kilogram level, such as graphite furnace atomic absorption spectrometry (GF-AAS), inductively coupled plasma atomic emission spectrometry (ICP-AES), and several other techniques—for instance, high-performance liquid chromatography (HPLC). Less available, but also very suitable techniques are nondestructive, so-called instrumental NAA (INAA) and measurement of proton-induced X-ray emission (PIXE).

However, for determination in blood and its derivatives and in urine, in which vanadium levels are usually below the microgram per liter level in occupationally nonexposed persons, the choice of suitable analytical methods is much limited. In view of the required detection limit (<10 pg/mL), the only techniques that can be considered are improved GF-AAS, isotope dilution mass spectrometry (IDMS), inductively coupled plasma mass spectrometry (ICP-MS), and radiochemical NAA (RNAA). In the case of GF-AAS, the direct determination of vanadium in urine or diluted serum is hardly feasible owing to an insufficient detection limit and the possibility of matrix interferences. However, a preconcentration procedure using extraction with *N*-benzoyl-*N*-(*o*-tolyl)hydroxyl amine from the wet-digested samples was successfully applied for vanadium determination in serum and urine (Ishida et al., 1989; Tsukamoto, 1990). IDMS looks like a promising analytical tool. However, this technique has been applied to vanadium determination in human serum only in one study (Fasset and Kingston, 1985) and the high mean value obtained (see Table 1) indicated that contamination problems might occur. Application of the usual ICP-MS is not suitable for the accurate determination of extremely low vanadium concentrations because of polyatomic interferences (Moens and Dams, 1995). Recently, however, high-resolution ICP-MS was demonstrated to be capable of this task (Moens et al., 1995) by vanadium determination in a second generation biological reference material (uncontaminated freeze-dried human serum) that yielded results in agreement with those obtained by RNAA (Byrne and Versieck, 1990; Kučera et al., 1992). While most of the above techniques may suffer from blank problems, the only blank-free method is NAA, provided that no preseparation step is used except for dry-ashing. RNAA with totally postirradiation separation is difficult to perform owing to the rather short half-life of the analytical radioisotope ^{52}V ($T_{1/2}$ = 3.75 min). The use of a high neutron fluence of 2×10^{15} to 1×10^{16} n/cm^2 for irradiation and a high-efficiency well-type HPGe detector for counting of the separated fractions minimizes the required amount of sample for analysis to 1–2 mL of blood, serum, or urine. The total sample decomposition and vanadium separation, with very high decontamination factors (10^7–10^8) for ^{24}Na, ^{38}Cl, and other interfering radionuclides, must be completed within 6–12 min. This has been mastered by only a few research groups (Byrne and Kosta, 1978; Cornelis et al., 1980, 1981; Byrne and Kučera, 1990; Heydorn, 1990; Byrne and Versieck, 1990; Kučera et al., 1992; Kučera et al., 1994). The separation time can be shortened by a few minutes if dry-ashing prior to irradiation is carried out (Byrne and Kučera, 1991). Therefore, this mode also results in a lower vanadium detection limit. Very high separation yield (>95%), which can accurately be determined by using the ^{48}V radiotracer, is achieved when, for instance, extraction of vanadium in the pentavalent state with N-benzoyl-N-phenyl hydroxylamine is employed. This separation can also be made very specific, so that the ^{52}V γ-rays are counted on a virtually "zero" spectral background (Byrne and Versieck, 1990). These features make the RNAA method superior for vanadium determination at the ultratrace level.

Other authors (Allen and Steinnes, 1978; Sabbioni and Maroni, 1983; Blotsky et al., 1989a; Blotsky et al. 1989b; Lavi and Alfassi, 1990) developed NAA with various preseparation procedures to overcome or circumvent the necessity of speedy operations with highly radioactive samples in RNAA. However, with this approach contamination and blank problems cannot be excluded. It has been shown by using nonparametric statistics that NAA with preseparation invariably gives high results (Heydorn, 1990). The only exception until now is the above-mentioned study performed in the Class-100 clean laboratory (Greenberg et al., 1990).

The possibilities of pursuing quality assurance of analytical procedures through employing comparative assays by applying adequate and independent analytical techniques have heretofore been limited because suitable techniques and well-elaborated procedures have been available in a few laboratories only.

Therefore, the use of reference materials (RMs) should have played a more important role. However, as with analytical techniques, the availability of suitable (i.e., matrix- and vanadium-level-matched) RMs decreases with decreasing vanadium levels. In a recent review (IAEA, 1995), 35 biological and environmental RMs with certified vanadium values higher than 1 mg/kg (and 29 RMs with noncertified values) are listed, while for vanadium concentrations below this level only 9 RMs (and 20 RMs with noncertified values) are available. Of the latter category, no biological RM exists with a certified vanadium value. For checking the accuracy of the determination of baseline vanadium values in blood, serum, and urine, there are only noncertified values available of 0.06 μg/kg and 0.65 μg/kg, dry weight, in NIST-SRM 1598 inorganic constituents in bovine serum and Versieck's second generation biological RM (freeze-dried human serum), respectively. This lack of low-level vanadium biological RMs is thus still a severe handicap for quality assurance of vanadium-related biological and biomedical studies, including establishing of vanadium reference values in human blood, serum, and urine.

3. SELECTED VANADIUM BASELINE LEVELS

The original spread of the published mean vanadium values in serum and plasma from 0.047 ng/mL to 420/ng mL (Versieck and Cornelis, 1980) can be significantly reduced toward lower values by taking into account that the highest values were obtained by analytical techniques lacking a sufficiently low detection limit or by techniques biased by unrecognized interferences, such as spectrophotometry or emission spectrography. Therefore, in the present review only such values in human blood, plasma or serum, and urine are discussed as have been obtained by analytical techniques capable of accurate vanadium determination at very low levels. However, it should be emphasized that not only the capability of an analytical technique itself, but also the availability in laboratories of the experience required to perform well-elaborated procedures in the optimum way are very important factors for accurate vanadium determination in human blood, serum, and urine (Heydorn 1990).

   Table 1 summarizes the mean vanadium values in human blood and serum that have been obtained by the potentially suitable analytical techniques. It is obvious that the lowest values were determined by RNAA procedures with minimum sample handling prior to irradiation and with proven accuracy by control analyses of suitable RMs. In this method, there are no detectable blank values (for instance, due to contamination by, or recoil effect of, impurities from irradiation polyethylene containers); the yield of (radio) chemical separation can be controlled by using the [48]V radiotracer; vanadium separation and counting of the analytical [52]V radioisotope can be performed in such a way that almost all neutron-induced radionuclides (namely, [24]Na, [42]K, [38]Cl, [56]Mn, [66]Cu, [80]Br) are removed and thus an almost zero background signal can

Table 1   Vanadium Levels in Blood and Serum (in ng/mL)

Tissue[a]	Ar. Mean ± sd or Median	N[b]	Analytical Method	Investigators
S	0.023 ± 0.014	4	NAA with preseparation in a Class 100 clean laboratory	Greenberg et al., 1990
S	0.031 ± 0.010	36w	RNAA[c]	Cornelis et al., 1980, 1981
S	0.064	22m		
B	0.042	17c	RNAA[c]	Kučera et al., 1992
B	0.056	11		
B	0.059	11	RNAA[c]	Byrne and Kučera, 1990
S	0.071	10	RNAA[c]	Byrne and Versieck, 1990
S	<0.08–0.24	64	GF-AAS with preconcentration	Ishida et al., 1989, Tsukamoto et al., 1990
B	0.35 ± 0.11	65	NAA with preseparation	Minoia et al., 1990
S	<0.5	Pooled sample	GF-AAS with preconcentration	Pyy et al., 1984
S	0.62 ± 0.03	415	GF-AAS, Zeeman correction	Minoia et al., 1990
S	0.67 ± 0.33	23	RNAA[d]	Simonoff et al., 1984, Simonoff et al., 1987
S	0.72 ± 0.21	8	RNAA	Heydorn, 1990
B	0.77 ± 0.09	7	RNAA[e]	Allen and Steinnes, 1978
B	0.94	12	NAA with preseparation	Sabbioni and Maroni, 1983
S	1.1	10	HPLC with colorimetric detection	Godin, 1990
S	<2	?	GF-AAS	Missenard et al., 1989
S	2.6 ± 0.3	7	IDMS	Fasset and Kingston, 1985
S	3.4 ± 0.6	10	GF-AAS, deuterium correction	Stroop et al., 1982
S	3.7 ± 1.6	20	NAA with preseparation	Lavi and Alfassi, 1990

[a] B, whole blood; S, serum.
[b] Number of subjects; w, women only; m, men only; c, children only.
[c] Minimum sample handling prior to irradiation; accuracy proved by suitable RMs.
[d] Contamination during sample preparation cannot be excluded.
[e] Elevated spectrum background due to incomplete removal of the ^{38}Cl radioisotope.

be achieved below the analytical peak of ^{52}V. Values very close to those determined by RNAA with minimum sample treatment were also obtained by NAA with preirradiation separation in a clean Class-100 laboratory and by an improved version of GF-AAS. The reason for the higher values (more than 0.5 ng/mL) also obtained by RNAA is most likely an insufficient prevention of contamination or incomplete removal of interfering radionuclides, which makes the signal-to-background ratio unfavorable. Also the other high results obtained by different techniques suffered most probably from contamination problems or unrecognized interferences.

Thus, a growing body of data appears to suggest that the baseline mean vanadium levels in blood and serum are in the range of 0.02 ng/mL to 0.1 ng/mL. No attempt should be made to interpret these levels as reference values, because they were determined in rather small population samples. The single large-scale study aimed at deriving reference values in tissues of Italian subjects (Minoia et al., 1990) yielded somewhat higher vanadium levels in blood and serum of 0.35 ± 0.11 ng/mL and 0.62 ± 0.03 ng/mL, respectively (Table 1). The question whether these values are elevated owing to specific geographical, environmental, and dietary factors or to unnoticed analytical problems cannot, however, be answered at present.

For urinary vanadium, the lowest mean value was again determined by RNAA with proven accuracy and with minimum sample handling prior to and during analysis (Table 2). Much more consistent results, compared with those for blood and serum, were also obtained by the GF-AAS method with various preconcentrations. Obviously, this is due to higher baseline vanadium concentrations in urine than in blood or serum. A methodological study on 20 subjects in which NAA with preirradiation separation, GF-AAS with de-

Table 2   Urinary Vanadium Levels (in ng/mL)

Ar. Mean $\pm$ sd or Median	N^a	Analytical Method	Investigators
0.212	21	RNAA[b]	Kučera et al., 1994
0.24 ± 0.14	42	GF-AAS with preconcentration	Ishida et al., 1989
0.36 ± 0.19	87	GF-AAS with preconcentration	Burrati et al., 1985
0.44 ± 0.15	11	GF-AAS with preconcentration	Buchet et al., 1982
0.8 ± 0.08	382	GF-AAS	Minoia et al., 1990
5.4 ± 2.8	20	NAA with preseparation	Sabbioni and Maroni, 1983
6.4 ± 5.8	20	GF-AAS with destruction of organic matrix	
5.9 ± 4.9	20	GF-AAS with preconcentration	

[a] Number of subjects.

[b] Minimum sample handling prior to irradiation, accuracy proved by suitable RMs.

struction of organic matter, and GF-AAS with preconcentration were employed resulted in about one order magnitude higher results (in the range of 5.4 ng/mL to 6.4 ng/mL) than those determined by this technique and RNAA by other researchers. However, a large variation of individual results suggests that sample contamination might be responsible for these high values.

It can be concluded that the most probable mean baseline urinary vanadium levels are in the range of 0.2 ng/mL to 0.4 ng/mL or even up to 0.8 ng/mL, the highest value and its 95% confidence interval being considered the reference value for Italian subjects (Minoia et al., 1990).

4. CONCLUSIONS

Scrutinizing of biological and analytical factors that may influence baseline vanadium levels in human blood, serum, and urine revealed that accurate determination of the true vanadium concentrations in these body fluids is still a demanding task. It appears that a choice of appropriate analytical methodology with the vanadium detection limit around 10 pg/mL, as well as sufficient experience in applying such well-elaborated analytical procedures, is of crucial importance for arriving at the true vanadium values. Of the available analytical techniques, RNAA seems to be superior for this purpose, because, apart from its inherent advantages, this method most frequently yielded results that are considered to be correct. Adequate care should also be paid to avoiding contamination during sampling and sample handling, especially if preconcentration procedures are employed. Among the presampling, biological factors, long-time residence in regions with polluted environment, mainly as a result of the burning of fuel oil and also coal, has been identified as possibly responsible for elevated vanadium levels in human blood, serum, and urine of occupationally nonexposed subjects.

From results obtained in rather small population samples it appears that baseline vanadium levels in blood and serum are most probably in the range of 0.02 ng/mL to 0.08 ng/mL, while about one order of magnitude higher values (i.e., 0.2 ng/mL to 0.4 ng/mL) may be expected in urine in most populations. More large-scale studies in well-described population samples are still needed, especially on samples collected in both environmentally polluted and unpolluted regions, to be able to confirm these tentatively proposed vanadium baseline values and to elucidate the possible influence of environmental factors, mostly air pollution.

REFERENCES

Allen, R. O., and Steinnes, E. (1978). Determination of vanadium in biological materials by radiochemical neutron activation analysis. *Anal. Chem.* **50,** 1553–1555.

Blotsky, A. J., Hamel, F. G., Stranik, A., Ebrahim, A., Sharma, R. B., Rack, E. P., and Solomon, S. S. (1989a). Determination of vanadium in biological tissue by anion exchange chromatography and neutron activation analysis. *J. Radioanal. Nucl. Chem. Articles* **131**, 319–329.

Blotsky, A. J., Duckworth, W. C., Ebrahim, A., Hamel, F. K., Rack, E. P., and Sharma, R. B. (1989b). Determination of vanadium in serum by pre-irradiation and post-irradiation chemistry and neutron activation analysis. *J. Radioanal. Nucl. Chem. Articles* **134**, 151–160.

Buchet, J. P., Knepper, E., and Lauwerys, R. (1982). Determination of vanadium in urine by electrothermal atomic absorption spectrometry. *Anal. Chim. Acta* **136**, 243–248.

Burrati, M., Pellegrino, O., Caravelli, G., Calzaferri, G., Bettinelli, M., Colombi, A., and Maroni, M. (1985). Sensitive determination of urinary vanadium by solvent extraction and atomic absorption spectroscopy. *Clin. Chem.* **150**, 53–58.

Byrne, A. R., and Kosta, L. (1978). Vanadium in foods and in human body fluids and tissues. *Sci. Total Environ.* **10**, 17–30.

Byrne, A. R., and Kučera, J. (1990). New data on levels of vanadium in man and his diet. In B. Momčilović (Ed.), *Trace Elements in Man and Animals 7 (TEMA-7)*. IMI, Zagreb, Ch. 25, pp. 18–20.

Byrne, A. R., and Kučera, J. (1991). Radiochemical neutron activation analysis of traces of vanadium in biological samples: A comparison of prior dry ashing with post-irradiation wet ashing. *Fresenius J. Anal. Chem.* **340**, 48–52.

Byrne, A. R., and Versieck, J. (1990). Vanadium determination in the ultratrace level in biological reference materials and serum by radiochemical neutron activation analysis. *Biol. Trace Elem. Res.* **26/27**, 529–540.

Cornelis, R., Versieck, J., Mees, L., Hoste, J., and Barbier, F. (1980). Determination of vanadium in human serum by neutron activation analysis. *J. Radioanal. Chem.* **55**, 35–43.

Cornelis, R., Versieck, J., Mees, L., Hoste, J., and Barbier, F. (1981). The ultratrace element vanadium in human serum. *Biol. Trace Elem. Res.* **3**, 257–263.

Dick, D. A. T., Dick, E. G., and Naylor, G. J. (1981). Plasma vanadium concentration in manic-depressive illness. *J. Physiol.* **310**, 24P.

Fasset, J. D., and Kingston, H. M. (1985). Determination of nanogram quantities of vanadium in biological material by isotope dilution thermal ionization mass spectrometry with ion counting detection. *Anal. Chem.* **57**, 2474–2478.

Godin, J. (1990). High-performance liquid chromatographic method for the determination of vanadium in serum. *J. Chromatog.* **532**, 445–448.

Greenberg, R. R., Kingston, H. M., Zeisler, R., and Woittiez, J. (1990). Neutron activation analysis of biological samples with a preirradiation separation. *Biol. Trace Elem. Res.* **26/27**, 17–25.

Gylseth, B., Leira, H. L., Steinnes, E., and Thomassen, Y. (1979). Vanadium in the blood and urine of workers in a ferroalloy plant. *Scand. J. Work Environ. Health* **5**, 188–194.

Heydorn, K. (1984). *Neutron Activation Analysis for Clinical Trace Element Research*. CRC Press, Boca Raton, FL.

Heydorn, K. (1990). Factors affecting the levels reported for vanadium in human serum. *Biol. Trace Elem. Res.* **26/27**, 541–551.

IAEA. (1995). *Survey of Reference Materials. Biological and Environmental Reference Materials for Trace Elements, Nuclides and Microcontaminants*. IAEA-TECDOC-854, Vol. 1. IAEA, Vienna.

Ishida, O., Kihura, K., Tsukamoto, Y., and Marumo, F. (1989). Improved determination of vanadium in biological fluids by electrothermal atomic absorption spectrometry. *Clin. Chem.* **35**, 127–130.

Iyengar, V. (1982). Presampling factors in the elemental composition of biological systems. *Anal. Chem.* **54**, 554A–558A.

Iyengar, G. V. (1987). Reference values for the concentrations of As, Cd, Co, Cr, Cu, Fe, I, Hg, Mn, Mo, Ni, Pb, Se, and Zn in selected human tissues and body fluids. *Biol. Trace Elem. Res.* **12**, 263–295.

Iyengar, V. G., Kollmer, W. E., and Bowen, H. J. M. (1978). *Elemental Composition of Human Tissues and Body Fluids.* Verlag Chemie, Weinheim.

Iyengar, V., and Woittiez, J. (1988). Trace elements in human clinical specimens: Evaluation of literature data to identify reference values. *Clin. Chem.* **34**, 474–481.

Kawai, T., Seiji, K., Watanabe, T., Nakatsuka, H., and Ikeda, M. (1989). Urinary vanadium as a biological indicator of exposure to vanadium. *Int. Arch. Occup. Health* **61**, 283–287.

Kiviluoto, M., Pyy, L., and Pakarinen, A. (1979). Serum and urinary vanadium of vanadium-exposed workers. *Scand. J. Work Environ. Health* **5**, 362–367.

Kiviluoto, M., Pyy, L., and Pakarinen, A. (1981). Serum and urinary vanadium of workers processing vanadium pentoxide. *Int. Arch. Occup. Environ. Health* **48**, 251–256.

Kučera, J., and Byrne, A. R. (1997). Quality assurance of neutron activation analysis of traces of vanadium in the workplace air, environment and human tissues. *Harmonization of Health-Related Environmental Measurements Using Nuclear and Isotopic Techniques (Proceedings of the IAEA Symposium).* IAEA-SM-344/27. Hyderabad, India, Nov. 4–7, 1996, IAEA, Vienna.

Kučera, J., Byrne, A. R., Mravcová, A., and Lener, J. (1992). Vanadium levels in hair and blood of normal and exposed persons. *Sci. Total. Environ.* **15**, 191–205.

Kučera, J., Lener, J., and Mňuková, J. (1994). Vanadium levels in urine and cystine levels in fingernails and hair of exposed and normal person. *Biol. Trace. Elem. Res.* **43/45**, 327–334.

Kučera, J., Soukal, L., and Horáková, J. (1993). Instrumental neutron activation analysis of Czechoslovak candidate reference material of municipal waste incinerator ash. *In Proceedings of the Third Meeting of the ECE Task Force HEMET (Heavy Metals Emission), Berlin, February 16–18, 1993,* pp. 93–97.

Lavi, N., and Alfassi, Z. B. (1988). Determination of trace amounts of titatinium and vanadium in human blood serum by neutron activation analysis: Coprecipitation with Pb(PDC)$_2$ or Bi(PDC)$_3$. *J. Radioanal. Nucl. Chem. Lett.* **126**, 361–374.

Lavi, N., and Alfassi, Z. B. (1990). Determination of trace amounts of cadmium, cobalt, chromium, iron, molybdenum, nickel, selenium, titanium, vanadium and zinc in blood and milk by neutron activation analysis. *Analyst* **115**, 817–822.

Minoia, C., Pietra, R., Sabbioni, E., Ronchi, A., Gatti, A., Cavalleri, A., and Manzo, L. (1992). Trace element reference values in tissues from inhabitants of the European Community. III. The control of preanalytical factors in the biomonitoring of trace elements in biological fluids. *Sci. Total Environ.* **120**, 63–79.

Minoia, C., Sabbioni, E., Apostoli, P., Pietra, R., Pozzoli, L., Gallorini, M., Nicolaou, G., Alessio, L., and Capodaglio, E. (1990). Trace element reference values in tissues from inhabitants of the European community, I. A study of 46 elements in urine, blood, and serum of Italian subjects. *Sci. Total Environ.* **95**, 89–105.

Minoia, C., Sabbioni, E., Ronchi, A., Gatti, A., Pietra, R., Nicolotti, A., Fortaner, S., Bulducci, C., Fonte, A., and Roggi, C. (1994). Trace element values in tissues from inhabitants of the European Community. IV. Influence of dietary factors. *Sci. Total Environ.* **141**, 181–195.

Missenard, C., Hansen, G., Kutter, D., and Kremer, A. (1989). Vanadium induced impairment of haem synthesis. *British J. Ind. Med.* **46**, 744–747.

Moens, L., and Dams, R. (1995). A comparison between two methods for trace and ultratrace element analysis. *J. Radioanal. Nucl. Chem. Articles* **192**, 29–38.

Moens, L., Vanhoe, H., Riondato, J., and Dams, R. (1995). Determination of trace and ultratrace elements in human serum via high resolution inductively coupled plasma mass spectrometry. *Proceedings of the Fifth COMTOX Symposium on Toxicology and Clinical Chemistry of Metals,* Vancouver, *July 10–13,* p. 2, Association of Clinical Scientists, Farmington, CT.

Obrusník, I., Stáková, B., and Blažek, J. (1989). Composition and morphology of stack emissions from coal and oil fuelled boilers. *J. Radioanal. Nucl. Chem. Articles* **133,** 377–390.

Philips, T. D., Nechay, T. D., and Heidelbaugh, N. D. (1983). Vanadium: Chemistry and kidney. *Federation Proc.* **42,** 2969–2973.

Pyy, L., Hakala, E. and Lajunen, L. J. (1984). Screening for vanadium in urine and blood serum by electrothermal atomic absorption spectrometry and d.c. plasma atomic emission spectrometry. *Anal. Chim. Acta,* **158,** 297–303.

Sabbioni, E., and Goetz, L. (1983). Mobilization of heavy metals from fossil-fuelled power plants. Potential ecological and biochemical implications, IV. Assessment studies of the European community situation., *Report EUR 6998/1983,* CEC, Luxembourg.

Sabbioni, E., Goetz, L., and Bignoli, G. (1984). Health and environmental implications of trace metals released from coal-fired power plants: an assessment study of the situation in the European Community. *Sci. Total Environ.* **40,** 141–154.

Sabbioni, E., Kučera, J., Pietra, R. and Vesterberg, O. (1996). A critical review on normal concentrations of vanadium in human blood, serum, and urine. *Sci. Total. Environ.* **188,** 49–58.

Sabbioni, E., and Maroni, M. (1983). A study on vanadium in workers from oil fired power plants. *Report EUR 9005 EN,* CEC, Joint Research Centre, Ispra Establishment, Italy.

Seiler, H. G. (1995). Analytical procedures for the determination of vanadium in biological materials. In Sigel, H., and Sigel, A., Eds. *Metal ions in biological systems. Vol. 31. Vanadium and its role in Life.* Marcel Dekker, Inc., New York, Basel.

Simonoff, M., Llabador, Y., Hamon, C., Berdeu, B., Simmonoff, G., Conri, C., Fleury, B., Couzigou P., and Lucena, A. (1987). Vanadium in depression and cirrhosis. *J. Radioanal. Nucl. Chem. Articles* **113,** 107–117.

Simonoff, M., Llabador, Y., MacKenzie Peers, A., and Simonoff G. (1984). Vanadium in human serum, as determined by neutron activation analysis. *Clin. Chem.* **30,** 1700–1703.

Stroop, S. D., Helinek, G., and Greene, H. L. (1982). More sensitive flameless atomic absorption analysis of vanadium in tissue and serum. Clin. Chem. **28,** 79–82.

Todaro, A., Bronzato, R., Buratti, M., and Colombi, A. (1991). Esposizione acuta a polveri conteneti vanadio: Effeti sulla salute e monitoraggio biologico in un gruppo di lavorati addetti alla manutenzione delle caldaie. *Med. Lav.* **82,** 141–147.

Tsukamoto, Y., Saka, S., Kumano, K., Iwanami, S., Ishida, O., and Marumo, F. (1990). Abnormal accumulation of vanadium in patients on chronic hemodialysis theraphy. *Nephron* **56,** 368–373.

Versieck, J., and Cornelis, R. (1980). Normal levels of trace elements in human blood plasma or serum. *Anal. Chim. Acta* **116,** 217–254.

Versieck, J., and Cornelis, R. (1989). *Trace Elements in Human Plasma or Serum.* CRC Press, Boca Raton, FL.

Vesterberg, O., Alessio, L., Brune, D., Gerhardsson, L., Herber, R., Kazantzis, G., Nordberg, G. F., and Sabbioni, E. (1993). International project for producing reference values for concentrations of trace elements in human blood and urine—TRACY. *Scand. J. Work Environ. Health* **19** (Suppl. 1), 19–26.

Waters, M. D. (1977). Toxicology of vanadium. *Adv. Mod. Toxicol.* **2,** 147–189.

WHO (1988). *Environmental Health Criteria 81. Vanadium.* World Health Organization, Geneva.

Zenz, C. (1980). Vanadium. In H. A. Waldron (Ed.), *Metals in the Environment.* Academic Press, London, New York, Toronto, Sydney, San Francisco, Chapter 10, pp. 301–313.

Zoller, W. H., Gladney, E. S., and Duce, R. A. (1974). Atmospheric concentration and sources of trace metals at the South Pole. *Science* **183,** 198–200.

6

VANADIUM AND METABOLIC PROBLEMS

Visith Sitprija and Somchai Eiam-Ong

Nephrology Unit, Department of Medicine, Faculty of Medicine, Chulalongkorn University Hospital, Bangkok, Thailand 10330

Vanadium in the Environment. Part 2: Health Effects, Edited by Jerome O. Nriagu.
ISBN 0-471-17776-8. © 1998 John Wiley & Sons, Inc.

---

1. BASIC CONSIDERATIONS

1.1. General Biochemical Characteristics of Vanadium

Vanadium, a group V element (molecular weight 50.942), is a greyish metal that occurs in the form of two natural isotopes, ^{50}V and ^{51}V (Clark, 1975). The metal belongs to the first transition series and forms oxidation states of -1, 0, $+2$, $+3$, $+4$, and $+5$ (Phillips et al., 1983). Vanadium compounds are mainly in valence states $+3$, $+4$ (tetravalent), and $+5$ (pentavalent) (Erdmann et al., 1984). In the presence of O_2, air, and oxidizing agents, or in oxygenated blood, vanadium is always in the $+5$ oxidation state. Oxidation state $+4$, however, is the most stable form and can occur when vanadium compounds are combined with reducing agents. (Chapter 4, Vol. 30)

1.2. Source and Distribution of Vanadium

Vanadium is widespread with various concentrations in all environments including rock, soil, water, air, plants, and animal tissue (WHO, 1988). Migration, diffusion, and concentration of vanadium in the biosphere occur as a net result of its extraction by living organisms from water, from food of both vegetable and animal origin, and from different types of rocks during their decomposition and the formation of soils. Vanadium concentrations in rocks depend on the pH of the rocks; the highest occur in basic rocks and the lowest in acid ones. The prevalence of vanadium approximately equals that of copper, lead, or zinc. Metallic vanadium does not occur in nature. About 70 vanadium compounds have been discovered (Chapter 1, Vol. 30). Many compounds are present in fossil fuels (oil, coal, shale). Power and heat-producing plants using fossil fuels appear to cause the most widespread discharge of vanadium into the environment. It has been estimated that combustion of coal and oil in power plants for electricity production could have mobilized about 10,000–20,000 tons of vanadium per year in the past decade (Sabbioni et al., 1984).

   The levels of vanadium in fresh water from different places are variable, depending on the difference in rainwater runoff from natural sources or in industrial effluent. In general, vanadium concentrations in drinking water are less than 10 $\mu g/L$ (Durfor and Becker, 1963). The range is between 1 and 30 $\mu g/L$, with an average of about 5 $\mu g/L$.

   Natural sources of airborne vanadium are continental dust and marine aerosols. It has been estimated that large cities may have annual average air levels of the order of 20–100 ng/m^3. Vanadium concentrations in rural air, less than 1 ng/m^3, are lower than those in urban air; which has markedly higher vanadium concentrations during the winter months than in the summer season. About 1 μg yearly or 250 ng daily of vanadium may enter the respiratory tract if one assumes an average air concentration of about 50 ng/m^3. Exposure to high concentrations of vanadium in the air may occur in working environments. The highest vanadium concentrations of 50–100 mg/m^3, or sometimes reaching 500 mg/m^3, have been observed in boiler cleaning.

It has been estimated that about 25% of soluble vanadium compounds may be absorbed by pulmonary route (ICRP, 1960). Animal studies have shown that the clearance time of soluble vanadium after exposure via inhalation depends on the time and nature of exposure to the compound. Following intratracheal instillation in rats, the clearance rate from lung of vanadium trioxide is more rapid than that of pentoxide or ammonium vanadate. Animal studies do not indicate significant accumulation in the lung. The concept of vanadium accumulation in the human lung with age thus remains unestablished.

All plants contain small amounts of vanadium. The concentration of vanadium in soil is about 10 times that in plants (Cannon, 1963). The aerial portions of plants have the lowest concentrations, while the root portions have nearly the same levels of vanadium as the soil in which the plants are grown. Absorption of vanadium from soil appears to be passive. Vanadium is present in all animals, but tissue levels in most vertebrates are very low. Marine species, especially invertebrate, have higher tissue levels of vanadium (Bertrand, 1950). The highest levels in land mammals occur in the liver and skeletal tissues (Schroeder, 1970).

The main source of vanadium intake for the general population is food. Available data show low levels of vanadium in most items of the human diet (Byrne and Kosta, 1978; Myron et al., 1977; Söremark, 1967). Grains have higher levels of vanadium than fruits and vegetables. Concentrations of vanadium in oils and fats and beef and pork are low, while those in the liver and kidneys of cows and pigs are higher. Higher levels are found in both flesh and internal organs of chicken and fish. Concentrations of vanadium tend to be higher in processed than in unprocessed foods (Myron et al., 1977; Myron et al., 1978). Studies in both humans and animals have shown that vanadium salts are poorly absorbed from the gastrointestinal tract, only 0.1–1% of the very soluble oxytartarovanadate being absorbed. Vanadium absorption occurs by a mechanism not well understood. It is possible that vanadium may share the same carrier as other elements for iron transport. These elements include cobalt, nickel, manganese, zinc, cadmium, and lead (Powell and Halliday, 1981; Barton et al., 1978). The range of vanadium concentrations in food is 0.1–10 μg/kg wet weight, with typical concentrations of about 1 μg/kg. Estimated daily intake ranges from 10 to 70 μg, the majority of estimates being below 30 μg. In general, the higher the solubility of vanadium compounds in water and biological media, the higher absorption and thus the more toxic the compound. The solubility, in decreasing order, of vanadium compounds in gastric juices is vanadyl sulfate, sodium vanadate, ammonium vanadate, vanadium pentoxide. In blood serum, the solubility orders are sodium vanadate, ammonium vanadate, vanadium pentoxide, and vanadyl sulfate.

Pentavalent vanadium is stable in aqueous solution over a wide range of pH (Chasteen, 1983; Nechay et al., 1986). Under physiological conditions at pH 7.4, the pentavalent state is in the form of an anion, vanadate: metavanadate ($H_2VO_4^-$ or VO_3^-), orthovanadate (VO_4^{3-}), or their isopolyanions ($V_{10}O_{28}^{6-}$). When the pH is above 13, vanadate exists mainly as orthovanadate and is analogous to inorganic phosphate (PO_4^{3-}). The pentavalent vanadium is reduced to the tetravalent form by relatively mild reducing agents. The tetravalent state at pH 7.4 is predominantly in the cationic form, vanadyl ion (oxovanadium, VO^{2+}). Vanadyl ion can be hydroxylated to vanadyl hydroxide, VO $(OH)_2$, which is sensitive to oxidation by oxygen in the air. Vanadyl ion easily forms complexes with other molecules including EDTA, ATP, catechol, catecholamines, acetoacetate, ribosides, hemoglobin, and serum transferrin. The tetravalent state of vanadium in these compounds is stabilized against oxidation and thus is the most stable oxidation state for vanadium. In summary, vanadium apparently exists as metavanadate in the plasma, and in vanadyl form inside cells (Robinson, 1981).

In humans absorbed vanadium is transported mainly in the plasma. The mean human serum concentration of vanadium is about or below 0.1–1 ng/ml (~2–20 nM) as determined by neutron activation analysis (Erdmann et al., 1984). It is estimated that 90% of total vanadate normally present in plasma, approximately 10^{-8} M, is bound to proteins. Free plasma vanadate concentration is about 10^{-9} M. In the absence of active transport, intracellular concentration of free metavanadate would be approximately 3×10^{-11} M, which is 30 times less than extracellular concentration. Regarding intracellular vanadyl ion, less than 1% is free, ranging from 10^{-10} to 10^{-9} M (<1% from 10^{-8} M to 4×10^{-7} M or 0.5–20 ng vanadium per gram) (Cantley et al., 1978). Such low concentrations of intracellular vanadyl ion appears to have no effect on enzymes or cellular functions. Vanadyl ion binds to various ligands mentioned earlier. Indeed, 90% of total intracellular vanadyl ion is trapped by the phosphates in several types of cells and is constituted, at least in part, as the basis for intracellular accumulation of vanadium. As stated earlier, bound vanadyl ion appears to be protected from oxidation to vanadate, which, as in unbound form, could otherwise spontaneously occur at the physiological range of intracellular pH. The binding ability of vanadate to protein is 100–400 times weaker than that of vanadyl ion to phosphate. As soon as vanadate enters the cell, vanadate is favorably reduced to vanadyl form, thus leading to the subsequent binding of vanadyl ion to phosphates. It has been suggested that intracellular excess of vanadate, which may not be reduced to the vanadyl state, would be responsible for toxic effects of vanadium, at least in cell cultures (Sabbioni et al., 1991).

The total-body pool of vanadium in human is about 100 μg, with a daily intake of 10–60 μg. It appears that vanadium concentrations are low in all tissues, though the liver, kidney, and lung often show higher levels than other tissues. These organs also are accumulators of the metal. Vanadium concentrations in the liver, kidney, and lung are in the range of 4.5–19, 3–7, and 10–130 μg/kg wet weight, respectively. Vanadium has a high affinity for nuclear and mitochondrial components, and also for Fe-containing proteins, such as transferrin and ferritin. Small amounts of vanadium are detected in the placenta, and vanadium passes through into the membranes rather than the fetus (Hackett and Kelman, 1983). Vanadium also is found in breast milk and saliva. It also can pass through the blood-brain barrier.

Since vanadium absorption in the gastrointestinal tract is very poor, ingested vanadium is mainly eliminated, unabsorbed, with the feces (Waters, 1977). The main route of excretion of absorbed vanadium is through the kidney. Vanadium levels in urine are of the order of 0.1–0.2 μg/L or about 12% of the amount intake. Correlation between exposure to vanadium and the levels in blood or those excreted in the urine are still uncertain. The ratio of amounts eliminated in the urine and feces is 5:1 (Talvitie and Wagner, 1954). It appears that the elimination of vanadium in the urine ceases before that in feces. This is likely due to its return to the intestine after internal resorption and excretion.

1.3. Biochemical Effects of Vanadium

Vanadium on one hand can inhibit a variety of enzymes; on the other hand the element can stimulate a number of enzymes (Boyd and Kustin, 1985; Crans et al., 1989; Dafnis and Sabatini, 1994; Erdmann et al., 1984; Nechay, 1984; Nechay et al., 1986). The former groups of enzymes include phosphoenzyme ion-transport ATPases, acid and alkaline phosphatases, H-ATPase, phosphotyrosyl protein phosphatase, dynein (contractile protein ATPase), myosin ATPase, phosphofructokinase, adenylate kinase, and choline esterase.

Regarding the phosphoenzyme ion-transport ATPase, also designated as the E_1/E_2 class of enzymes, the simplest single criterion for membership in the group appears to be an aspartyl residue at the active site of phosphorylation. The family includes the following enzymes: Na-K-ATPase, Ca-ATPase, and H-K-ATPase. Vanadium is a strong inhibitor of Na-K-ATPase and H-K-ATPase (Eiam-Ong et al., 1995a; Jandhyala and Hom, 1983), both of which have important roles in metabolic problems discussed later in the chapter. Located in the cell membrane of most types of cells, Na-K-ATPase expresses its function as the sodium pump, moving three intracellular Na^+ out of the cell and two extracellular K^+ into the cell. In the cells that have polarity, including renal tubular epithelial cell, Na-K-ATPase resides in the basolateral membrane (Eiam-Ong et al., 1995b).

Vanadate inhibits Na-K-ATPase from the cytoplasmic side. It enters the cell and forms a stable inactive complex with the phosphorus site in the E_2 conformation of the enzyme (Huang and Askari, 1981; Jørgensen, 1983). Inhibition by vanadate requires a divalent cation such as Mg^{2+}. Inhibition is facilitated by K^+ and is antagonized by Na^+. Vanadyl ion inhibits Na-K-

ATPase by a mechanism different from that of vanadate. It appears that vanadate is a more potent inhibitor of the enzyme than vanadyl ion.

H-K-ATPase is basically found in the epithelium of the gastric mucosa. The pump drives protons in an electroneutral pattern. The H-K-ATPase is inhibited by vanadate, omeprazole, and SCH_{28080} but not by ouabain, the specific inhibitor of the Na-K-ATPase (Eiam-Ong et al., 1995a). An H-K-ATPase similar to that found in the gastric mucosa has been demonstrated in the mammalian nephron (Doucet and Marsy, 1987). In the kidney, H-K-ATPase activity is exclusively detected in the terminal segments.

Vanadate also can stimulate a variety of enzymes, including adenylate cyclase, glyceraldehyde-3-phosphate dehydrogenase, NADPH oxidase, tyrosine phosphorylase, glycogen synthase, lipoprotein lipase, phosphoglucomutase, glucose-6-phosphate dehydrogenase, and cytochrome oxidase (Erdmann et al., 1984; Nechay, et al., 1986). Studies both in vivo and in vitro have shown that vanadium enhances lipid peroxidation but decreases activities of antioxidant enzymes, especially catalase and glutathione peroxidase, in various organs including liver and kidney (Donaldson et al., 1985; Russanov et al., 1994; Younes et al., 1984). It is postulated that the peroxidating properties of vanadium are the proximate cause of vanadium toxicity (Boyd and Kustin, 1985; Byczkowski and Sorenson, 1984; Crans et al., 1989; Cros et al., 1992; Erdmann et al., 1984; Wennig and Kirsch, 1988).

1.4. Vanadium Toxicity

Despite present knowledge indicating the role of vanadium as an essential element for chicks and rats, conclusive evidence that vanadium is essential for other species, including humans, is lacking (Schroeder et al., 1963). Vanadium is an ultratrace element in humans. Heretofore, there has been no definite proof that vanadium deficiency reproducibly and consistently impairs a biological function in any animals.

Studies in animals have shown that vanadium is better tolerated by small animals, including the rat and mouse, than by larger animals, such as the rabbit and horse (Hudson, 1964). The toxicity of vanadium is low when given orally, moderate when inhaled, and high when injected. As a rule, the toxicity of vanadium increases as the valency increases, pentavalent vanadium being the most toxic. In general, toxic effects in humans and animals under natural conditions occur infrequently. Vanadium toxicity in humans is almost always related to industrial processes (Schroeder et al., 1963), while in animals the reported effects of a natural vanadium toxicity occurs in using contaminated phosphate in diets for chicks and laying hens (Berg, 1963; Berg et al., 1963).

In the past, vanadium compounds were widely used as therapeutic agents (for anemia, chlorosis, tuberculosis, dental caries, and diabetes mellitus), as an antiseptic, as a spirochetocide, and as a tonic. Because of the poor absorption from the gastrointestinal tract, vanadium has no serious toxicity when ingested. The lethal dose, administered in a soluble form directly into the circulation, for a 70-kg person is about 30 mg of vanadium pentoxide (Hudson, 1964).

Vanadium toxicity can be classified into local and systemic effects. Local effects generally result from skin or inhalation exposure. Intoxication by vanadium of various metabolic functions or organ systems stems from the biochemical effects stated above. Since vanadium affects various enzymes, it can cause diverse effects on different organ systems. In the nervous system, vanadium has a selective effect on adrenergic pathways (lowering the level of noradrenaline), inhibits calcium transport, impairs protein synthesis, and causes physiological disturbance of the central nervous system (impaired conditioned reflexes and neuromuscular excitability). Nonspecific neurological signs and symptoms, including headache, weakness, nausea, vomiting, and tinnitus, have been reported (WHO, 1988). Patients with manic-depressive illness have high levels of vanadium in plasma and hair (Dick et al., 1982). The level falls to normal with recovery of the patients.

Regarding the cardiovascular system, vanadium causes vasoconstriction in various organs (spleen, kidney, and intestine), and induces physiological cardiovascular changes (occurrence of arrythmias and extrasystole, prolongation of the QRS-T interval, and decrease in the height of the P and T waves of the EKG) (Carmignani et al., 1991; Hudson, 1964). The reported clinical manifestations include palpitation of the heart at rest and on exercise, and transient coronary insufficiency. Vanadium can increase both systolic and diastolic blood pressure, the mechanism of which is basically mediated by increasing total peripheral resistance (Boscolo et al., 1994). The effect of vanadium on renin depends on the dose. Exposure to 10–40 ppm of vanadium results in an increase of plasma renin activity and kininase I, kininase II, and kallikrein activities. In animals, long-term inhalation exposure to vanadium could cause fatty changes in the myocardium as well as perivascular swelling.

Human data are not available on the effects of vanadium on the liver. Animal studies have shown fatty changes with partial cell necrosis in the liver after exposure to vanadium. The toxic effects observed appear to correlate with the concentration of vanadium in the liver.

Pulmonary effects of vanadium consist of asthmatic reactions in conjunction with nonspecific bronchial hyperreactivity, a dose-dependent decline in forced expiratory volume in one second (FEV_1) and forced vital capacity (FVC). Regarding the immune system, vanadium depresses antibody-forming cells, and enhances DNA synthesis in splenic leukocytes (Sharma et al., 1981). Vanadium increases resistance to *E. coli* endotoxin but decreases resistance to *Listeria* lethality. In animals, vanadium causes morphological changes in spermatozoa, desquamation of spermatogenic epithelium in the seminal tubules, absence of fertilization of the female, and decrease in the number of fetuses (Roshchin et al., 1980). Some evidence has suggested that vanadium may have teratogenic effects. Available data, however, do not show carcinogenic effects from vanadium.

Indeed, the most impressive and consistent effect of vanadate is functional alterations of mammalian kidney. Vanadate, regardless of the route by which it is administered, causes profound renal artery vasoconstriction (Day et al., 1980; Higashi and Bello-Reuss, 1980; Inciarte et al., 1980; Larsen et al., 1979; Lopez-Novoa et al., 1982) and results in a rise in intracellular calcium concentration in the renal artery (Benabe et al., 1984). Vanadate has a natriuretic effect throughout the entire nephron. This effect is reversible and is secondary to inhibition of the basolateral Na-K-ATPase (Balfour et al., 1978; Kumar and Corder, 1980). The inhibition induced by vanadate, however, is most profound when the element is present in the lumen. It is believed that vanadate is filtered into the lumen, and then enters the renal tubular cells from the luminal side (Edwards and Grantham, 1983a). The cortex appears to be able to accumulate more vanadium, and increased potassium concentration could augment vanadium binding, leading to more inhibition of Na-K-ATPase activity.

Vanadium, by inhibiting Na-K-ATPase, can inhibit proximal tubule absorption of various substances, including phosphate, bicarbonate, glucose, and water. Vanadium also inhibits paraaminohippuric acid (PAH) secretion in the S_2 portion of the proximal tubule (Smith et al., 1982). In the distal nephron, vanadium inhibits the action of vasopressin in the collecting tubule (Edwards and Grantham, 1983b). Vanadate does not abolish the potassium adaptation response to diets high in potassium for long periods (Higashino et al., 1983).

Vanadium, administered intraperitoneally to rats for 10 days, has been reported to cause hypokalemic distal renal tubular acidosis almost identical to the "classical" form observed in humans (Dafnis et al., 1992). Vanadium is detected in highest concentration in the urine; this is followed by cortex and medulla, and the least is observed in the blood. Both H-K-ATPase and Na-K-ATPase activities are inhibited in the collecting tubule.

Vanadium toxicity may be divided into acute and chronic forms (WHO, 1988). The acute form is characterized by a latent period that varies with the concentration of vanadium, the individual sensitivity of the subject, and the properties of the specific vanadium compounds. The most rapidly toxic form is vanadium chloride. The more soluble salts of vanadium pentoxide have a more rapid action than the vanadium oxides. Acute vanadium effects are divided into "mild," "moderate," and "severe" degrees. Mild toxicity is characterized by rhinitis with a profuse and often bloody discharge, sneezing, conjunctivitis, an itching and burning sensation in the throat, varying body temperature, and diarrhea. These symptoms disappear 2–5 days after cessation of exposure. The additional findings observed in moderate vanadate toxicity include bronchitis with expiratory dyspnea and bronchospasm, vomiting and diarrhea, rashes and eczema with itching papules and dry patches. The hallmark features of severe toxicity are bronchitis and bronchopneumonia, disorders of the nervous system (severe neurotic states and tremor of the fingers and hands), and more prominent signs and symptoms of the kind found in the milder forms.

Regarding chronic vanadium intoxication, the deleterious effects of the element on the respiratory system are still unestablished. Wheezing is more common among exposed workers than among unexposed ones. Lung function tests and chest radiography have not revealed persistent lung damage (Kiviluoto, 1980). Cardiovascular manifestations include accentuated second sound on the pulmonary artery and an attenuated first sound on the apex cords, sinus arrhythmia, bradycardia, coronary spasm, and various electrocardiographic changes. Other abnormal findings are enlarged liver, deterioration in liver function test, a tendency towards anemia, leukopenia and basophilic granulation of leukocytes, and decreases in blood sulfhydryl groups, vitamin C, and cholesterol.

Of interest, the kidneys are a critical organ for vanadium poisoning. Kidney damage, characterized by grave dystrophic changes in the epithelium of the convoluted tubules and disturbed tubular secretion, occurs immediately after the start of exposure to low vanadium doses in both acute and chronic intoxication. These changes, however, are irreversible despite the discontinuation of exposure.

2. ROLE OF VANADIUM IN METABOLIC PROBLEMS IN NORTHEASTERN THAILAND

2.1. General Considerations

Northeastern Thailand is the largest part of the country and covers nearly one-third of the country (Fig. 1). The northeastern region of Thailand is geographically different from the central part of the country. Despite lying in the tropical zone, it is an arid plateau where soil is more sandy and less fertile. In summer, which lasts for 3–4 months, precipitation is negligible and the weather is hot and dry. The central region is a plain; there is much more

Figure 1. Map of northeastern Thailand and neighboring countries.

rainfall and the soil is more fertile. Northeastern Thailand is noted for having the biggest population in the nation, which is of low socioeconomic status. Income per capita is lowest in northeastern Thailand. The northeastern inhabitants are ethnically related more closely to the neighboring countries, namely, Laos and Cambodia, while the central population are ethnic central Thais or Thai-Chinese. Several endemic health problems occur in northeastern Thailand. Besides nutritional deficiency and infectious diseases, metabolic problems in this particular area include distal renal tubular acidosis (dRTA), renal stone disease, sudden unexplained death syndrome (SUDS), malnutrition-related diabetes mellitus (MRDM), and hypokalemic periodic paralysis (Sitprija et al., 1991; Sitprija et al., 1993).

2.1.1. Distal Renal Tubular Acidosis (dRTA)

The syndrome of classical (Type I) dRTA consists of hypokalemia, hyperchloremic metabolic acidosis, inability to lower the final urine pH to below 5.5, nephrocalcinosis and nephrolithiasis, and osteomalacia or renal rickets (Eiam-Ong and Kurtzman, 1995; Eiam-Ong et al., 1995a; Eiam-Ong et al., 1995b). The basic mechanism of classical dRTA has been established to be a failure of the proton pump in the collecting tubule, which leads to a failure of proton secretion. The collecting tubule, the main site of renal acidification, secretes proton by two proton pumps, H-ATPase and H-K-ATPase. Traditionally, the defect in the H-ATPase pump has been considered the proximate cause of classical dRTA. The discovery that H-K-ATPase in the renal tubule is similar to that of the gastric mucosa provides new insight into classical dRTA. Theoretically, the decrease in collecting tubule H-K-ATPase activity could lead to both acidification defects and urinary potassium loss and hence could cause hypokalemic hyperchloremic metabolic acidosis resembling classical dRTA.

Recently, there is a report that primary renal tubular acidosis is endemic in northeastern Thailand (Nilwarangkur et al., 1990). Initial estimation has suggested that the disease may affect as many as 450,000 people. The symptom of muscle weakness commonly occurs in the summer, while the disease may be relatively asymtomatic during the other seasons. Patients have hypokalemic hyperchloremic metabolic acidosis. The ratio of female to male patients is 3.3:1. The mean age is 39 years, ranging from 18 to 76 years. Hypokaliuria and hypocitraturia are the consistent findings in all victims. The finding of "hypokaliuria" in dRTA patients in northeastern Thailand is unique and, thus, is contradictory to "kaliuresis," which is generally observed in dRTA patients reported in the Western literature (Batlle et al., 1981a; Batlle et al., 1981b; Sebastian et al., 1971). Follow-up studies by the same group of investigators in five villages in northeastern Thailand have shown that the prevalence of dRTA is 2.8% of the population with the ratio of female to male of about 1.6 to 1 (Nimmannit et al., 1996). Studies by Na_2SO_4 or Na_3PO_4 tests have shown that proton pump failure is the proximate cause of dRTA in most of these patients (Vasuvattakul et al., 1987). At present, the underlying mechanism of this proton pump failure is still not established.

That environmental factors, especially vanadium, would be the underlying etiology of dRTA in northeastern Thailand stems from two crucial pieces of evidence. First, patients with dRTA in northeastern Thailand have decreased gastric acidity (Sitprija et al., 1988). Gastric analysis was performed by a standard method a few weeks following alkaline treatment when hypokalemia had been corrected, with clinical recovery. Alkaline treatment was discontinued for 24 h before the test. A baseline gastric acidity was obtained by gastric fluid collection during four periods of 15-min each. This was followed by intramuscular injection of pentagastrin (6 μg/kg) with continued gastric fluid collection for 1 h at 15-min intervals. The data have shown that the patients with dRTA have low basal gastric acidity and decreased response to pentagastrin stimulation, which was indicated by the lower values of basal and maximal acid secretions (Table 1). The study thus provides a hypothesis that dRTA in northeastern Thailand might be a generalized disease with the defect in hydrogen ion transport in both renal tubules and gastric parietal cells. The defect may reside in H-K-ATPase, which is inhibited by H-K-ATPase inhibitor. Vanadium is one of the possible inhibitors.

The second piece of evidence is that water buffalo in northeastern Thailand also exhibit muscular weakness, hypokalemia, and increased urine pH in the summer (Chaiyabutr et al., 1983; Sitprija et al. 1990). The animals respond well to potassium administration. This is interpreted as a heat stress reaction especially in animals with poor development of sweat glands such as buffalo. It is not known whether they have urinary acidification defect. Environmental factors, however, are likely to play an important role in this circumstance.

2.1.2. Renal Stone Disease

In the northeastern Thailand, renal stone disease is a common community health problem (Chatudompan, 1987; Halstead and Valyasevi, 1967; Nimmannit et al., 1996). The kidney stone is generally composed of mixed calcium oxalate and phosphate (Nirdnoy et al., 1987). A recent community survey at

Table 1   Gastric Acidity Data of Patients with Distal Renal Tubular Acidosis in Northeastern Thailand[a]

Gastric Acidity (mmol/h)	Patient (7)	Control (7)	P
Basal	1.1 ± 0.6	2.7 ± 1.8	NS
Maximal	7.4 ± 2.7	13.4 ± 3.4	<0.05
Peak	9.6 ± 3.4	16.6 ± 5.2	<0.05

[a] Data are expressed as mean ± SD. Number in parentheses indicates sample size. NS = not significant; $P < 0.05$ = statistically significant.

Source: Adapted from Sitprija et al. (1988).

one rural district has reported the prevalence of definitive past or present stone-bearing cases to be 3.76 per 1,000 population (Sriboonlue et al., 1992). By extrapolation, nearly 60,000 resident northeasterners of Thailand have renal stone disease. The ratio of male to female victims is 2 : 1; the peak age of patients is in the range of 41–50 years.

Most former renal stone patients are from the low-income laboring rural class. The epidemiological circumstances of these patients from northeastern Thailand seem to be different from those reported in the Western Hemisphere, where former renal stone patients are predominantly found in affluent social classes (Scott, 1987). Therefore, it is likely that the pathogenesis of renal stone disease in Thailand is different from that in the Western world.

Biochemical studies have demonstrated that metabolic disorders such as hyperuricemia, hyperuricosuria, hypophosphaturia, and hypercalcicuria are uncommon findings in former renal stone patients from northeastern Thailand (Sriboonlue et al., 1991). Indeed, most of the victims have normal urinary calcium, magnesium, phosphate, uric acid, and oxalate. Unique laboratory features observed in this disorder include hypokalemia, cellular potassium depletion, hypokaliuria, and moderate to marked hypocitraturia (Sriboonlue et al., 1991; Sriboonlue et al., 1993). Furthermore, hypocitraturia, is found in every subject. These data seem to suggest that, with respect to the pathogenesis of renal stone disease in northeastern Thailand, a decrease in urinary inhibitory activity is likely to play a more dominant role than an increase in urinary aggregatory activity. In the low-citrate circumstance, urine is more likely to have decreased inhibitory activity against the crystallization of calcium oxalate and calcium phosphate, leading to an increase in the propensity for the spontaneous nucleation and crystal growth of calcium salts.

A recent study has shown that increased potassium intake by potassium supplements to these patients can normalize the serum potassium and can increase urine potassium (Sriboonlue et al., 1993). Urine citrate, however, still remains low, indicating the existence of decreased cellular potassium uptake and the persistence of low intracellular potassium and intracellular acidosis. These findings confirm the presence of circulating substances acting as an inhibitor of cellular potassium uptake.

Indeed, some authors have suggested that renal stone disease is pathogenetically related to distal renal tubular acidosis (Nimmannit et al., 1996). These authors have revealed that villagers with acidification defect, shown by the short acid loading test, have 2.4 times the chance of having renal stone and/ or nephrocalcinosis. In other studies, urinary acidification by the long acid loading test, which is more reliable than the short one, is normal in renal stone patients from northeastern Thailand (Tosukhowong et al., 1991; Tungsanga et al., 1992).

2.1.3. *Sudden Unexplained Death Syndrome*

Sudden unexplained death syndrome (SUDS) is defined as the sudden death of a person at least 2 years old when postmortem does not reveal a demonstrable cause. The victim must have been born in, or have had at least one parent born in, a South-East Asian country (Cambodia, Laos, Thailand, Vietnam, or the Philippines). A probable case of SUDS is one that fulfills the above criteria without a postmortem examination (Parrish et al., 1987). SUDS has been observed among residents of this region as well as among natives who have migrated to Western countries. In Thailand, sudden unexplained death in a young adult is called *lai-tai,* which is translated as "deadly nightmare." In a recent survey by the Thai Ministry of Public Health, the incidence of SUDS in 20- to 59-year-old native Thais is highest in northeastern Thailand (Chokvivat et al., unpublished data). Moreover, about 80% of more than 100 Thai laborers reporting SUDS in Singapore during 1987–1989 were ethnic Thai–Lao from northeastern Thailand, and those who are from the northeastern region had the highest rate of SUDS (Goh et al., 1990).

Follow-up study has identified the cases of SUDS in healthy villagers, 20–49 years old, who died suddenly without explanation (Tungsanga and Sriboonlue, 1993). Live healthy villagers who were age- and sex-matched with the dead served as controls. The study has shown that 30 of 31 subjects were male, with a mean age of 38 ± 8 years. The incidence of SUDS is 38 per 100,000 men, 20–49 years old, per year. The peak risk is at 45–49 years. About 75% of SUDS cases have an annual income per household of less than the mean per capita income in Thailand. There are more SUDS cases in the hot season than any other season. The onset is nocturnal in 84% of cases. The incidence of SUDS among other family members is more frequently associated with victims than with controls. A history of muscle soreness, malaise, and recent hard labor are seen as frequently in SUDS victims as in controls. In witnessed cases, symptoms usually last for a few minutes prior to death. Common symptoms or signs are respiratory manifestations (groaning, choking, or coughing) and muscular spasticity or paralysis. Of interest is the fact that the most consistent but yet unexplained finding is an association between the onset of SUDS and a preceding rest period.

The sudden onset of the attacks and the short symptomatic intervals till death in witnessed cases suggest a cardiovascular event as a proximate cause. Ventricular fibrillation has been observed in previous studies of Southeast Asian immigrants (Otto et al., 1984), but there is no evidence of coronary heart disease (CHD) as an important factor. None of the victims undergo the preceding chest pain that is typical of angina or myocardial infarction, and more than 50% are less than 40 years old, as compared with less than 6% of Thais who experience acute myocardial infarction (Chaithiraphan et al., 1984). These patients also are comparable in age to those Thai workers who die in Singapore and are free of significant CHD. It appears unlikely that these subjects die from CHD.

The causes of SUDS are still bewildering. Hypokalemia and potassium deficiency have been proposed as a risk factor for SUDS in northeastern Thailand (Nimmannit et al., 1990; Nimmannit et al., 1991), but to date no strong supporting evidence for this hypothesis has been published. A low thiamine intake, and consumption of raw, fermented fish products containing thiaminase, among northeastern Thais might contribute to thiamine deficiency (Changbumrung et al., 1989; Tanphaichitr et al., 1990). The association of a prolonged QT interval in the electrocardiogram and poor thiamine intake with a history of nonfatal seizure-like episodes during sleep has been observed among Laotian refugees in Thailand (Munger et al., 1991). These findings suggest a contributing role of thiamine deficiency in SUDS. However, such an association of thiamine deficiency and SUDS is not found in a recent survey of Thai workers in Singapore. Thiamine deficiency combined with hypokalemia may be important as cofactors for SUDS.

2.1.4. Malnutrition-Related Diabetes Mellitus

The prevalence of diabetes mellitus is high in both urban and rural areas in Thailand. In Western countries, almost all patients are more than 40 years old and obese. In developing countries, especially in the tropics, a significant proportion of patients are undernourished and are less than 35 years old. A recent study in northeastern Thailand has shown several types of diabetes in the young (Kiatsayompoo et al., 1993). These include non-insulin-dependent diabetes mellitus (NIDDM) (38.4%), malnutrition-related diabetes mellitus (MRDM) (36.4%), maturity onset diabetes in the young, insulin-dependent diabetes mellitus (IDDM) (9.9%), and secondary diabetes (2.6%). MRDM can be further classified into two groups: 61.82% of patients with MRDM have fibrocalculous pancreatic disease (FCPD) and the rest (38.18%) have protein-deficient pancreatic diabetes (PDPD). There are more females than males among IDDM and NIDDM patients (the ratios are 2:1 and 2.6:1, respectively), but there are fewer females than males among MRDM patients (the ratio is 0.7:1). The percentage of MRDM patients is higher in northeastern Thailand than in other parts of the country, but it is similar to the percentage in tropical countries, including Jamica, Uganda, Nigeria, India, Indonesia, and Bangladesh (Hugh-Jones, 1955; Olurin, 1969; Shaper, 1960; Tulloch, 1961; Viswanathan, 1980).

Most of the MRDM patients are farmers, which is different from the occupations typical of NIDDM and IDDM patients. Compared with other groups, MRDM patients have higher incidence of abdominal pain and cataract. The histories of heavy alcohol and raw cassava consumption are, however, not different from those of other diabetic groups. The pathogenesis of MRDM is still unknown.

2.2. Blood, Tissue, and Urinary Parameters in Healthy Northeastern Villagers

It appears that a state of potassium depletion (as illustrated by hypokalemia, low red cell potassium, and hypokaliuria) and hypocitraturia are also observed in northeastern villagers who do not suffer from these disorders (Sriboonlue et al., 1991; Sriboonlue et al., 1993). The low urine citrate has been used as an index to indicate intracellular acidosis and cellular potassium depletion. In a recent study of cadavers of subjects who had died from accidents and had been apparently healthy before death, potassium contents in muscle and kidney obtained from northeastern villagers were found to be significantly lower than those obtained from central region dwellers (muscle potassium content: 295 ± 9 vs. 339 ± 10 mmol/kg dry weight; kidney potassium content: 176 ± 6 vs. 196 ± 15 mmol/kg dry weight) (Lelamali et al., unpublished data). The low serum potassium and the low urine potassium likely reflect decreased potassium intake, since potassium consumption is low among the villagers. A recent survey among healthy rural northeastern Thais has shown that their daily potassium intake is as low as 15–30 mmol (Puwastien et al., unpublished data). Potassium loss through excessive sweating due to heat, however, can contribute to the state of potassium depletion.

Follow-up study has shown that Na-K-ATPase activity of the erythrocyte membrane of the villagers is only 50% of that of the central region population (52 ± 4 vs. 98 ± 5 nmol P_i/mg/h, Na-K-ATPase activity; $P < 0.001$) (Tosukhowong et al., 1992). By contrast, the intracellular erythrocyte sodium levels of the villagers are significantly higher than those of the central region dwellers (13.6 ± 3.1 vs. 8.1 ± 1.4 mmol/L; $P < 0.001$). Erythrocyte sodium inversely correlated with erythrocyte Na-K-ATPase activity. The relationships among erythrocyte sodium, potassium depletion state, and Na-K-ATPase activity, however, are not easily established. Theoretically, increased intracellular sodium would stimulate membrane Na-K-ATPase activity (Clausen and Kjeldsen, 1987). Hence, the greater the intracellular sodium, the greater the Na-K-ATPase activity. The inverse relationship between erythrocyte sodium and Na-K-ATPase activity in northeastern villagers appears contradictory to this physiological view. Thus, it is likely that there is a quantitative or qualitative diminution in the function of Na-K-ATPase leading to secondary retention of intracellular sodium.

The average salt intake among healthy residents in northeastern Thailand is as low as 73 ± 2 mEq/day as opposed to 200 ± 10 mEq/day among central region dwellers (Puwastien et al., unpublished data). It is known that high salt intake can decrease Na-K-ATPase activity (Quintanilla et al., 1988), suggesting that this factor is not the cause of the observed reduction in Na-K-ATPase activity. Potassium depletion can increase erythrocytic membrane Na-K-ATPase activity, the quantity of the Na-K-ATPase enzyme, and membrane permeability of sodium influx (Chan and Sanstone, 1969; Clausen and Kjeldsen, 1987; Hoffmann and Smith, 1970). Again, the detection of decreased Na-K-ATPase activity despite potassium depletion indicates that potassium depletion is the result, not the proximate cause, of the deficiency in Na-K-ATPase. Thus, the potassium-depleted state of subjects in northeastern Thailand is caused by low intake and by low cellular uptake due to decreased Na-K-ATPase activity. The high erythrocyte sodium level is the result of decreased Na-K-ATPase activity and potassium depletion, or may be caused by a primary abnormality in membrane permeability for sodium influx.

A recent study has been performed to elucidate whether decreased Na-K-ATPase activity is hereditary or acquired (Tosukhowong et al., 1996). Decreased Na-K-ATPase activity may be caused by a reduction in the pump numbers or function. In this regard, an assay of ouabain-binding capacity can be used to quantitatively determine the number of Na-K-ATPase units on cell membranes (Cheng et al., 1984; Deluise and Flier, 1985; Hoffman, 1969; Schmalzing et al., 1981). As illustrated in Table 2, northeastern newborns have higher erythrocyte sodium levels than central region newborns. Other parameters, however, are not different between the two groups. Northeastern villagers have the lowest Na-K-ATPase activities and fewest erythrocyte ouabain-binding sites (OBS) but have the highest erythrocyte sodium. The erythrocyte Na-K-ATPase/OBS ratio, an expression of Na-K- ATPase activity equalized for the number of Na-K-pump units, also is lowest in the northeastern villagers. It is of interest that the abnormalities of these parameters in northeastern migrant workers, who move to work in the central region for at least 1 year, are less severe than in villagers who remain in the northeastern region. Thus, the available data suggest that northeastern villagers in Thailand tend to have lower erythrocyte Na-K-ATPase activity than central region dwellers and that this condition probably is acquired after birth. The defect in Na-K-ATPase is likely associated with low numbers of Na-K-pump units and an incapacity of the pump to express activity. There might be circulating Na-K-ATPase inhibitors and metabolic disturbances that cause attenuation of Na-K-ATPase function and synthesis in northeastern villagers, and such substances may have an environmental origin.

2.3.   Vanadium Levels in Northeastern Thailand

All the above defects, detected in both healthy and diseased subjects in northeastern Thailand, indicate that there might be one or more Na-K-ATPase inhibitors in these people and the substances are likely to be acquired from the environment. The inhibitors could depress various enzymes, including Na-K-ATPase and H-K-ATPase. Vanadium, as stated earlier, is a strong inhibitor of both H-K-ATPase and Na-K-ATPase. To elucidate this hypothesis, a recent study has shown that soil vanadium concentrations in the northeastern region are higher than those in the central part of the country. (Table 3) (Sitprija et al., 1993) There is no significant difference in the vanadium concentration in water obtained from the two parts of the country, although the value in villages tends to be higher. Urine vanadium levels in villagers is higher than in central region dwellers. It should be noted that these values are strikingly high compared with the normal value of 0.2–0.4 μg/day (based on a 24-h urine of 2,000 ml) reported by the World Health Organization. The normal value, however, must be interpreted with caution, since it could vary among different geographical areas. Irrespective of the varying data, the values obtained in northeastern villagers are on the high side. The values of vanadium content of the kidney and lung of patients who live in the villages and die from various

Table 2 Erythrocyte Sodium, Na-K-ATPase Activity, Ouabain-Binding Sites (OBS), Na-K-ATPase Activity/OBS in Different Thai Populations[a]

	Group[b]				
	1 (30)	2 (31)	3 (30)	4 (19)	5 (25)
Erythrocyte					
Sodium (mmol/L)	9.8 ± 0.35	11.8 ± 0.24^{1}	$7.9 \pm 0.35^{1,2}$	$11.6 \pm 0.88^{1,3,5}$	$8.4 \pm 0.43^{1,2}$
Na-K-ATPase activity (nmol P_i/mg · h)	100 ± 5	106 ± 8	118 ± 7	$63 \pm 4^{1,2,3,5}$	93 ± 8^{3}
OBS (sites per cell)	441 ± 11	445 ± 21	484 ± 14	397 ± 16^{3}	443 ± 23
Na-K-ATPase activity/OBS	0.23 ± 0.01	0.25 ± 0.02	0.24 ± 0.01	$0.16 \pm 0.01^{1,2,3}$	0.21 ± 0.01

[a] Data are expressed as mean ± SEM. The superscript number indicates the population group that is significantly different at $P < 0.05$.

[b] Group 1, central newborns; Group 2, northeastern newborns; Group 3, central region dwellers; Group 4, northeastern villagers; Group 5, northeastern migrant workers. Number in parentheses indicates sample size.

Source: Adapted from Tosukhowong et al. (1996).

Table 3 Vanadium Concentration in Soil, Water, and Tissue[a]

	Northeastern Areas	Central Areas	P
Soil vanadium concentration			
(μg/g dry weight of soil)			
Surface soil	39.3 ± 12.7 (26)	12.7 ± 9.1 (10)	<0.05
Below surface			
10 cm	41.9 ± 29.6 (26)	18.2 ± 10.8 (10)	<0.05
50 cm	46.9 ± 22.9 (26)	25.6 ± 9.8 (10)	<0.05
100 cm	42.8 ± 18.2 (26)	28.3 ± 9.7 (10)	<0.05
Water vanadium concentration			
(μg/L)			
6 m in depth	1.09 ± 2.01 (31)	Not determined	
36 m in depth	0.71 ± 1.89 (31)	0.77 ± 0.89 (20)	NS
Tissue vanadium concentration			
(μg/kg)			
Liver	27.8 ± 19.9 (15)	14.0 ± 3.4	NS
Kidney	16.4 ± 6.7 (15)	9.8 ± 1.9	<0.05
Lung	37.5 ± 15.8 (15)	19.6 ± 9.4	<0.05

[a] Data are expressed as mean ± *SD*. Numbers in parentheses indicate size of samples. NS = nonsignificant.

Source: Adapted from Sitprija et al. (1993).

nonrenal diseases are higher than those of patients who live only in the central region and die of diseases not related to the kidney. The liver vanadium concentrations are found to be higher but not statistically different from those of the central region patients. In normal persons, as described earlier, tissue vanadium is usually less than 10 μg/kg of tissue.

Indeed, the findings of high vanadium concentrations in various environmental sections of northeastern Thailand are expected from the point of view of paleontology. Northeastern Thailand is a high plateau that has existed since the Mesozoic era, while the other parts of the country were under sea level. The soil displays the character of sedimentary rock similar to that of the Colorado Valley in the United States and is rich in vanadium.

Fossils of many ancient reptiles including dinosaurs have been discovered in northeastern Thailand. From an evolutionary view point, the high vanadium content would have been beneficial for the survival of reptiles, because urinary acidification is not needed. Indeed, the end product of protein metabolism in reptiles is uric acid. Acid urine would cause precipitation of uric acid in the renal tubules. Failure to acidify urine by inhibition of H-K-ATPase would make uric acid more soluble and would be appropriate for survival of the animals. Reptiles have cloacas for this uricosuric reason. In alligators, urine is alkaline and acid excretion is in the form of ammonium bicarbonate (Ventura et al., 1989). Moreover, inhibition of Na-K-ATPase would favor sodium entry into the cell and raise intracellular osmolality, which would be suitable for the possibly high extracellular osmolality due to water loss from the high temperature of the ancient world.

The high tissue and urine vanadium contents are interpreted as indicating high vanadium intake of the villagers, presumably from prolonged consumption of natural products of the land. As stated above, vanadium absorption from the gastrointestinal tract is very poor and may be closely related to iron transport. In the presence of iron deficiency the absorption of iron and metals closely related to iron could be enhanced. Since iron deficiency is common in northeastern Thailand, absorption of vanadium could be increased in these villagers.

2.4.   Possible Role of Vanadium in Metabolic Problems in Northeastern Thailand

Taken together, available data have suggested that vanadium might play an important role in the pathogenesis of these metabolic problems in northeastern Thailand. To define the specifically pathogenetic role of vanadium in more detail, each metabolic syndrome deserves separate description.

2.4.1.   Role of Vanadium in Distal Renal Tubular Acidosis

H-K-ATPase is present in the intercalated cells of the collecting tubules and in gastric parietal cells (Eiam-Ong et al., 1995a). The demonstration of decreased gastric acidity in patients with distal renal tubular acidosis and hypokalemia is consistent with decreased H-K-ATPase activity. Kaliuresis from decreased H-K-ATPase activity would result in hypokalemia. Thus, in this circumstance, hypokalemia is the result of the deficiency of ATPase enzyme. Hypokalemia caused by other etiologies including low intake and loss through sweating also affects both H-K-ATPase and H-ATPase. On the one hand, hypokalemia is a potent stimulator of H-K-ATPase; but the presence of vanadium could inhibit H-K-ATPase activity despite hypokalemia. On the other hand, hypokalemia, irrespective of the causes, could directly inhibit aldosterone secretion (Eiam-Ong et al., 1995a).

Since H-ATPase activity is regulated by aldosterone, diminished aldosterone concentration would depress H-ATPase activity. Therefore, activities of H-K-ATPase and H-ATPase are both decreased, leading to impairment in urinary acidification. Obviously, symptomatic cases of dRTA, like the top of an iceberg, are much fewer than asymptomatic subjects, who have a subtle degree of deficiency in both proton pumps. The magnitude of the reduction in both enzyme activities would vary among individuals depending on various factors including host response, duration and amount of vanadium intake, gender, and probably genetic variations. Despite the high tissue concentration of vanadium in the autopsy material, the villagers have normal serum electrolytes and acid urine as assessed from the medical records (Sitprija et al., 1993).

In subjects from both the northeastern and the central region in whom the urinary vanadium concentration has been studied, the random urine pH is acid (pH 5.5–6.5) on the dipstick test. Differentiation between the two groups cannot be made. Indeed, some aggravating factors would exist. In this regard, it appears that seasonal factors might play such a role. Symptoms of dRTA usually occur in the hot season, when potassium loss via sweating is increased (Nilwarangkur et al., 1990). It is estimated that the amount of potassium loss through sweating may be 5–15 mmol/day, averaging 10 mmol/day, during the summer time (Sriboonlue et al., 1991). This amount of potassium loss is substantially higher than that reported in previous work from Western countries (4–8 mmol/L of sweat) (Kohler, 1987). This could aggravate hypokalemia, leading to an acutely symptomatic attack of dRTA. In accord with this hypothetical scenario, previous animal studies have demonstrated some valuable data. Inhibition of H-K-ATPase by omeprazole, another inhibitor of H-K-ATPase, in normal animal experiments does not alter urine acidification, but hydrogen ion secretion is decreased when the animals are potassium-depleted (Wingo, 1989). Omeprazole has no effect on urine acidification in human subjects with normal serum potassium (Howden et al., 1986).

Potassium depletion and hypokalemia can also impair renal acidification by other mechanisms. Hypokalemia can enhance prostaglandin production, which could directly decrease medullary collecting tubule (MCT) proton secretion (Hays et al., 1986). Furthermore, chronic potassium depletion may cause tubulointerstitial nephropathy (Tolins et al., 1987). Studies in humans have shown that potassium depletion for at least 1 month can produce a characteristic vacuolar lesion in the epithelial cells of the proximal tubule and, occasionally, the distal tubule (Relman and Schwartz, 1956; Schwartz and Relman, 1967; Torres et al., 1990). Hypokalemia can reduce the intracellular pH of the proximal tubular cell, which in turn stimulates NH_3 production and NH_3 accumulation in the interstitium. Since NH_3 can activate the complement system, tubular injury may develop (Tolins et al., 1987). Consequently, interstitial fibrosis, tubular atrophy, and cyst formation, especially in the renal medulla, could occur. A defect in NH_3 transportation from the medullary interstitium to the lumen of MCT would decrease luminal NH_4^+ concentration and reduce net acid excretion. In this regard, some authors have suggested that classical dRTA may actually represent a disorder that consists of a chronic persistent reduction in intracellular pH in the proximal tubular cell, resulting in increased NH_4^+ production (Halperin et al., 1994). Another hypothesis of the pathogenic role of ammonia in dRTA is that the permeability of NH_3 and/or NH_4^+ is altered in the collecting duct (Kurtz et al., 1990).

A recent study has been performed to examine whether tubulointerstitial pathology could occur in the kidney of northeastern villagers with potassium depletion (Lelamali et al., unpublished data). A pathological study of kidney tissue obtained from northeastern villagers who had died from accidents and had been apparently healthy before death has shown that there was no tubulointerstitial pathology detected in the kidney. This occurs despite a state of potassium depletion in the kidney tissue, which would facilitate ammonia production.

The role of vanadium as the proximate cause of dRTA in northeastern Thailand, however, is negated in some degree by the fact that vanadium also inhibits Na-K-ATPase, which affects renal potassium excretion. Potassium is secreted by the principal cells of the cortical collecting tubule in a two-step process that involves uptake from the blood into cells via the basolateral Na-K-ATPase and passive diffusion from the cell to the lumen through the apical membrane K channel (Alpern and Rector, 1996). Therefore, urinary potassium excretion is influenced by basolateral Na-K-ATPase as well as apical K channels in principal cells. Vanadium thus has opposing effects on renal potassium secretion. Kaliuresis caused by decreased H-K-ATPase would be offset by potassium retention that had resulted from decreased Na-K-ATPase. This, and probably the effect of low potassium intake, might explain why dRTA patients in northeastern Thailand have "hypokaliuria" instead of "kaliuresis."

Hypokalemia observed in classical dRTA is generally believed to be the result of intact potassium secretory capacity in the face of high levels of aldosterone, secondary to variable degrees of volume contraction and an increase in distal delivery of sodium with poorly reabsorbed anions (Sebastian et al., 1971). The increase in urinary potassium excretion occurs in response to Na_2SO_4, suggesting that the capacity of the distal nephron to generate a lumen-negative potential difference and the capacity to secrete potassium are intact in this condition (Batlle et al., 1981a; Batlle et al., 1981b).

2.4.2. *Role of Vanadium in Renal Stone Disease*

Hypocitraturia is generally observed in the healthy villagers and renal stone patients (Sriboonlue et al., 1992; Sriboonlue et al., 1993). Indeed, potassium depletion can decrease urinary citrate excretion. Theoretically, decreased intracellular potassium and increased intracellular sodium can increase intracellular H^+ by depressing Na-H-antiport. Decreased intracellular pH increases tubular reabsorption of citrate, leading to decreased urinary citrate (Hamm, 1990). Inhibition of Na-K-ATPase by vanadium could be responsible for potassium depletion and hypocitraturia. The low urinary citrate could be the cause of renal stone diseases, since citrate is a stone inhibitor.

2.4.3. *Role of Vanadium in Sudden Unexplained Death Syndrome*

The causes of sudden unexplained death are perhaps multiple, including thiamine deficiency, genetic variations, cardiac conduction system defect, potassium depletion, and unidentified toxic factors (Sitprija et al., 1991). The syndrome could be due to decreased Na-K-ATPase induced by vanadium with cardiac muscle cell depolarization similar to a digitalis effect. Hypokalemia and stress or increased sympathetic activity can cause ventricular fibrillation and sudden death in a manner similar to the effect upon digitalized subjects (Podrid et al., 1990; Rosen et al., 1975). Cardiac arrhythmia and prolongation of the QT interval can occur during vanadium exposure (Roshchin, 1967).

2.4.4.   *Role of Vanadium in Malnutrition-Related Diabetes Mellitus*

Potassium depletion, resulting from various causes including vanadium, appears to play an important role in the pathogenesis of MRDM. The association between potassium deficiency and glucose intolerance is well established. There is a deficit in insulin release in response to hyperglycemia in hypokalemic patients. This suppression of insulin secretion is prostaglandin-mediated (Robertson, 1983). Infusion of prostaglandins in conjunction with glucose has been shown to reduce insulin secretion by β cells. Administration of acetylsalicylic acid increases insulin release and glucose use despite hypokalemia (Gaynor et al., 1982). The pathogenesis of malnutrition-related diabetes mellitus in northeastern Thailand needs further exploration. Indeed, vanadium has been used to treat diabetes mellitus. The insulin-mimetic effect of vanadium is mediated by activating Na-K-ATPase and stimulating both glucose transport and glucose oxidation (Erdmann et al., 1984; WHO, 1988).

The roles of vanadium in the pathogenesis of metabolic syndromes in northeastern Thailand can be summarized in Figure 2.

The metabolic syndromes found in northeastern Thailand are also observed in Laos and in the southern part of China, both of which are geographically close to northeastern Thailand. Vanadium content is noted for being high in the soil of these countries. Due to the lack of a systematized data collecting system, there are no epidemiological details of the metabolic syndromes in these two places.

Besides potassium depletion, low protein intake caused by the low socioeconomic status of the northeastern villagers could contribute to the pathogenesis of these metabolic problems. A recent survey has shown that daily dietary protein of healthy northeastern villagers is as low as 40–50 g (Puwastien et al., unpublished data). Low protein intake can increase prostaglandin production, leading to reduction in acid secretion (Benabe et al., 1989). Furthermore, low protein consumption could decrease ATPase activity (Benabe et al., 1989; Benabe and Martinez-Maldonado, 1991; Senay and Marver, 1986). As in the case of potassium depletion, increased prostaglandin production by low protein intake and the consequently impaired insulin release could contribute to malnutrition-related diabetes mellitus.

Despite all the above-described evidence that suggests a cause–effect relationship between vanadium and these metabolic syndromes, one should keep in mind that all the observed phenomena may be epiphenomena. Indeed, both healthy and diseased villagers have potassium depletion and hypocitraturia, the severities of which are not different between the two groups. People in both groups have similar life-styles and have lived all their lives in northeastern rural areas. It appears that the environmental factors affecting both groups of villagers are quite similar. In this regard, genetic factors might play an important role in the expression of these syndromes and warrant further investigation.

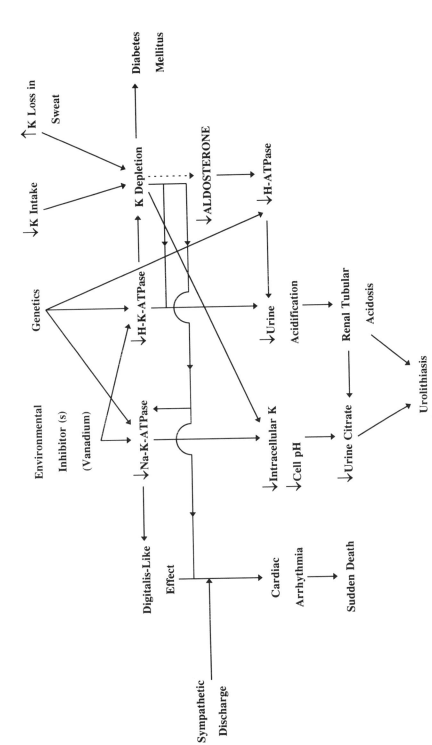

Figure 2. Diagram showing the possible mechanisms involved in the pathogenesis of the metabolic syndromes in northeastern Thailand. Hypokalemia decreases aldosterone secretion (dotted line), assuming that blood volume is normal.

113

3. FUTURE DIRECTION

Tropical diseases reflect the interaction of the human body with various external environmental factors, including the tropical climate, socioeconomy, nutrition, access to medical care, and genetic variations. Available data to date have suggested that a state of vanadium and potassium depletion could play an important role in the pathogenesis of metabolic problems in northeastern Thailand. Chronic vanadium ingestion and prolonged low potassium intake could cause impairment in functions or/and structures of various ATPases and also could mediate numerous metabolic disturbances in northeastern villagers. Low-protein-calorie malnutrition and heat would have some contributory roles, as well. Further clinical and laboratory studies need to be performed to explain these metabolic problems in more detail. It is crucial to gain a greater knowledge of tissue ATPase levels, biochemical disturbances at cellular levels, and so forth. In this regard, kidney tissue should be the main target of these investigations. It also is essential to obtain more information regarding the role of genetic variations in these metabolic syndromes. Methodological and technical advances in the present and the upcoming century would help us to unfold the complete picture of these metabolic syndromes and to completely elucidate their pathogenesis.

ACKNOWLEDGMENTS

The authors would like to thank Ms. Tipwan Tongthamrongrat for her excellent typographical assistance.

REFERENCES

Alpern, R. J. and Rector, F. C. Jr. (1996). Renal acidification mechanism. In B. M. Brenner (Ed.), *The Kidney.* 5th ed. W. B. Saunders Company, Philadelphia, pp. 408–471.

Balfour, W. E., Grantham, J. J., and Glynn, I. M. (1978). Vanadate-stimulated natriuresis. *Nature* **275,** 768.

Barton, J. C., Conrad, M. E., Nuby, S., and Harrison, L. (1978). Effects of iron on the absorption and retention of lead. *J. Lab. Clin. Med.* **92,** 536–547.

Batlle, D. C., Mozes, M. F., Manaligod, J., Arruda, J. A. L., and Kurtzman, N. A. (1981a). The pathogenesis of hyperchloremic metabolic acidosis associated with kidney transplantation. *Am. J. Med.* **70,** 786–796.

Batlle, D. C., Sehy, J. T., Roseman, M. K., Aruda, J. A. L., and Kurtzman, N. A. (1981b). Clinical and pathophysiologic spectrum of acquired distal renal tubular acidosis. *Kidney Int.* **20,** 389–396.

Benabe, J. E., Aroz, F., Cordova, H., and Martinez-Maldonado, M. (1989). High protein diet increases renal concentration in the rabbit : Role of medullary prostaglandins. *Kidney Int.* **35,** 493.

Benabe, J. E., Cruz-Soto, M. A., and Martinez-Maldonado, M. (1984). Critical role of extracellular calcium in vanadate-induced renal vasoconstriction. *Am. J. Physiol.* **246,** F317–F322.

Benabe, J. E., and Martinez-Maldonado, M. (1991). Renal effects of dietary protein excess and deprivation. *Semin. Nephrol.* **11,** 76–85.

Berg, L. R. (1963). Evidence of vanadium toxicity resulting from the use of certain commercial phosphorus supplements in chick rations. *Poult. Sci.* **42,** 760–769.

Berg, L. R., Bearse, G. E., and Merill, L. H. (1963). Vanadium toxicity in laying hens. *Poult. Sci.* **42,** 1407–1411.

Bertrand, D. (1950). Survey of contemporary knowledge of biochemistry. II. The biogeochemistry of vanadium. *Bull. Am. Mus. Natl. Hist.* **94** (7), 407–455.

Boscolo, P., Carmignani, M., Volpe, A. R., Felaco, M., Rosso, G. D., Porcelli, G., and Giuliano, G. (1994). Renal toxicity and arterial hypertension in rats chronically exposed to vanadate. *Occup. Environ. Med.* **51,** 500–503.

Boyd, D. W. and Kustin, K. (1985). Vanadium: A versatile biochemical effector with an elusive biological function. *Adv. Inorg. Chem.* **6,** 311–365.

Byczkowski, J. Z. and Sorenson, J. R. J. (1984). Effects of metal compounds on mitochondrial function: A review. *Sci. Total Environ.* **37,** 133–162.

Byrne, A. R. and Kosta, L. (1978). Vanadium in foods and human body fluids and tissues. *Sci. Total Environ.* **10,** 17–30.

Cannon, H. L. (1963). The biogeochemistry of vanadium. *Soil. Sci.* **96** (3), 196–204.

Cantley, L. C. Jr., Cantley, L. G., and Josephson, L. (1978). A characterization of vanadate interactions with the (Na,K)-ATPase: Mechanistic and regulatory implications. *J. Biol. Chem.* **253,** 7361–7368.

Carmignani, M., Boscolo, P., Volpe, A. R., Togna, G., Masciocco, L., and Preziosi, G. (1991). Cardiovascular system and kidney as specific targets of chronic exposure to vanadate in the rat: Functional and morphological findings. *Arch. Toxicol.* **14**(Suppl), 124–127.

Chaithiraphan, S., Ngam U-kos, P., Laothavorn, P., Watanachai, K., and Kochaseni, S. (1984). Acute myocardial infarction: A collaborative study of 1,541 cases from four medical centers in Thailand. *J. Med. Assoc. Thailand* **67,** 382–390.

Chaiyabutr, N., Changpongsang, S., Loypetjra, P., and Pichaicharnarong, A. (1983). Renal function studies in normal and heat stressed swamp buffaloes. *Proc. Fifth World Conf. Animal Production.* Tokyo, pp. 763–764.

Chan, P. C. and Sanstone, W. R. (1969). The influence of low-potassium diet on rat-erythrocyte-membrane adenosine triphosphatase. *Arch. Biochem. Biophys.* **134,** 48–52.

Changbumrung, S., Tungtrongchitr, R., and Hongtong, K., (1989). Food patterns and habits of people in an endemic area for liver fluke infection. *J. Nutr. Assoc. Thailand* **23,** 133–146.

Chasteen, N. D. (1983). The biochemistry of vanadium. *Struct. Bonding* **53,** 105–135.

Chatudompan, S. (1987). Prevalence of urinary stones in Khon Kaen province. In S. Nimmannit, L. Ong-Aj-Yooth, and A. Tantiwong (Eds.), *Proceedings of the First National Symposium on Urinary Stone Disease and Renal Tubular Acidosis, Khon Kaen, Thailand.* Medical Media, Bangkok, Thailand., pp. 9–12.

Cheng, J. T., Kahn, T., and Kaji, D. M. (1984). Mechanism of alteration of sodium, potassium pump of erythrocytes from patients with chronic renal failure. *J. Clin. Invest.* **74,** 1811–1820.

Clark, R. J. H. (1975). *The chemistry of vanadium, niobium, and tantalum.* Pergamon Press, Oxford, pp. 491–535.

Clausen, T. and Kjeldsen, K. (1987). Effect of potassium deficiency on Na, K homeostasis and Na^+, K^+-ATPase in muscle. *Curr. Top. Membr. Transp.* **28,** 403–419.

Crans, D. C., Bunch, R. L., and Theisen, L. A. (1989). Interaction of race levels of vanadium (IV) and vanadium (V) in biological systems. *J. Am. Chem. Soc.* **111,** 7597–7607.

Cros, G., Mongold, J. J., Serrano, J.-J., Ramanadham, S., and Mc Neill, J. H. (1992). Effects of vanadyl derivertives on animal models of diabetes. *Mol. Cell. Biochem.* **109,** 163–166.

Dafnis, E. and Sabatini, S. (1994). Biochemistry and pathophysiology of vanadium. *Nephron* **67,** 133–143.

Dafnis, E., Spohn, M., Kurtzman, N. A., and Sabatini, S. (1992). Vanadate causes hypokalemic distal renal tubular acidosis. *Am. J. Physiol.* **262,** F499–F453.

Day, H., Middendorf, D., Lukert, B., Heinz, A., and Grantham, J. (1980). The renal response to intravenous vanadate in rats. *J. Lab. Clin. Med.* **96,** 382–395.

Deluise, M. and Flier, J. S. (1985). Evidence for coordinate genetic control of Na-K-pump density in erythrocyte and lymphocyte. *Metabolism* **34,** 771–776.

Dick, D. A. T., Naylor, G. J., and Dick, E. G. (1982). Plasma vanadium concentration in manic-depressive illness. *Psychol. Med.* **12,** 533–537.

Donaldson, J., Hemming, R., and Labella, F. (1985). Vanadium exposure enhances lipid peroxidation in the kidney of rats and mice. *Can. J. Physiol. Pharmacol.* **63,** 196–199.

Doucet, A. and Marsy, S. (1987). Characterization of K-ATPase activity in distal nephron: Stimulation by potassium depletion. *Am. J. Physiol.* **253,** F418–F423.

Durfor, C. N., and Becker, E. (1963). *Public water supplies of the 100 largest cities in the United States, 1962.* Water-Supply Paper No. 1812. U.S. Geological Survey, Washington, D.C.

Edwards, R. M., and Grantham, J. J. (1983a). Effect of vanadate on fluid absorption and PAH secretion in isolated proximal tubules. *Am. J. Physiol.* **244,** F367–F375.

Edwards, R. M., and Grantham, J. J. (1983b). Inhibition of vasopressin action by vanadate in the cortical collecting tubule. *Am. J. Physiol.* **244,** F722–F777.

Eiam-Ong, S., and Kurtzman, N. A. (1995). Renal tubular acidosis. In S. G. Massry and R. J. Glassock (Eds.), *Textbook of Nephrology.* 3d ed. Chapter 23, Part 5. William and Wilkins, Boston, pp. 457–468.

Eiam-Ong, S., Laski, M. E., and Kurtzman, N. A. (1995a). Disease of renal ATPase, *Am. J. Med. Sci.* **309,** (1), 13–25.

Eiam-Ong, S., Tungsanga, K., Tosukhowong, P., and Sitprija, V. (1995b). Renal ATPase-associated disorders. *Nephrology* **1,** 181–190.

Erdmann, E., Werdan, K., Krawietz, W., Schmitz, W., and Scholz, H. (1984). Vanadate and its significance in biochemistry and pharmacology. *Biochem. Pharmacol.* **33**(7), 945–950.

Gaynor, M. L., Ferguson, E. R., and Knochel, J. P. (1982). The effect of prostaglandin synthesis inhibitors on glucose tolerance in potassium deficiency. *Clin. Res.* **30,** 571.

Goh, K. T., Chao, T. C., and Chew, C. H. (1990). Sudden nocturnal deaths among Thai construction workers in Singapore (Letter). *Lancet* **335,** 1154.

Hackett, P. L., and Kelman, B. J. (1983). Availability of toxic trace metals to the conceptus. *Sci. Total Environ.* **28,** 433–442.

Halperin, M. L., Carlisle, E. J. F., Donnelly, S., Karnel, K. S., and Vasuvattakul, S. (1994). Renal tubular acidosis. In R. G. Narins (Ed.), *Clinical Disorders of Fluid and Electrolyte Metabolism.* 5th ed. McGraw-Hill, New York, pp. 875–910.

Halstead, S. B. and Valyasevi, A. (1967). Studies of bladder stone disease in Thailand. III: Epidemiologic studies in Ubol province. *Am. J. Clin. Nutr.* **20,** 1329–1339.

Hamm, L. L. (1990). Renal handling of citrate. *Kidney Int.* **38,** 728–735.

Hays, S., Kokko, J. P., and Jacobson, H. R. (1986). Hormonal regulation of proton secretion in rabbit medullary collecting duct. *J. Clin. Invest.* **78,** 1279–1286.

Higashi, Y., and Bello-Reuss, E. (1980). Effects of sodium orthovanadate on whole kidney and single nephron function. *Kidney Int.* **18,** 302–308.

Higashino, H., Bogden, J. D., Lavenhar, M. A., Baumm, J. W. Jr, Hirotsu, T., and Aviv, M. (1983). Vanadium, Na-K-ATPase, potassium adaptation in the rat. *Am. J. Physiol.* **244,** F105–F111.

Hoffman, J. F. (1969). The interaction between tritiated ouabain and the Na-K pump in red blood cells. *J. Gen. Physiol.* **54,** 343S–350S.

Hoffman, W. W., and Smith, R. A. (1970). Hypokalemic periodic paralysis studies in vitro. *Brain* **93,** 445–474.

Howden, C. W., Beastall, G. H., and Reid, J. L. (1986). An investigation into the effects of omeprazole on renal tubular function and endocrine function in man. *Scand. J. Gastroenterol.* **21,** 169–170.

Huang, W-H., and Askari, A. (1981). Simultaneous bindings of ATP and vanadate to (Na^+-K^+)-ATPase. *J. Biol. Chem.* **259,** 13287–13291.

Hudson, T. G. F. (1964). *Vanadium: Toxicity and Biological Significance.* Elsevier Science Publishers, Amsterdam, New York, Oxford, 384 pp.

Hugh-Jones, P. (1955). Diabetes in Jamica. *Lancet* **1,** 891–897.

ICRP (1960). *Report of Committee II on Permissible Dose for Internal Radiation (1959). Recommendations of the International Commission on Radiological Protection.* ICRP Publication No. 2. Pergamon Press, Oxford.

Inciarte, D. J., Steffen, R. P., Dobbins, D. E., Swindall, B. J., Johnston, J., and Haddy, F. T. (1980). Cardiovascular effects of vanadate in the dog. *Am. J. Physiol.* **239,** H47–H56.

Jandhyala, B. S. and Hom, G. J. (1983). Minireview: Physiological and pharmacological properties of vanadium. *Life Sci.* **33,** 1325–1340.

Jørgensen, P. L. (1983). A conformational change in the α-subunit, and cation transport by Na-K-ATPase. *Ciba Found. Symp.* **95,** 253–272.

Kiatsayompoo, S., Lueprasitasakul, W., Bhuripanyo, P., and Graisopa, S. (1993). Diabetes mellitus in the young in Srinagarind hospital. *J. Med. Assoc. Thai.* **76** (5), 247–251.

Kiviluoto, M. (1980). Observations on the lungs of vanadium workers. *Br. J. Ind. Med.* **37,** 363–366.

Kohler, H. (1987). Fluid metabolism in exercise. *Kidney Int.* **32,** (Suppl 21), S92–S96.

Kumar, A. and Corder, C. N. (1980). Diuretic and vasoconstrictor effects of sodium orthovanadate on the isolated perfused rat kidney. *J. Pharmacol. Exp. Ther.* **213,** 85–90.

Kurtz, I., Dass, P. D., and Cramer, S. (1990). The importance of renal ammonia metabolism to whole body acid-base balance: A reanalysis of the pathophysiology of renal tubular acidosis. *Miner. Electrolyte Metab.* **16,** 331–338.

Larsen, J. A., Thomsen, O. Q., and Hansen, O. (1979). Vanadate-induced oliguria in the anesthetized cat. *Acta Physiol. Scand.* **106,** 495–496.

Lopez-Novoa, J. M., Mayol, V., and Martinez-Maldonado, M. (1982). Renal actions of orthovanadate in the dog. *Proc. Soc. Exp. Biol. Med.* **170,** 418–426.

Munger, R. G., Prineas, R. J., and Crow, R. S. (1991). Prolonged QT interval and risk of sudden death in South-East Asian men. *Lancet* **338,** 280–288.

Myron, D. R., Givand, S. H., and Nielsen, F. H. (1977). Vanadium content of selected foods as determined by flameless atomic absorption spectroscopy. *J. Agric. Food. Chem.* **25**(2), 297–299.

Myron, D. R., Zimmerman, T. J., Shuler, T. R., Klevay, L. M., Lee, D. E., and Nielsen, F. H. (1978). Intake of nickel and vanadium by humans. A survey of selected diets. *Am. J. Clin. Nutr.* **31**(3), 527–531.

Nechay, B. R. (1984). Mechanisms of action of vanadium. *Annu. Rev. Pharmacol. Toxicol.* **24,** 508–524.

Nechay, B. R., Nanninga, L. B., Nechay, P. S. E., Post, R. L., Grantham, J. J., Macara, I. G., Kubena, L. F., Timothy, D. P., and Nielson, F. H. (1986). Role of vanadium in biology. *Fed. Proc.* **45,** 123–132.

Nilwarangkur, S., Nimmanit, S., Chaovakul, V., Susaengrat, V., Ong-aj-yooth, S., Vasuvattakul, S., Pidetcha, P, and Malasit, P. (1990). Endemic primary distal renal tubular acidosis in Thailand. *Q. J. Med.* **14,** 289–301.

Nimmannit, S., Malasit, P., Chaovakul, V., Susaengrat, W., and Nilwarangkur, S. (1990). Potassium and sudden unexplained nocturnal death syndrome (Letter). *Lancet* **336,** 116–117.

Nimmannit, S., Malasit, P., Chaovakul, V., Susaengrat, W., Vasuvattakul, S., and Nilwarangkur, S. (1991). Pathogenesis of sudden unexplained nocturnal death (lai tai) and endemic distal renal tubular acidosis. *Lancet* **338,** 930–932.

Nimmannit, S., Malasit, P., Susaengrat, W., Ong-Aj-Yooth, S., Vasuvattakul, S., Pidetcha, P., Vasuvattakul, S., and Nilwarangkur, S. (1996). Prevalence of endemic distal renal tubular acidosis and renal stone in the northeast of Thailand. *Nephron* **72,** 604–610.

Nirdnoy, M., Gojaseni, P., and Watanachote, D. (1987). A comparison of chemical and infrared spectroscopic analyses of urinary stones. In S. Nimmannit, L. Ong-Aj-Yooth, and A. Tantiwong (Eds), *Proceedings of the First National Symposium on Urinary Stone Disease and Renal Tubular Acidosis, Khon Kaen, Thailand.* Medical Media Press, Bangkok, Thailand, pp. 45–49.

Olurin, E. O. (1969). Pancreatic calcification: A report of 45 cases. *Br. Med. J.* **4,** 534–539.

Otto, C. M., Tauxe, R. V., and Cobb, L. A. (1984). Ventricular fibrillation causes sudden death in Southeast Asian immigrants. *Ann. Intern. Med.* **100,** 45–47.

Parrish, R. G., Tucker, M., Ing, R., Encamacion, C., and Eberhardt, M. (1987). Sudden unexplained death syndrome in Southeast Asian refugees: A review of CDC surveillance. *MMWR. CDC. Surveil. Summ.* **36,** 43–53.

Phillips, T. D., Nechay, B. R., and Heidelbaugh, N. D. (1983). Vanadium: Chemistry and the kidney. *Fed. Proc.* **42,** 2969–2973.

Podrid, P. J., Fuchs, T., and Candinas, R. (1990). Role of the sympathetic nervous system in the genesis of ventricular arrhythmia. *Circulation* **82,** 1103–1113.

Powell, L. W. and Halliday, J. W. (1981). Iron absorption and iron overload. *Clin. Gastroenterol.* **10,** 707–735.

Quintanilla, A. P., Weffer, M. I., Koh, H., Rahman, M., Motini, A., and del Greco, F. (1988). Effect of high salt intake on sodium, potassium-dependent adenosine triphosphatase activity in the erythrocytes of normotensive men. *Clin. Sci.* **75,** 167–170.

Relman, A. S. and Schwartz, W. B. (1956). The nephropathy of potassium depletion: A clinical and pathological entity. *N. Engl. J. Med.* **255,** 195–203.

Robertson, P. (1983). Hypothesis: PGE, carbohydrate homeostasis, and insulin secretion. *Diabetes* **32,** 231–234.

Robinson, K. A. (1981). Concerning the form of biochemically active vanadium. *Proc. R. Soc. Lond. B. Biol. Sci.* **212,** 65–84.

Rosen, M. R., Wit, A. L., and Hoffman, B. F. (1975). Electrophysiology and pharmacology of cardiac arrhythmias: Cardiac arrhythmia and toxic effects of digitalis. *Am. Heart. J.* **89,** 391–399.

Roshchin, A. V. (1967). Toxicology of vanadium compounds used by modern industry. *Hyg. Sanit.* **32,** 345–352.

Roshchin, A. V., Ordzhonikidze, E. K., and Shalganova, I. V. (1980). Vanadium-toxicity, metabolism, carrier state. *J. Hyg. Epidemiol. Microbiol. Immunol.* **24**(4), 377–383.

Russanov, E., Zaporowska, H., Ivancheva, E., Kirkova, M., and Konstantinova, S. (1994). Lipid peroxidation and antioxidant enzymes in vanadate-treated rats. *Comp. Biochem. Physiol.* **107**(3), 415–421.

Sabbioni, E., Goetz, L., and Bignoli, G. (1984). Health and environmental implications of trace metals released from coal fired plants: An assesment study of the situation in the European Community. *Sci. Total Environ.* **40,** 141–154.

Sabbioni, E., Pozzi, G., Pintar, A., Casella, L., and Garattini, S. (1991). Cellular retention, cytotoxicity and morphological transformation by vanadium (IV) and vanadium (V) in BALB/3T3 cell lines. *Carcinogenesis* **12,** 47–52.

Schmalzing, G., Pfaff, E., and Breyer-Pfaff, U. (1981). Red cell ouabain binding sites, Na, K-ATPase and intracellular Na^+ as individual characteristics. *Life Sci.* **29,** 371–381.

Schroeder, H. A. (1970). *Vanadium.* Air Quality Monograph No. 70–73. American Petroleum Institute, Washington, D.C.

Schroeder, H. A., Balassa, J. J., and Tipton, I. H. (1963). Abnormal trace metals in man: Vanadium. *J. Chronic. Dis.* **16**, 1047–1071.

Schwartz, W. B., and Relman, A. S. (1967). Effect of electrolyte disorders on renal structure and function. *N. Engl. J. Med.* **276**, 383–387.

Scott, R. (1987). Prevalence of calcified upper urinary tract stone disease in a random population survey. *Br. J. Urol.* **59**, 111–117.

Sebastian, A., Mc Sherry, E., and Morris, R. C. Jr. (1971). Renal potassium wasting in renal tubular acidosis (RTA). Its occurrence in types 1 and 2 RTA despite sustained correction of systemic acidosis. *J. Clin. Invest.* **50**, 667–678.

Senay, F. D. Jr. and Marver, D. P. (1986). Effect of dietary protein on medullary thick ascending limb Na^+-K^+-ATPase (Abstract). *Clin. Res.* **34**, 609.

Shaper, A. G. (1960). Chronic pancreatic disease and protein malnutrition. *Lancet* **2**, 1223–1234.

Sharma, R. P., Bourcier, D. R., Brinkerhoff, C. R., and Christensen, S. A. (1981). Effects of vanadium on immunologic functions. *Am. J. Ind. Med.* **2**(2), 91–99.

Sitprija, V., Eiam-Ong, S., Suvanapha, R., Kullavanijaya, P., and Chinayon, S. (1988). Gastric hypoacidity in distal renal tubular acidosis. *Nephron* **50**, 395–396.

Sitprija, V., Tungsanga, K., Eiam-Ong, S., Leelhaphunt, N., and Sriboonlue, P. (1990). Renal tubular acidosis, vanadium, and buffaloes. *Nephron* **54**, 97–98.

Sitprija, V., Tungsanga, K., Eiam-Ong, S., Tosukhowong, P., Leelhaphunt, N., Sriboonlue, P., and Prasongwatana, V. (1991). Metabolic syndrome caused by decreased activity of ATPase. *Semin. Nephrol.* **11**, 453–464.

Sitprija, V., Tungsanga, K., Tosukhowong, P., Leelhaphunt, N., Kruerklai, D., Sriboonlue, P., and Saew, O. (1993). Metabolic problems in northeastern Thailand: Possible role of vanadium. *Miner. Electrolyte Metab.* **19**, 51–56.

Smith, J. H., Braselton, W. E., Tonsager, S. R., Mayor, G. H., and Hook, J. B. (1982). Effects of vanadate on organic ion accumulation in rat renal cortical slices. *J. Pharmacol. Exp. Ther.* **220**, 540–546.

Söremark, R. (1967). Vanadium in some biological specimens. *J. Nutr.* **92**, 183–190.

Sriboonlue, P., Prasongwatana, V., Chata, K., and Tungsanga, K. (1992). Prevalence of upper urinary tract stone disease in a rural community of northeastern Thailand. *Br. J. Urol.* **69**, 240–244.

Sriboonlue, P., Prasongwatana, V., Tungsanga, K., Tosukhowong, P., Phantumvanit, P., Bejraputra, O., and Sitprija, V. (1991). Blood and urinary aggregator and inhibitor composition in control and renal-stone patients from northeastern Thailand. *Nephron* **59**, 591–596.

Sriboonlue, P., Tungsanga, K., Tosukhowong, P., and Sitprija, V. (1993). Seasonal changes in serum and erythrocyte potassium among renal stone formers from northeastern Thailand. *Southeast Asian J. Trop. Med. Public Health* **24**(2), 287–292.

Talvitie, N. A., and Wagner, W. D. (1954). Studies in vanadium toxicology. II. Distribution and excretion of vanadium in animals. *Arch. Ind. Hyg.* **9**, 414–422.

Tanphaichitr, V., Lerdvuthisopon, N., Dhanamitta, S., and Valyasevi, A. (1990). Thiamine status in Northeastern Thais. *Thai. J. Intern. Med.* **6**, 43–46.

Tolins, J. P., Hostetter, M. K., and Hostetter, T. H. (1987). Hypokalemic nephropathy in the rat: Role of ammonia in chronic tubular injury. *J. Clin. Invest.* **79**, 1447–1453.

Torres, V. E., Young, W. F. Jr., Offord, K. P., and Hattery, R. R. (1990). Association of hypokalemia, aldosteronism, and renal cysts. *N. Engl. J. Med.* **322**, 345–351.

Tosukhowong, P., Chotikasatit, C., Tungsanga, K., Sriboonlue, P., Prasongwattana, V., Pansin, P., and Sitprija, V. (1992). Abnormal erythrocyte Na-K-ATPase activity in Northeastern Thai population. *Southeast Asian J. Trop. Med. Public Health* **23**, 526–530.

Tosukhowong, P., Tungsanga, K., Kittinantavorakoon, C., Chaitachawong, C., Pansin, P., Sriboonlue, P., and Sitprija, V. (1996). Low erythrocyte Na-K-pump activity and number in northeast Thailand adults: Evidence suggesting an acquired disorder. *Metabolism* **45**(7), 804–809.

Tosukhowong, P., Tungsanga, K., Sriboonlue, P., Prasongwatana, V., Bavornpadoongkitti, S., Laornuan, S., and Sitprija, V. (1991). Hypocitraturia and hypokaliuria in renal stone formers from northeastern Thailand: Uncommon association with distal renal tubular acidosis. *J. Nephrol.* **4**, 227–232.

Tulloch, J. A. (1961). J-type diabetes. *Lancet* **1**, 119–121.

Tungsanga, K., and Sriboonlue, P. (1993). Sudden unexplained death syndrome in North-east Thailand. *Int. J. Epidemiol.* **22**(1), 81–87.

Tungsanga, K., Sriboonlue, P., Bavornpadoongkitti, S., Tosukhowong, P., and Sitprija, V. (1992). Urinary acidification in renal stone patients from northeastern Thailand. *J. Urol.* **147**, 325–328.

Vasuvattakul, S., Nimmanit, S., and Nilwarangkur, S. (1987). Pathogenesis of primary distal renal tubular acidosis. In S. Nimmannit, L. Ong-Aj-Yooth, and A. Tantiwong (Eds.), *Proceedings of the First National Symposium on Urinary Stone Disease and Renal Tubular Acidosis, Khon, Kaen, Thailand.* Medical Media, Bangkok, Thailand, pp. 193–199.

Ventura, S. C., Northrup, T. E., Schneider, G., Cohen, J. J., and Garella, S. (1989). Transport and histochemical studies of bicarbonate handling by the alligator kidney. *Am. J. Physiol.* **256**, F239–F245.

Viswanathan, M. (1980). Pancreatic diabetes in India: An overview, in Secondary Diabetes. *Diabetes* **29**, 105–116.

Waters, M. D. (1977). Toxicology of vanadium. *Adv. Mod. Toxicol.* **2**, 147–189.

Wenning, R. and Kirsch, N. (1988). Vanadium. In H. G. Seiler, H. Siegel, and A. Siegel (Eds.), *Handbook on Toxicity of Inorganic Compounds.* Marcel Dekker, New York, pp. 749–765.

Wingo, C. S. (1989). Active proton secretion and potassium absorption in the rabbit outer medullary collecting duct. Functional evidence for proton-potassium activated adenosine triphosphatase. *J. Clin. Invest.* **84**, 361–365.

WHO (World Health Organization). (1988). *Vanadium in Environmental Health, Criteria 81.* Finland, World Health Organization, Geneva, 170 pp.

Younes, M., Albrecht, M., and Siegers, C. P. (1984). Lipid peroxidation and lysosomal enzyme release induced by vanadate in vitro. *Res. Commun. Chem. Pathol. Pharmacol.* **43**, 487–495.

7

VANADIUM AND ITS SIGNIFICANCE IN ANIMAL CELL METABOLISM

Halina Zaporowska and Agnieszka Ścibior

Department of Cell Biology, Institute of Biology,
Maria Curie-Skłodowska University, Akademicka 19, 20-033
Lublin, Poland

1. INTRODUCTION

Vanadium is an essential trace element for animals owing to its physiological and biochemical activities (Nielsen and Uthus, 1990; French and Jones, 1993). The metabolism and toxicology of vanadium in laboratory animals and humans are well established. However, the effects of vanadium on cell metabolism are considerably less known, although many reports on this subject have appeared in recent years (Cortizo and Etcheverry, 1995; Wenzel et al., 1995;

Vanadium in the Environment. Part 2: Health Effects, Edited by Jerome O. Nriagu.
ISBN 0-471-17776-8. © 1998 John Wiley & Sons, Inc.

Jonas and Henquin, 1996). Therefore, the aim of this chapter is a discussion of vanadium's effect on some selected metabolic processes in animal cells.

2. ACCUMULATION OF VANADIUM IN THE CELL AND ITS METABOLISM

Vanadium is known to be present in the tissues of several animals, including humans and other mammals. Bones, kidneys, liver, and spleen contain larger quantities of vanadium. The smallest amounts are found in the brain (Edel and Sabbioni, 1989; Lener et al., 1989; Al-Bayati et al., 1991; Ramanadham et al., 1991). It was shown that vanadium accumulation in tissues is directly related to the dose administered. No significant differences were observed in the retention of ^{48}V in rats ingesting tetravalent or pentavalent vanadium ions (Edel and Sabbioni, 1988). Cell cultures incubated in the medium supplemented with different vanadium compounds accumulated this element, above all in the nucleus and mitochondria (Bracken et al., 1985; Stern et al., 1993; Sit et al., 1996). Investigations of intracellular vanadium accumulation were also carried out on liver and kidney cells of laboratory animals that had previously received suitable intravenous or intratracheal doses of ^{48}V. It was shown that in hepatocytes and kidney cells most vanadium was accumulated in the nucleus, mitochondria, and cytosol (Edel and Sabbioni, 1988; Edel and Sabbioni, 1989).

In body fluids and extracellularly, vanadium exists predominantly in the $5+$ oxidation state as vanadate (VO_3^-). Vanadate enters the cell through nonspecific anion channels (Heinz et al., 1982). In the cell it undergoes bioreduction from pentavalent form to tetravalent vanadyl (VO^{2+}). Vanadium may be reduced by cellular glutathione, catechols, cysteine, NADH, NADPH, and L-ascorbic acid (Legrum, 1986; Kretzschmar and Bräunlich, 1990; Sabbioni et al., 1993). Shi and Dalal (1990) demonstrated that glutathione reductase (GR; EC 1.6.4.2) can also fulfill the function of vanadate reductase. Moreover, Bandwar and Rao (1995) have recently shown in vitro vanadate reduction by various hydroxy-containing compounds such as pentoses, hexoses, glycols, and ethanolamines. Thus, the intracellular form of vanadium may depend on the ratio of the whole pool of reductants to the cell oxidants (Fig. 1). In the cell, vanadate can affect the phosphorylation and dephosphorylation processes of proteins (Tessier et al., 1989; Chao et al., 1992; Pugazhenthi and Khandelwal, 1992; Imbert et al., 1994; Yamaguchi et al., 1995), regulate many enzyme activities (Ueki et al., 1992; Wenzel et al., 1995), and participate in the generation of free radicals (Carmichael, 1990; Kalyani et al., 1992; Ding et al., 1994; Shi et al., 1996).

Protein phosphorylation by kinases and dephosphorylation by phosphatases are key reactions involved in the regulation of the cell metabolism, proliferation, and differentiation. Vanadate has been shown to inhibit phosphotyrosine phosphatase activity and to stimulate tyrosine kinase, leading to an accumula-

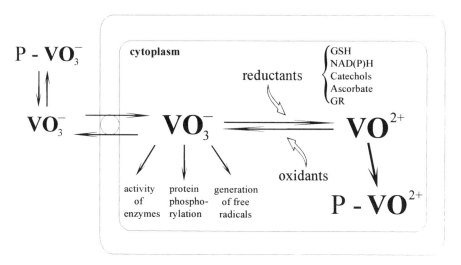

Figure 1. Vanadium transport to the cell and its effect on intracellular processes. P, protein; GSH, reduced glutathione; GR, glutathione reductase. Adapted from Heinz et al. (1982) and Sabbioni et al. (1993); figure changed and supplemented.

tion of phosphotyrosine in various cellular proteins. For example, Lerea et al. (1989) have reported that vanadate added to electropermeabilized human platelets consistently caused a time- and dose-dependent increase in the content of phosphotyrosine in proteins of 50 and 38 kDa. This effect probably reflected the inhibition of protein tyrosine phosphatase (PTPase). Pugazhenthi and Khandelwal (1992) have also demonstrated in vitro that ortho- and meta-vanadate stimulated phosphorylation of six rat liver cytosolic proteins, whereas vanadyl sulfate had the opposite, inhibitory effect. This stimulatory effect of vanadate on the phosphorylation of several rat liver proteins occurred probably by a protein kinase C-mediated mechanism. Chao et al. (1992) have demonstrated that vanadate stimulated tyrosine phosphorylation of numerous cellular proteins in intact Kupffer cells in a time- and concentration-dependent manner. However, Zor et al. (1993) reported that treatment of mice macrophages with vanadate plus H_2O_2 led to a dose-dependent activation of protein kinase C (PKC) and protein tyrosine kinase (PTK) and to inhibition of protein tyrosine phosphatase (PTPase). Of particular significance also is the fact that vanadate causes protein tyrosine phosphorylation of insulin receptors, leading to down-regulation of the insulin receptor (Torossian et al., 1988).

Vanadate ion shows great reactivity in vitro. The result of the reaction with ADP and ATP is AMPV instead of ADP and $AMPV_2$ or ADPV as analogues of ATP (Tracey et al., 1988; Geraldes et al., 1989). Ion VO^{2+} can also form complexes with other biologically important compounds such as glutathione, nucleic acids (DNA, RNA), amino acids, peptides, and proteins (Crans et al., 1989; Lord and Reed, 1990; Ferrer et al., 1991). Of the combinations of vanadyl with proteins the best known are complexes with hemoglobin, transferrin, ferritin, lactoferrin, calcineurin, and calmodulin (Sabbioni et al., 1980; Ahmed et al., 1987; Parra-Diaz et al., 1995; Saponja and Vogel, 1996). ESR spectroscopy studies suggested that VO^{2+} binds to calmodulin with $4:1$ stoichiometry (Nieves et al., 1987). It has been shown that VO^{2+} also binds to carbohydrates.

Vanadate is structurally similar to phosphate and can act as a phosphate analogue in biological systems. For example, in the reaction with glucose it forms glucose-6-vanadate (Gresser and Tracey, 1990). Moreover, vanadate promotes cleavage of phosphoenolpyruvate with liberation of phosphate and formation of vanadoenolpyruvate complex (Aureliano et al., 1994).

Do identical complexes also occur in vivo under vanadium influence and what are the consequences of this process for the cell (and organism)? A full answer cannot be given to this question on the basis of the present studies. Moreover, the in vitro effects of vanadium cannot be extrapolated a priori in vivo and vice versa.

Dissolution of vanadium compounds in aqueous media leads to, apart from monovanadate ($H_2VO_4^-$), a variety of oligomeric forms, particularly divanadate ($H_2V_2O_7^{2-}$), cyclic tetravanadate ($V_4O_{12}^{4-}$), cyclic pentavanadate ($V_5O_{15}^{5-}$), and decavanadate ($V_{10}O_{28}^{6-}$ or $HV_{10}O_{28}^{5-}$). Therefore, vanadate aqueous solutions consist of complex mixtures of different oligomers (Amado et al., 1993; Correia et al., 1994; Lobert et al., 1994). The ratio of these vanadate oligomers in solution depends on several factors, particularly on the buffer type used, pH, total concentration of oxyanion, and ionic strength. Vanadium ions exist as hydrated monomers at micromolar concentrations near neutral pH. At concentrations higher than 0.1 mM, at neutral pH, vanadate begins to polymerize. For example, in organic buffers a low pH and high vanadate concentration favors oligomer formation, especially tetra- and decavanadate (Correia et al., 1994). Thus, the results of the investigations conducted in vitro depended on the vanadate oligomer species formed in the studies (Sandiraseg-arane and Gopalakrishnan, 1994).

3. EFFECT ON PHOSPHOINOSITIDE METABOLISM AND SIGNAL TRANSDUCTION

Every cell, tissue, and organ, and the whole organism, can function only under conditions of correct signal transduction. Transmission of information between cells takes place owing to so-called extracellular (primary) and intracellular (secondary) messengers. The primary messengers are hormones, neurotransmitters, and growth factors. Among the secondary messengers, the best known are cyclic nucleotides (e.g., cAMP, cGMP), inositol-1,4,5-trisphosphate (IP_3), 1,2-diacylglycerol (DAG), and some ions (e.g., Ca^{2+}) (Berridge, 1987).

In vitro investigations have shown that vanadium can regulate the activity of secondary messengers. It has been shown that vanadium activates adenylate cyclase (AC; EC 4.6.1.1), which catalyzes the reaction of ATP transformation to cAMP. At concentrations $>10^{-5}$ M vanadate has been found to enhance adenylate cyclase from various sources. For instance, vanadate increased the activity of AC in the heart muscle of cats in the presence of 300 μM NH_4VO_3 or VCl_3 (Schmitz et al., 1982) and in platelets after incubation with 100 μM NH_4VO_3 (Ajtai et al., 1983), but vanadyl sulfate inhibited this enzyme in rat plasmalemma cells isolated from adrenal tissue (Hayashi and Kimura, 1986).

The effect of vanadium on AC activity is explained variously. In the opinion of some authors orthovanadate can form complexes with GDP, which, "imitating" GTP, binds with protein G, and in this way activates AC (Dehaye and Grosfils, 1993). Other authors think that this enzyme can have a specific place for oxyions (e.g., for oxycation VO^{2+}), which, after incorporation with enzyme, regulate AC activity (Mittag et al., 1993). The effect of AC activation is increased cAMP level. An increase in cAMP was observed in different tissues (Schmitz et al., 1982; Ajtai et al., 1983).

The formation of two other secondary messengers (IP_3 and DAG) is also an enzymatic process. Binding a primary messenger to a receptor (R) at the cell surface leads to G-protein activation, which then stimulates specific plasma-membrane phospholipase C (PLC; EC 3.1.4.3). This enzyme catalyzes hydrolysis of membrane-associated phosphatidylinositol-4,5-bisphosphate (PIP_2) to IP_3 and DAG (Fig. 2). Then IP_3 can go to cytosol, where it plays the main role in mobilizing intracellular Ca^{2+} levels. For example, IP_3 can bind with specific receptors of endoplasmic reticulum (ER). Stimulation of these receptors leads to mobilization of Ca^{2+} ions from ER to cytosol (Berridge, 1987; Barańska, 1992; Pawełczyk, 1996). Moreover, IP_3 can be metabolized via two separate pathways. One pathway depends on a series of dephosphorylation reactions leading to IP_2, IP, and inositol (I), whereas, in the other pathway IP_3 is transformed to PI and PIP, and PIP_2 is resynthesized (Fig. 2). The amounts of PI, PIP, and PIP_2 are the result of the two opposite reactions. PI

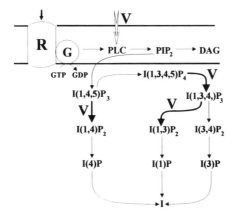

Figure 2. Effect of vanadium on phosphoinositide metabolic pathways (abbreviations as in text). Solid arrows, inhibitory effect of vanadium; open arrows, stimulatory effect of vanadium. Adapted and elaborated from Berridge (1987), Barańska (1992), and Bencherif and Lukas (1992).

and PIP kinases catalyze PIP and PIP_2 synthesis, while phosphomonoesterases dephosphorylate PIP and PIP_2 to PI and PIP, respectively (Berridge, 1987).

Addition of 100 μM vanadate alone did not cause any significant increase in PIP and PIP_2 levels in GH_4C_1 cell membranes. However, when added together with 40 μM tamoxifen (antiestrogen used widely in the therapy for breast cancer), it synergistically enhanced the formation of PIP and PIP_2 (Friedman, 1993).

Zick and Sagi-Eisenberg (1990) showed that treatment of different cell lines—rat hepatoma (Fao), murine muscle (BC3H-1), Chinese hamster ovary (CHO) and rat basophilic leukemia (RBL)—with a combination of 3 mM H_2O_2 and 1 mM sodium orthovanadate (Na_3VO_4) markedly stimulated protein tyrosine phosphorylation, which was accompanied by an increase in IP_3 formation. However, Morita et al. (1992) reported that Na_3VO_4 alone, but at a concentration twice as high, increased IP_3 content in the fat pads of rats. Randazzo et al. (1992) showed inositol phosphate (PI) accumulation in NIH 3T3 cells incubated in lithium and 100 and 1,000 μM orthovanadate. Increased inositol phosphate level was also observed in vitro in vanadate-treated canine trachelis muscle (Lee et al., 1994). In the concentration range from 0.1 to 0.5 mM sodium metavanadate ($NaVO_3$) stimulated ^{32}P-orthophosphate incorporation into phosphatidylinositol (PI) and phosphatidic acid (PA) in brain microvessels. At concentrations higher than 0.5 mM the stimulatory effect of vanadate decreased (Catalán et al., 1991).

Increased IP_3 level in the examined vanadium-treated cell lines resulted on the one hand from PLC activation by vanadium (Dehaye and Grosfils 1993) and, on the other from inhibition of dephosphorylation inositol phosphates (Bencherif and Lukas, 1992). It is shown that vanadium inhibits in vitro hydrolysis of inositol-1,4,5-trisphosphate [$I(1,4,5)P_3$] to inositol-1,4-bisphosphate [$(1,4)P_2$]; inositol-1,3,4-trisphosphate [$I(1,3,4)P_3$] to inositol-1,3-bisphosphate [$I(1,3)P_3$]; and inositol-1,3,4,5-tetrakisphosphate [$I(1,3,4,5)P_4$] to inositol-1,3,4-trisphosphate [$I(1,3,4)P_3$]. The mechanism of these effects consists of direct inhibition of inositol polyphosphate-5-phosphomonoesterase (IPP-5-PME) and inositol polyphosphate-4-phosphomonoesterase (IPP-4-PME). Thus, in the cells studied vanadium increased the accumulation of $I(1,4,5)P_3$, $I(1,3,4)P_3$, and $I(1,3,4,5)P_4$ (Bencherif and Lukas, 1992; Morita et al., 1992; Dehaye and Grosfils, 1993).

4.   EFFECT ON Ca^{2+} ACCUMULATION AND TRANSPORT IN THE CELL

The discussed effect of vanadium on phosphoinositide metabolism suggests indirect influence of this element on signal transduction in the cell. IP_3 plays an important role in the regulation of Ca^{2+} ion concentration in the cell. Therefore, vanadium, which increased the IP_3 level in cytoplasm, disturbed homeostasis and caused indirect release of Ca^{2+} ions to cytosol from their intracellular stores, particularly from the ER. The increase of Ca^{2+} ion inflow through the plasma membrane is probably less significant, because vanadium compounds are the potent inhibitors of several ATPases, including Na$^+$, K$^+$-ATPase, Ca^{2+}-ATPase, and Ca^{2+}, Mg^{2+}-ATPase (Stern et al., 1993; Janiszewska, et al., 1993), as well as calcium transport through isolated adipocyte plasma membrane and endoplasmic reticulum (Delfert and McDonald, 1985). However, vanadium salts stimulate Ca^{2+} influx into cultured epidermal carcinoma cells. Macara (1986) demonstrated that vanadate (200 μM) activated ^{45}Ca^{2+} influx into human epidermal carcinoma A431 cells to a degree similar to that produced by epidermal growth factor (EGF).

Gullapalli et al. (1989) investigated the effect of vanadate on redistribution of subcellular calcium in rat hepatocytes. They showed that administration of vanadium to the rat disturbed the hepatic subcellular calcium status and caused mobilization of calcium ions from mitochondria. Hepatic mitochondria isolated from vanadate-treated animals showed a 50% decrease in their calcium content, and the cytosolic fraction showed a corresponding increase.

It is difficult to explain unambiguously the mechanism of vanadium action on mobilization of Ca^{2+} ions. Its effect via increased IP$_3$ level is surely not the only pathway. According to Gullapalli et al. (1989), vanadate appears to act also at the level of the plasma membrane, involving the activation of α-adrenergic receptors. In the studies of Gullapalli et al. phenoxybenzamine, an antagonist to α-adrenergic receptors, effectively inhibited vanadate-induced Ca^{2+} mobilization. Thus, vanadate can mimic α-adrenergic agonists in vivo.

Richelmi et al. (1989) demonstrated that incubation of isolated rat hepatocytes with vanadate (0.25, 0.5, and 1 mM) resulted in progressive accumulation of Ca^{2+} in intracellular compartments. Vanadate-induced Ca^{2+} accumulation was related to inhibition of the plasma membrane Ca^{2+}-extruding system, but did not involve either enhanced activity of the Na$^+$/Ca^{2+} exchange or increased permeability of the plasma membrane. As shown by these authors, most of the Ca^{2+} accumulated in response to vanadate treatment was sequestered by mitochondria with a small contribution of the endoplasmic reticulum. Richelmi et al. (1989) also proposed a scheme for vanadate-induced intracellular Ca^{2+} accumulation and cytotoxicity. According to this scheme Ca^{2+} sequestration in mitochondria leads to mitochondrial membrane damage, release of Ca^{2+} into the cytosol, increase of cytosolic Ca^{2+} level, and subsequent cytoxicity.

Besides hepatocytes, vanadate causes the cytosolic free Ca^{2+} level to increase in other cell types also: rat erythrocytes (David-Dufilho et al., 1993), HL60 granulocytes (Bianchini et al., 1993), rat cardiomyocytes (Shah et al., 1995), smooth muscle cells (Sandirasegarane and Gopalakrishnan, 1995).

An addition of low concentrations of vanadate (10–200 μM) to cultured aortic smooth muscle cells (ASMC) produced a rapid and concentration-dependent increase in cytosolic Ca^{2+} level. The authors concluded that the increased cytosolic Ca^{2+} level in rat ASMC in the presence of vanadium may be due to Ca^{2+} influx or Ca^{2+} mobilization from IP$_3$-sensitive and/or IP$_3$-

insensitive storage pools, and possibly also from a vanadate-sensitive Ca^{2+} pool (Sandirasegarane and Gopalakrishnan, 1995).

Orthovanadate, which is effective in mobilizing intracellular Ca^{2+} in intact cells, has no effect when added alone to permeabilized cells. However, in the presence of GTP vanadate acted synergistically and stimulated accumulation of inositol phosphates and Ca^{2+} release from the IP_3-sensitive pool (Muldoon et al., 1987).

David-Dufilho et al. (1993) investigated the influence of Na_3VO_4 (0.5– 1 mM) on $^{45}Ca^{2+}$ accumulation in rat erythrocytes. The authors reported that vanadate increased accumulation of $^{45}Ca^{2+}$ in normotensive rat erythrocytes, but not in cells from spontaneously hypersensive rats (SHR). Moreover, it was shown that erythrocytes from SHR are characterized by higher Ca^{2+} values and low sensitivity to vanadate ions.

However, the effect of vanadium on Ca^{2+} transport in the cell is not fully known. This problem requires further investigations and observations.

Recent studies have shown that vanadate (10 μM), a potent inhibitor of tyrosine phosphatase, can markedly potentiate sodium butyrate-induced *apoptosis* in human promyelocytic leukemia cell line HL-60. On the other hand no significant differences in the relative levels of increased tyrosine phosphorylation of pp37 and pp97 proteins were observed in the cells treated with sodium butyrate (BuONa) in the presence of vanadate as compared to the cells treated with BuONa alone (Chang and Yung, 1996).

5. SUMMARY

Vanadium is a transitional metal widely distributed in nature. It stimulates or inhibits many enzymes in vivo and in vitro (Bourgoin and Grinstein, 1992; Ueki et al., 1992; Wenzel et al., 1995).

Vanadium is a well-known pro-oxidant. It oxidizes NADH and other intracellular reducing agents and generates active oxygen forms (Kalyani et al., 1992). The participation of vanadate in generation of free radicals is reportedly one of the mechanisms by which this element exerts its cellular effects. Moreover, other pro-oxidants can act synergistically with vanadium in some of its actions on animal cell metabolism. However, other mechanisms of vanadium cytoxicity are also plausible.

One of more interesting effects of vanadate is its ability to mimic and potentiate the effects of growth factors such as insulin (Brichard and Henquin, 1995) and epidermal growth factor (EGF) on intact cells (Macara, 1986; Stern et al., 1993). It was also shown that vanadate can regulate the activity of secondary messengers and in this way affect signal transduction in the cell. For example, vanadate activates PLC-coupled G-protein and consequently stimulates IP_3 synthesis in different cells (Zick and Sagi-Eisenberg, 1990; Tertrin-Clary et al., 1992). On the other hand, vanadium also inhibits dephosphorylation inositol phosphates (Bencherif and Lukas, 1992). The result of the cooperation of these two mechanisms of vanadium action is increased IP_3 level, which then leads to mobilization of Ca^{2+} from intracellular stores and the subsequent increase in the intracellular Ca^{2+} level. This phenomenon is very important in the stimulation of many cellular processes, but excess Ca^{2+} can be also toxic (Richelmi et al., 1989). Activation of adenylate cyclase by vanadium leads to increased cAMP level (Schmitz et al., 1982), but inhibition of cAMP production by vanadate and absence of effect on cAMP level were also described (Madsen et al., 1994). Thus, vanadium can affect many cellular processes regulated by cAMP. Another pathway of vanadium action on cell metabolism is inhibition of phosphotyrosine phosphatase activity and stimulation of tyrosine kinase, leading to an accumulation of phosphotyrosine in many cellular proteins (Jonas and Henquin, 1996). The above-mentioned mechanisms of vanadium action on cell metabolism are not the only ones. This metal can act genotoxically and mutagenically and induce gene expression in many cultures in vitro.

REFERENCES

Ahmed, R. H., Nieves, J., Kim, L., Echegoyen, L., and Puett, D. (1987), Vanadyl binding to a testicular S-100-like protein and to calmodulin: Electron paramagnetic resonance spectra of VO^{2+}-protein complexes. *J. Prot. Chem.* **6**, 431–439.

Ajtai, K., Tuka, K., and Birò, E. N. A. (1983). The activation of human platelet adenylate cyclase by vanadate. *Thromb. Res.* **29**, 371–376.

Al-Bayati, M. A., Raabe, O. G., Giri, S. N., Knaak, and J. B. (1991). Distribution of vanadate in the rat following subcutaneous and oral routes of administration. *J. Am. Coll. Toxicol.* **10**, 233–241.

Amado, A. M., Aureliano, M., Ribeiro-Claro, P. J. A., and Teixeira-Dias, J. J. C. (1993). Combined Raman and ^{51}V NMR spectroscopic study of vanadium(V) oligomerization in aqueous alkaline solutions. *J. Raman Spectrosc.* **24**, 699–703.

Aureliano, M., Leta, J., Madeira, V. M. C., and de Meis, L. (1994). The cleavage of phosphoenolpyruvate by vanadate. *Biochem. Biophys. Res. Commun.* **201**, 155–159.

Bandwar, R. P., and Rao, C. P. (1995). Relative reducing abilities *in vitro* of some hydroxy-containing compounds, including monosaccharides, towards vanadium(V) and molybdenum(VI). *Carbohydr. Res.* **277**, 197–207.

Barańska, J. (1992). Rozpad fosfolipidów a przekazywanie informacji w komórce. *Monografia Pol. Tow. Biochem. Warszawa,* 1992.

Bencherif, M., and Lukas, R. J. (1992). Vanadate amplifies receptor-mediated accumulation of inositol trisphosphates and inhibits inositol tris- and tetrakis-phosphatase activities. *Neurosci. Lett.* **134**, 157–160.

Berridge, M. J. (1987). Inositol trisphosphate and diacyloglycerol: Two interacting second messengers. *Annu. Rev. Biochem.* **56**, 159–93.

Bianchini, L., Todderud, G., and Grinstein, S. (1993). Cytosolic $[Ca^{2+}]$ homeostasis and tyrosine phosphorylation of phospholipase $C\gamma2$ in HL60 granulocytes. *J. Biol. Chem.* **268**, 3357–3363.

Bourgoin, S., and Grinstein, S. (1992). Peroxides of vanadate induce activation of phospholipase D in HL-60 cells. Role of tyrosine phospholylation. *J. Biol. Chem.* **267**, 11908–11916.

Bracken, W. M., Sharma, R. P., and Elsner, Y. Y. (1985). Vanadium accumulation and subcellular distribution in relation to vanadate induced cytotoxicity *in vitro*. *Cell Biol. Toxicol.* **1**, 259–267.

Brichard, S. M., and Henquin, J. C. (1995). The role of vanadium in the management of diabetes. *Trends in Pharmacological Sciences* **16**, 265–270.

Carmichael, A. J. (1990). Vanadyl-induced Fenton-like reaction in RNA. An ESR and spin trapping study. *FEBS Lett.* **261**, 165–170.

Catalán, R. E., Martínez, A. M., Aragonés, M. D., Miguel, B. G., Diaz, G., and Hernández, F. (1991). Pertussis toxin-insensitive regulation of phosphatidylinositol hydrolysis by vanadate in brain microvessels. *Biochem. Int.* **25**, 985–993.

Chang, S. T., and Yung, B. Y. M. (1996). Potentiation of sodium butyrate-induced apoptosis by vanadate in human promyelocytic leukemia cell line HL-60. *Biochem. Biophys. Res. Commun.* **221**, 594–601.

Chao, W., Liu, H., Hanahan, J. and Olson, M. S. (1992). Protein tyrosine phosphorylation and regulation of the receptor for platelet-activating factor in rat Kupffer cells. *Biochem. J.* **288**, 777–784.

Correia, J. J., Lipscomb, L. D., Dabrowiak, J. C., Isern, N., and Zubieta, J. (1994). Cleavage of tubulin by vanadate ion. *Arch. Biochem. Biophys.* **309**, 94–104.

Cortizo, A. M., and Etcheverry, S. B. (1995). Vanadium derivatives act as growth factor-mimetic compounds upon differentiation and proliferation of osteoblast-like UMR106 cells. *Mol. Cell Biochem.* **145**, 97–102.

Crans, D. C., Bunch, R. L., and Theisen, L. A. (1989). Interaction of trace levels of vanadium(IV) and vanadium(V) in biological systems. *J. Am. Chem. Soc.* **111**, 7597–7607.

David-Dufilho, M., Pernollet, M. G., Morris, M., Astarie-Dekequer, C., and Devynck, M. A. (1993). Erythrocyte Ca^{2+} handling in the spontaneously hypertensive rat; effect of vanadate ions. *Life Sci.* **54**, 267–274.

Dehaye, J. P., and Grosfils, K. (1993). Interaction of vanadate with isolated rat parotid acini. *Gen. Pharmacol.* **24**, 479–488.

Delfert, D. M., and McDonald, J. M. (1985). Vanadyl and vanadate inhibit Ca^{2+} transport system of the adipocyte plasma membrane and endoplasmic reticulum. *Arch. Biochem. Biophys.* **241**, 665–672.

Ding, M., Gannett, P. M., Rojanasakul, Y., Liu, K., and Shi, X. (1994). One-electron reduction of vanadate by ascorbate and related free radical generation at physiological pH. *J. Inorg. Biochem.* **55**, 101–112.

Edel, J., and Sabbioni, E. (1988). Retention of intratracheally instilled and ingested tertavalent and pentavalent vanadium in the rat. *J. Trace Elem. Electrolytes Health Dis.* **2**, 23–29.

Edel, J., and Sabbioni, E. (1989). Vanadium transport across placenta and milk of rats to the fetus and newborn. *Biol. Trace Elem. Res.* **22**, 265–275.

Ferrer, E. G., Williams, P. A. M., and Baran, E. J. (1991). A spectrophotometric study of the VO^{2+}-glutathione interactions. *Biol. Trace Elem. Res.* **30**, 175–183.

French, R. J., and Jones, P. J. H. (1993). Role of vanadium in nutrition: Metabolism, essentiality and dietary considerations. *Life Sci.* **52**, 339–346.

Friedman, Z. Y. (1993). Tamoxifen and vanadate synergize in causing accumulation of polyphosphoinositides in GH_4C_1 membranes. *J. Pharmacol. Exp. Ther.* **267**, 617–623.

Geraldes, C. F. G. C., and Castro, M. M. C. A. (1989). Multinuclear NMR study of the interaction of vanadate with mononucleotides, ADP, and ATP. *J. Inorg. Biochem.* **37**, 213–232.

Gresser, M. J., and Tracey, Y. S. (1990). Vanadate as phosphate analogs in biochemistry. In N. D. Chasteen (Ed.), *Vanadium in Biological Systems. Physiology and Biochemistry.* Kluwer Academic Publishers, The Netherlands, pp. 63–79.

Gullapalli, S., Shivaswamy, V., Ramasarma, T., and Kurup, C. K. R. (1989). Redistribution of subcellular calcium in rat liver on administration of vanadate. *Mol. Cell, Biochem.* **90**, 155–164.

Hayashi, Y., and Kimura, T. (1986). The effects of vanadium compounds on the activation of adenylate cyclase from rat adrenal membrane. *Biochim. Biophys. Acta* **869**, 29–36.

Heinz, A., Rubinson, K. A., and Grantham, J. J. (1982). The transport and accumulation of oxyvanadium compounds in human erythrocytes in vitro. *J. Lab. Clin. Med.* **100**, 593–612.

Imbert, V., Peyron, J. F., Farahi Far, D., Mari, B., Auberger, P., and Rossi, B. (1994). Induction of tyrosine phosphorylation and T-cell activation by vanadate peroxide, an inhibitor of protein tyrosine phosphatases. *Biochem. J.* **297**, 163–173.

Janiszewska, G., Lachowicz, L., Jaskólski, D., and Gromadzińska, E. (1993). Vanadium inhibition of human parietal lobe ATPases. *Int. J. Biochem.* **26**, 551–553.

Jonas, J. C., and Henquin, J. C. (1996). Possible involvement of a tyrosine kinase-dependent pathway in the regulation of phosphoinsitide metabolism by vanadate in normal mouse islets. *Biochem. J.* **315**, 49–55.

Kalyani, P., Vijaya, S., and Ramasarma, T. (1992). Characterization of oxygen free radicals generated during vanadate-stimulated NADH oxidation. *Mol. Cell. Biochem.* **111**, 33–40.

Kretzschmar, M., and Bräunlich, H. (1990). Role of glutathione in vanadate reduction in young and mature rats: Evidence for direct participation of glutathione in vanadate inactivation. *J. Appl. Toxicol.* **10**, 295–300.

Lee, S. H., Hwang, T. W., and Jung, J. S. (1994). Differential actions of AlF^{-4} and vanadate on canine trachealis muscle. *Pflugers Arch. Eur. J. Physiol.* **427**, 295–301.

Legrum, W. (1986). The mode of reduction of vanadate ($+V$) to oxovanadium ($+V$) by glutathione and cysteine. *Toxicology* **42**, 281–289.

Lener, J., Mravcová, A., and Babický, A. (1989). Retention and distribution of vanadium in rats. *Physiol. Bohemoslov.* **38**, 366.

Lerea, K. M., Tonks, N. K., Krebs, E. G., Fischer, E. H., and Glomset, J. A. (1989). Vanadate and molybdate increase tyrosine phosphorylation in a 50-kilodalton protein and stimulate secretion in electropermeabilized platelets. *Biochemistry* **28**, 9286–9292.

Lobert, S., Isern, N., Hennington, B. S., and Correia, J. J. (1994). Interaction of tubulin and microtubule proteins with vanadate oligomers. *Biochemistry* **33**, 6244–6252.

Lord, K. A., and Reed, G. H. (1990). Vanadyl(IV) complexes with pyruvate kinase: Activation of the enzyme and electron paramagnetic resonance properties of ternary complexes with the protein. *Arch. Biochem. Biophys.* **281**, 124–131.

Macara, I. G. (1986). Activation of $^{45}Ca^{2+}$ influx and $^{22}Na^+/H^+$ exchange by epidermal growth factor and vanadate in A431 cells is independent of phosphatidylinositol turnover and is inhibited by phorbol ester and diacylglycerol. *J. Biol. Chem.* **261**, 9321–9327.

Madsen, K. L., Porter, V. M., and Fedorak, R. N. (1994). Vanadate reduces sodium-dependent glucose transport and increases glycolytic activity in LLC-PK1 epithelia. *J. Cell. Physiol.* **158**, 459–466.

Mittag, T. W., Guo, W., and Taniguchi, T. (1993). Interaction of vanadate and iodate oxyanions with adenylyl cyclase of ciliary processes. *Biochem. Pharmacol.* **45**, 1311–1316.

Morita, T., Motoyashiki, T., Tsuruzono, Y., Kanagawa, A., Tominaga, N., and Ueki, H. (1992). Rapid increase of inositol 1,4,5-trisphosphate content in isolated rat adipose tissue by vanadate. *Chem. Pharm. Bull.* **40**, 2242–2244.

Muldoon, L. L., Jamieson, G. A., and Villereal, M. L. (1987). Calcium mobilization in permeabilized fibroblasts: Effect of inositol trisphosphate, orthovanadate, mitogens, phorbol ester, and guanosine triphosphate. *J. Cell. Physiol.* **130**, 29–36.

Nielsen, F. H., and Uthus, E. O., (1990). The essentiality and metabolism of vanadium. In N. D. Chasteen (Ed.), *Vanadium in Biological Systems. Physiology and Biochemistry.* Kluwer Academic Publishers, The Netherlands, pp. 51–62.

Nieves, J., Kim, L., Puett, D., and Echegoyen, L. (1987). Electron spin resonance of calmodulin-vanadyl complexes. *Biochemistry* **26**, 4523–4527.

Parra-Diaz, D., Wei, Q., Lee, E. Y. C., Echegoyen, L., and Puett, D. (1995) Binding of vanadium(IV) to the phosphatase calcineurin. *FEBS Lett.* **376**, 58–60.

Pawełczyk, T. (1996). Izoenzymy fosfolipazy C specyficznie hydrolizujacej fosfatydyloinozytole i regulacja ich aktywności. *Post. Biochem.* **42**, 290–298.

Pugazhenthi, S., and Khandelwal, R. L. (1992). Vanadate increases protein kinase C-induced phosphorylation of endogenous proteins of liver in vitro. *Biochem. Int.* **26**, 241–247.

Ramanadham, S., Heyliger, C., Gresser, M. J., Tracey, A. S., and McNeill, J. H. (1991). The distribution and half-life for retention of vanadium in the organs of normal and diabetic rats orally fed vanadium(IV) and vanadium(V). *Biol. Trace Elem. Res.* **30**, 119–124.

Randazzo, P. A., Olshan, J. S., Bijivi, A. A., and Jarett, L. (1992). The effect of orthovanadate on phosphoinositide metabolism in NIH 3T3 fibroblasts. *Arch. Biochem. Biophys.* **292**, 258–265.

Richelmi, P., Mirabelli, F., Salis, A., Finardi, G., Berte, F., and Bellomo, G. (1989). On the role of mitochondria in cell injury caused by vanadate-induced Ca^{2+} overload. *Toxicology* **57**, 29–44.

Sabbioni, E., Pozzi, G., Devos, S., Pintar, A., Casella, L., and Fischbach, M. (1993). The intensity of vanadium(V)-induced cytotoxicity and morphological transformation in BALB/3T3 cells is dependent in glutathione-mediated bioreduction to vanadium(IV). *Carcinogenesis* **14**, 2565–2568.

Sabbioni, E., Rade, J., and Bertolero, F. (1980). Relationships between iron and vanadium metabolism: The exchange of vanadium between transferrin and ferritin. *J. Inorg. Biochem.* **12**, 307–315.

Sandirasegarane, L., and Gopalakrishnan, V. (1994). Limitations of the radioreceptor assay of inositol 1,4,5-trisphosphate in vanadate-treated cell suspensions. *Biochem. J.* **298**, 511–512.

Sandirasegarane, L., and Gopalakrishnan, V. (1995). Vanadate increases cytosolic free calcium in rat aortic smooth muscle cells. *Life Sci.* **56**, 169–174.

Saponja, J. A., and Vogel, H. J. (1996). Metal-ion binding properties of the transferrins: A vanadium-51 NMR study. *J. Inorg. Biochem.* **62**, 253–270.

Schmitz, W., Scholz, H., Erdmann, E., Krawietz, W., and Werdan, K. (1982). Effect of vanadium in the +5, +4 and +3 oxidation states on cardiac force of contraction, adenylate cylase and $(Na^+ + K^+)$-ATPase activity. *Biochem. Pharmacol.* **31**, 3853–3860.

Shah, K. R., Matsubara, T., Foerster, D. R., Xu, Y. J., and Dhalla, N. S. (1995). Mechanisms of inotropic responses of the isolated rat hearts to vanadate. *Int. J. Cardiol.* **52**, 101–113.

Shi, X., and Dalal, N. S. (1990). Glutathione reductase functions as vanadate(V) reductase. *Arch. Biochem. Biophys.* **278**, 288–290.

Shi, X., Wang, P., Jiang, H., Mao, Y., Ahmed, N., and Dalal, N. (1996). Vanadium(IV) causes 2′-deoxyguanosine hydroxylation and deoxyribonucleic acid damage via free radical reactions. *Ann. Clin. Lab. Sci.* **26**, 39–49.

Sit, K. H., Paramanantham, R., Bay, B. H., Wong, K. P., Thong, P., and Watt, F. (1996). Induction of vanadium accumulation and nuclear sequestration causing cell suicide in human Chang liver cells. *Experientia* **52**, 778–785.

Stern, A., Yin, X., Tsang, S-S., Davison, A., and Moon, J. (1993). Vanadium as modulator of cellular regulatory cascades and oncogene expression. *Biochem. Cell Biol.* **71**, 103–112.

Tertrin-Clary, C., DeLa Llosa-Hermier, M. P., Roy, M., Chenut, M. C., Hermier, C., and DeLa Llosa, P. (1992). Activation of phospholipase C by different effectors in rat placental cells. *Cell. Signal.* **4**, 727–732.

Tessier, S., Chapdelaine, A., and Chevalier, S. (1989). Effect of vanadate on protein phosphorylation and on acid phosphatase activity in the canine prostate. *Mol. Cell. Endocrinol.* **64**, 87–94.

Torossian, K., Freedman, D., and Fantus, I. G. (1988). Vanadate down-regulates cell surface insulin and growth hormone receptors and inhibits insulin receptor degradation in cultured human lymphocytes. *J. Biol. Chem.* **263**, 9353–9359.

Tracey, A. S., Gresser, M. J., and Liu, S. (1988). Interaction of vanadate with uridine and adenosine monophosphate. Formation of ADP and ATP analogues. *J. Am. Chem. Soc.* **110,** 5869–5874.

Ueki, H., Okuhama, R., Sera, M., Inoue, T., Tominaga, N., and Morita, T. (1992). Stimulatory effect of vanadate on $3', 5'$-cyclic guanosine monophosphate-inhibited low Michaelis–Menten constant $3', 5'$-cyclic adenosine monophosphate phosphodiesterase activity in isolated rat fat pads. *Endocrinology* **131,** 441–446.

Wenzel, U. O., Fouqueray, B., Biswas, P., Grandaliano, G., and Choudhury, G. G. (1995). Activation of mesangial cells by the phosphatase inhibitor vanadate. *J. Clin. Invest.* **95,** 1244–1252.

Yamaguchi, M., Oishi, H., Araki, S., Saeki, S., Yamane, H., Okamura, N., and Ishibashi, S. (1995). Respiratory burst and tyrosine phosphorylation by vanadate. *Arch. Biochem. Biophys.* **323,** 382–386.

Zick, Y., and Sagi-Eisenberg, R. (1990). A combination of H_2O_2 and vanadate concomitantly stimulates protein tyrosine phosphorylation and polyphosphoinositide breakdown in different cell lines. *Biochemistry* **29,** 10240–10245.

Zor, U., Ferber, E., Gergely, P., Szücs, K., Dombrádi, V., and Goldman, R. (1993). Reactive oxygen species mediate phorbol ester-regulated tyrosine phosphorylation and phospholipase A_2 activation: Potentiation by vanadate. *Biochem. J.* **288,** 777–888.

8

HEMATOLOGICAL EFFECTS OF VANADIUM ON LIVING ORGANISMS

Halina Zaporowska and Agnieszka Ścibior

Department of Cell Biology, Institute of Biology, Maria Curie-Skłodowska University, Akademicka 19, 20-033 Lublin, Poland

1. INTRODUCTION

In mammals including humans the quantity of vanadium in blood is low, usually of the order of 0.2–0.5 ng/g (Byrne and Versieck, 1990). Most of it is found in the blood plasma (Edel and Sabbioni, 1989), where it is bound by transferrin. Exchange of this element may also occur between plasma transferrin and liver ferritin (Sabbioni et al., 1980; Chasteen et al., 1986).

Vanadium in the Environment. Part 2: Health Effects, Edited by Jerome O. Nriagu.
ISBN 0-471-17776-8. © 1998 John Wiley & Sons, Inc.

In certain diseases the vanadium level in human blood may vary greatly. A decrease in its level was noted in atherosclerosis, gastric ulcer, and liver cirrhosis (Nozdrjuchina et al., 1974), whereas a considerable rise was observed in subjects occupationally exposed to vanadium (Missenard et al., 1989; Corrigan et al., 1992), in chronic renal diseases (Bello-Reuss et al., 1979), in hemodialyzed patients (Hosokawa and Yoshida, 1990; Romero, 1994), and in some psychic diseases (Conri et al., 1986; Naylor et al., 1987). Some authors suggest a relationship between high vanadium concentration in the environment and endemic occurrence of goiter (Barannik and Michaliuk, 1970) and renal tubular acidosis in humans (Sitprija et al., 1990).

2. INFLUENCE OF VANADIUM ON THE RED BLOOD CELL SYSTEM

Intoxication with vanadium compounds in most experimental animals cause distinct changes in the red blood cell system (Table 1). In many animals a decrease in the erythrocyte count and hemoglobin level was observed (Gulko 1956; Seljankina, 1961; Roshchin, 1967; Chakraborty et al., 1977; Zaporowska and Wasilewski, 1989a, 1989b, 1992; Zaporowska et al., 1993). These changes were accompanied by increased percentage of reticulocytes and polychromatophilic erythrocytes in the peripheral blood (Zaporowska and Wasilewski, 1989a, 1992). This increased entering of immature blood cells into the circulation with a simultaneous decrease in the erythrocyte count may suggest a disorder in erythropoiesis and erythrocyte maturation in circumstances of vanadium intoxication. However, other authors did not observe such changes in the erythrocyte count and hemoglobin level in animals (Platonow and Abbey, 1968; Hansard et al., 1978; Novakova et al., 1981; Sokolov, 1983; Kubena et al., 1985; Uthus and Nielsen, 1990; Dai and McNeill, 1994; Dai et al., 1995) or in humans exposed to vanadium (Missenard, 1989). No statistically significant differences in hematocrit between controls and vanadium-treated or untreated diabetic rats were observed by Domingo et al. (1991).

Blalock and Hill (1987) found an increase in the hemoglobin level and hematocrit index in broiler chickens fed a diet containing 10, 20, and 40 ppm vanadium. An increase in the erythrocyte count, hemoglobin level, and hematocrit index was observed in quails after oral administration of 10 mg V/ kg b.w./day during 4–8 weeks (Table 1). The increase in the latter parameters without changes in the red blood cell indices—mean corpuscular hemoglobin (MCH), mean corpuscular volume (MCV), and mean corpuscular hemoglobin concentration (MCHC)—suggest that only relative changes result from thickening of blood cells (Kaczanowska and Zaporowska, 1990).

Many articles describe the hematological consequences of excess vanadium in laboratory animals and most of their data are quoted in this chapter. The authors of the cited articles used various laboratory animals, different vanadium compounds, and frequently different routes of administration and

Table 1 Effect of Vanadium Compounds on the Red Blood Cell System: A Summary of Various Studies

Species	Vanadium Compounds	Dose	Route	Dosing Period	Results	References
Rabbits	V_2O_5	$0.5-2.2$ mg/m^3	Inhalation	$1.5-3$ months	Increased RBC count, increased Hb level, increased reticulocyte count	Gulko (1956)
		$0.5-2.2$ mg/m^3	Inhalation	$4-5$ months	Decreased RBC count, decreased Hb level, decreased reticulocyte count	
Rats	V_2O_5 NH_4VO_3	0.5 and 1 mgV/kg b.w.	Intragastric	21 days	Decreased RBC count, decreased Hb level	Seljankina (1961)
Rats	V_2O_3	$40-75$ mg/m^3	Inhalation	2 h/day for 9–12 months	Decreased Hb level	Roshchin (1967)
Rabbits	V_2O_5	$8-18$ mg/m^3	Inhalation	2 h/day for 9–12 months		
Rats	V_2O_5	$3-4$ mg V/kg b.w.	Per os	3 weeks	Decreased Hb level	Chakraborty et al. (1977)

Table 1. *(Continued)*

Species	Vanadium Compounds	Dose	Route	Dosing Period	Results	References
Rats	NH_4VO_3	5–6 mg V/kg b.w./day	Drinking water	4 weeks	Decreased RBC count, decreased Hb level	Zaporowska et al. (1993)
		13 mg V/kg b.w./day		4 weeks	Decreased RBC count, decreased Hb level	Zaporowska and Wasilewski (1992)
		20–23 mg V/kg b.w./day		4–8 weeks	Decreased RBC count, decreased Hb level	Zaporowska and Wasilewski (1989a)
Calves	NH_4VO_3	1, 3, 5, and 7.5 mg V/kg b.w.	Per os	1 month	No changes	Platonow and Abbey (1968)
Sheep	NH_4VO_3	10, 100, and 200 ppm V	In diet	84 days	No changes	Hansard et al. (1978)
Rabbits	Not reported	0.5 mg V/kg b.w.	Drinking water	4 months	No changes	Novakova et al. (1981)
Leghorn chicks	$Ca_3(VO_4)_2$	50 mg V/kg diet	In diet	28 days	No changes	Kubena et al. (1985)

Species	Compound	Dose	Route	Duration	Effect	Reference
Rats	NH_4VO_3	1 μg V/g diet	In diet	8 weeks	No changes	Uthus and Nielsen (1990)
Rats	V_2O_5 V_2O_5	0.1 and 1 mg/m^3 10 and 20 mg/m^3	Inhalation Inhalation	24 days 24 days	No changes Increased Hb level	Sokolov (1983)
Broiler chicks	$VOCl_2$	10, 20, and 40 ppm V	In diet	1–2 weeks	Increased Hb level, increased Ht	Blalock and Hill (1987)
Quail	NH_4VO_3	10 mg V/kg b.w./day	Per os	4 weeks	Increased RBC count, increased Hb level, increased Ht	Kaczanowska and Zaporowska (1990)
Rats	$VOSO_4$	34–155 mg/kg/day	Drinking water	12 months	No changes	Dai and McNeill (1994)
Rats	NH_4VO_3 $VOSO_4$ Bis(maltolato) oxovanadium(IV)	0.19 mmol V/kg/day 0.15 mmol V/kg/day 0.18 mmol V/kg/day	Drinking water	12 weeks 12 weeks 12 weeks	No changes No changes No changes	Dai et al. (1995)

dosing periods. The authors of many articles published only the vanadium concentration in the drinking water (Novakova et al., 1981) or in the diet (Hansard et al., 1978; Kubena et al., 1985; Blalock and Hill, 1987; Uthus and Nielsen, 1990) of the experimental animals. Therefore, the vanadium doses consumed by the investigated laboratory animals were not given precisely. The authors did not use a daily monitoring of the vanadium solution intake or the food with increased vanadium level consumed during the experimental period. This was necessary because in animals repeatedly intoxicated with vanadium per os a statistically significant decrease in food and water intake was observed, as compared with controls (Zaporowska and Wasilewski, 1989a, 1992; Dai et al., 1995). The calculation of daily vanadium doses consumed by the experimental animals should be based on their body weights, fluid (or food) intake, and the concentrations of vanadium compounds in drinking water (or in the diet). However, the fact should be taken into consideration that the average intake of water (or food) by control rats is usually higher than in vanadium-treated animals. Therefore, the calculations based on the average water intake by control rats lead to increased vanadium doses consumed by intoxicated animals. Hence a checklist and comparison of the results (Table 1) are rather difficult. Moreover, it can be seen that the animals have a very different susceptibility to vanadium. A comparison of the experimental results of the authors who studied hematological parameters in the same species of laboratory animals during vanadium intoxication is also difficult. The age of the animals was very often different. Besides, the composition of the diet used is very important. Thus, the age of laboratory animals, the composition of the diet (particularly the content of antioxidants), and the resistance of these animals to stress are of great significance in toxicological investigations (Turturro et al., 1993; Vogel, 1993).

It has also been demonstrated that sodium chloride may to a certain extent act protectively in intoxication with vanadium. For example, intravenous administration of vanadium induces marked hemodynamic changes in rats, which can be attenuated by a suitable sodium level in the diet (Lopez-Novoa and Garrido, 1986). The protective action of sodium chloride in vanadium intoxication was also described by Hill (1990) in broiler chickens. The author demonstrated that increasing the dietary supplement of NaCl from 0.5 to 2% resulted in amelioration of vanadium toxicity and improved the growth rate of the birds. Investigations carried out in our laboratory aimed at determining the effect of four chosen vanadium concentrations (0.01, 0.05, 0.15, and 0.30 mg/ml) given to rats in their drinking water with sodium chloride (80 mM solution) on certain hematological parameters. The observed changes of blood diagnostic parameters (RBC count, WBC count, hemoglobin level, and hematocrit index) went in the same direction (Zaporowska, 1996) as in our earlier investigations, when the rats received solely vanadium (Zaporowska and Wasilewski, 1989a, 1992; Zaporowska et al., 1993). Thus, in these conditions sodium chloride did not exhibit a distinct protective effect against vanadium toxicity.

The influence of vanadium on erythrocyte maturation is also confirmed by the results of experiments in vitro. The reticulocyte is the last immature cell in the erythropoietic series. This cell contains a nonlysosomal, ATP-dependent system that eliminates various proteins during maturation and passage of reticulocytes to mature erythrocytes. It has been demonstrated that at 5–100 μM concentration vanadium inhibits proteolysis in reticulocyte extracts (Tanaka et al., 1984). English et al. (1983) demonstrated that an addition of 10–20 μM ammonium vanadate to a culture of Friend murine erythroleukemia cells inhibits the differentiation process of these cells, causing at the same time a decrease in hemoglobin synthesis. Other authors observed a significant increase in the number of erythrocyte precursors in the spleen of mice intoxicated with a vanadium dose of 10 mg/kg body weight for a period of 6 weeks (Cohen et al., 1986). This may account for increased hemopoietic activity in the spleen of these animals.

Trummert and Boehm (1957) described macrocytosis in rabbits intoxicated with vanadium. In the studies of other authors this phenomenon was not observed. Moreover, Vives-Corrons et al. (1981) noted a decrease in deformability of human erythrocytes after incubation with vanadium.

Heme synthesis is probably affected by vanadium. In persons exposed to vanadium by inhalation, increased excretion with urine of the precursors of this compound: δ-aminolevulinic acid, porphobilinogen and coproporphyrins (Garlej et al., 1975), as well as a higher protoporphyrin level in blood serum was observed (Missenard et al., 1989). Under natural conditions, however, interaction of several factors occurs. Therefore, the described changes should be considered to be the result of vanadium action (concentration of which was probably the highest) with other elements. Therefore, additional laboratory experiments concerning the influence of vanadium on the successive stages of heme synthesis and the activity of enzymes catalyzing them could supplement the present knowledge in this respect. Moreover, in vanadium intoxication an increase in methemoglobin level in blood was observed (Garlej et al., 1975).

3. EFFECT ON RED BLOOD CELL METABOLISM

Vanadium penetrates the erythrocyte through its membrane by means of the anion transport system. In the cell it undergoes reduction from pentavalent form to tetravalent vanadyl (Heinz et al., 1982). Vanadium reduction may occur with the participation of glutathione, cysteine, NADH, NADPH, and L-ascorbic acid (Sakurai et al., 1981; Legrum, 1986; Kretzschmar and Bräunlich, 1990). Macara et al. (1980) reported that in human erythrocytes glutathione is more effective in vanadate reduction to vanadyl than NADH and NADPH. Shi and Dalal (1990) demonstrated that glutathione reductase can fulfill the function of vanadate reductase. The formed vanadyl ion may bind with the cell proteins (e.g., with hemoglobin) as well as with nucleotides and carbohydrates. Vanadyl oxidation is also possible. Thus, the intracellular form of vanadium depends on the ratio of the whole reductant pool to the cell oxidants (see Zaporowska and Ścibior, this volume, Ch. 7).

Vanadate affects the level of NADH and NADPH in intact red cells. Human red blood cells incubated with sodium orthovanadate at a concentration of 0.5 mM showed little oxidation of NADH and NADPH. In these cells little NADPH oxidation by vanadate occurs despite inhibition of intracellular superoxide dismutase and the presence of a sufficient concentration of L-ascorbic acid, which can reduce vanadate to vanadyl (Yoshino et al., 1988).

Vanadium causes changes in the permeability of the erythrocyte membrane, and this may be the cause of metabolism disturbances in these cells. It has been demonstrated that this element inhibits active and passive Ca^{2+} ion transport (Rossi et al., 1981; Varečka et al., 1986) and increases selective permeability of K^+ ions of the erythrocyte membranes (Fuhrmann et al., 1984). Sodium orthovanadate at a concentration of 1 mmol induces Ca^{2+}-dependet Na^+ influx in red cells. The extent of the Ca^{2+}-dependent Na^+ influx induced by vanadate is very small in human red blood cells but clearly visible in guinea pig red blood cells (Varečka et al., 1994). Moreover, vanadium has an inhibitory influence on the activity of Na^+, K^+-ATPase, Ca^{2+}-ATPase, Ca^{2+}, Mg^{2+}-ATPase, PTPase, phosphoglyceromutase, phosphoglucomutase, adenyl kinase, and 6-phosphogluconate dehydrogenase. On the other hand, it has a stimulating effect on the activity of NADH oxidase, adenyl cyclase, and 2,3-bisphosphoglycerate phosphatase (Table 2).

Studies in recent years have demonstrated that free radicals and lipid peroxides are formed in the presence of vanadium in vitro and in vivo. The tendency of malondialdehyde (MDA) level to increase in the blood of animals intoxicated with vanadium (Zaporowska et al., 1993) indicates peroxidative changes in the blood cell membrane. Incubation of human erythrocytes with vanadium in vitro also produces peroxidative changes in the erythrocyte membrane and an increase in MDA level (Heller et al., 1987; Zaporowska and Słotwińska, 1996). This leads to the occurrence of holes and lysis of the cell, which may explain the hemolytic action of vanadium compounds observed in vitro (Hansen et al., 1985; Hamada, 1994) and in vivo (Hogan, 1990).

MDA, as the product of lipid peroxidation, may react with the amino acids of proteins, forming bonds of the Schiff base type. This may lead to aggregation of membrane proteins, to increased rigidity of the plasmalemma, and in consequence to diminished ability of erythrocyte deformation (Vives-Corrons et al., 1981; Chatterjee et al., 1988). MDA, like vanadium, may cause inhibition of membrane Na^+,K^+-ATPase activity and intracellular accumulation of Na^+ (Pfafferott et al., 1982). A further consequence of direct action of vanadium (inhibition of Na^+,K^+-ATPase) and its indirect influence (increased MDA level) may be increased cell volume. Hence in hematological investigations the decrease in the erythrocyte count need not always be associated with distinct (and statistically significant) decrease of hematocrit. After a certain time changes in the erythrocyte membrane may be so pronounced that they will lead to hemolysis. Moreover, cells with reduced deformability are more

Table 2 Effect of Vanadium on Some Selected Enzyme Activities in Red Blood Cells

Enzyme	Source	Effect	Conditions	Reference
Adenylate cyclase	Turkey RBC	Stimulation	In vitro	Krawietz et al. (1982)
Adenylate kinase	Human RBC	Inhibition	In vitro	Vives-Corrons et al. (1981)
Na^+,K^+–ATPase	Human RBC	Inhibition	In vitro	Cantley et al. (1978)
				Beaugé et al. (1980)
				Schrier et al. (1986)
Ca^{2+}-ATPase	Human RBC	Inhibition	In vitro	Barrabin et al. (1980)
				Bond and Hudgins (1980)
				Varečka and Carafoli (1982)
Ca^{2+}, Mg^{2+}-ATPase	Human RBC	Inhibition	In vitro	O'Neal et al. (1979)
				Schrier et al. (1986)
2,3-Bisphosphoglycerate phosphatase	Human RBC	Stimulation	In vitro	Mendz et al. (1990)
Glucose-6-phosphate dehydrogenase	Human RBC	No changes	In vitro	Vives-Corrons et al. (1981)
	Rat RBC	No changes	In vivo	Zaporowska and Wasilewski (1992)
Lactate dehydrogenase	Rat RBC	No changes	In vivo	Zaporowska and Wasilewski (1992)
Phosphoglycerate mutase	Human RBC	Inhibition	In vitro	Vives-Corrons et al. (1981)
Phosphoglucomutase	Human RBC	Inhibition	In vitro	Vives-Corrons et al. (1981)
6-Phosphogluconate dehydrogenase	Human RBC	No changes	In vitro	Vives-Corrons et al. (1981)
	Human RBC	Inhibition	In vitro	Crans et al. (1990)
NADH oxidase	Rat RBC	Stimulation	In vitro	Vijaya et al. (1984)

frequently retained in the reticuloendothelial system of the spleen and eliminated earlier from blood circulation (Kogawa et al., 1976). Thus, the observed decrease in the erythrocyte count in animals intoxicated with vanadium may be the result both of the hemolytic activity of this element and the reduced survival time of erythrocytes.

Human red blood cells undergo both ATP-dependent and ATP-independent transitions from smooth biconcave discocytic forms to crenated (echinocytic) states. Vanadate can cause shape change in intact red blood cells. At concentrations >2 μM, vanadate blocked ATP-dependent shape changes completely (Patel and Fairbanks, 1986). However, orthovanadate at micromolar concentrations (2–100 μM) had no effect in vitro on RBC shape, deformability, osmotic fragility, or metabolism (Schrier et al., 1986). Xu et al. (1991) demonstrated that ammonium metavanadate (3–100 μM) transformed intact human red blood cells from normal discocytes to echinocytes and more rapidly in Mg^{2+}-depleted cells. These results suggest that vanadate-induced shape changes are associated with inhibition of Mg^{2+}-ATPase activity localized in the plasma membrane of the red blood cells.

Investigations in vitro also demonstrated the influence of vanadium on oxygen transport by hemoglobin. Erythrocytes have a characteristic metabolic pathway—the Rapoport–Luebering cycle, which is a branch of glycolysis at the level of 1,3-diphosphoglyceric acid (1,3-DPG). In this cycle 2,3-diphosphoglyceric acid (2,3-DPG) is formed from 1,3-DPG in a reaction catalyzed by the enzyme phosphoglyceromutase (Fig. 1). The formed compound (2,3-DPG) regulates the affinity of hemoglobin to oxygen (Beutler, 1990; Staniszewska, 1993). Thus, the efficiency of oxygen transport by hemo-

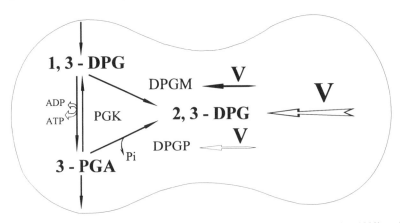

Figure 1. Rapoport–Lubering cycle in erythrocytes (Beutler, 1990; Staniszewska, 1993) and the effect of vanadium on this pathway. 1,3-DPG, 1,3-diphosphoglyceric acid; 2,3-DPG, 2,3-diphosphoglyceric acid; 3-PGA, 3-phosphoglyceric acid; DPGM, diphosphoglyceromutase (EC 2.7.5.3); DPGP, diphosphoglycero phosphatase (EC 3.1.3.13); PGK, phosphoglycerate kinase (EC 2.7.2.3); P_i, inorganic phosphate. Single-point open arrow, stimulation of enzyme activity; solid arrow, inhibition of enzyme activity; double-point open arrow, stimulation of nonenzymatic hydrolysis of 2,3-DPG.

globin is dependent on the 2,3-DPG concentration in erythrocytes, which binds with hemoglobin molecules and in this way depresses its affinity to oxygen.

Human erythrocytes incubated in vitro with vanadium compounds have a decreased 2,3-DPG level, and this causes an increased affinity of hemoglobin to oxygen (Vives-Corrons et al., 1981; Ninfali et al., 1983). The studies of Stankiewicz et al. (1987) and Mendz et al. (1990) indicated a stimulating influence of vanadium on 2,3-diphosphoglycerophosphatase activity leading to a decrease in the 2,3-DPG level in blood cells, and thus to increased hemoglobin affinity to oxygen. However, Vives-Corrons et al. (1981) demonstrated an inhibitory effect of vanadium on diphosphoglyceromutase activity in human erythrocytes. Thus, the decreased 2,3-DPG level in erythrocytes incubated with vanadium ions is the result both of inhibition of diphosphoglyceromutase activity and stimulation of 2,3-diphosphoglycerophosphatase activity (Fig. 1). It has also been demonstrated that this element enhances nonenzymatic 2,3-DPG hydrolysis to 3-phosphoglycerate (3-PGA) and inorganic phosphorus (Stankiewicz 1989).

Vanadium also affects glucose metabolism in the red blood cell. Human erythrocytes when incubated with glucose in the presence of 0.5 mM NH_4VO_3 increase glucose utilization as well as pyruvate and lactate synthesis (Ninfali et al., 1983). However, at higher concentrations vanadium was observed to inhibit glycolysis. Benabe et al. (1985) explain this by oxidation of the $-SH$ groups of 3-phosphoglycerate dehydrogenase, which leads to depression of the activity of this enzyme and in consequence to a reduced glycolysis rate.

Sodium orthovanadate (4 μM) markedly stimulated in vitro the growth of human blood erythroid burst-forming units, and this effect appears to be very similar to the effect of stem cell factor on these cells but not to the effects of interleukin-3 or erythropoietin (Dai and Krantz, 1992).

4. INFLUENCE OF VANADIUM ON THE WHITE BLOOD CELL SYSTEM

In poisonings with vanadium changes are also noted in the leukocyte system (Table 3). Roshchin (1967) found leukopenia in rabbits during their intoxication with V_2O_3 by inhalation and a shift of the white blood cell picture to the left. Minden (1968) reported eosinophilia in workers occupationally exposed to vanadium compounds. Novakova et al. (1981) did not observe changes in the leukocyte count in rabbits after a 4-month period of oral vanadium intoxication. Sokolov (1983), however, noted an increase in the leukocyte count in rats intoxicated with V_2O_3 by inhalation.

In our investigations (Zaporowska and Wasilewski, 1992), no significant changes were observed in the leukocyte system of rats given for drinking solely an aqueous solution of ammonium metavanadate at a concentration of 0.3 mg V/ml for a period of 2, 4, and 8 weeks. After 12 weeks of intoxication, however, the neutrophil counts increased significantly (Zaporowska and Wasi-

Table 3 Effect of Vanadium Compounds on the White Blood Cell System: A Summary of Various Studies

Species	Vanadium Compounds	Dose	Route	Dosing Period	Results	References
Rats	V_2O_3	40–75 mg/m^3	Inhalation	2 h/day for 9–12 months	Decreased WBC count, shift of the WBC, picture to the left	Roshchin (1967)
Rabbits	Not reported	0.5 mg V/kg b.w.	Drinking water	4 months	No changes	Novakova et al. (1981)
Rats	V_2O_5	0.1 and 1 mg/m^3 10 and 20 mg/m^3	Inhalation Inhalation	24 days 24 days	No changes Increased WBC count	Sokolov (1983)
Rats	NH_4VO_3	5–6 mg V/kg b.w./ day 13 mg V/kg b.w./day	Drinking water	4 weeks 4 weeks	No changes Increased WBC count	Zaporowska et al. (1993) Zaporowska and Wasilewski (1992)

Animal	Compound	Dose	Route	Duration	Effect	Reference
Quail	NH_4VO_3	20–23 mg V/kg b.w./day		4–8 weeks	No changes	Zaporowska and Wasilewski (1989a)
	NH_4VO_3	10 mg V/kg b.w./day	Per os	4 and 8 weeks	Decreased WBC count	Kaczanowska and Zaporowska (1990)
Rats	$VOSO_4$	34–155 mg/kg/day	Drinking water	12 months	No changes	Dai and McNeill (1994)
Rats	NH_4VO_3	0.19 mmol V/kg/day	Drinking water	12 weeks	No changes	Dai et al. (1995)
	$VOSO_4$	0.15 mmol V/kg/day		12 weeks	No changes	
	Bis(maltolato) oxovanadium(IV)	0.18 mmol V/kg/day		12 weeks	No changes	

lewski, 1989b). In animals drinking 0.15 mg V/ml ammonium metavanadate solution over 4 weeks, a statistically significant increase in the leukocyte count was found that resulted from the increased number of both neutrophils and lymphocytes (Zaporowska and Wasilewski, 1992).

The present investigations have also demonstrated the influence of vanadium on phagocyte metabolism. Rabbit alveolar macrophages cultured in vitro with vanadium added to the medium exhibited depressed viability, reduced phagocytosis, and a decrease in acid phosphatase and β-glucuronidase activity (Waters et al., 1974). Vanadium had also a cytotoxic effect on murine peritoneal macrophages. The cells isolated from mice intoxicated earlier with vanadium showed a lowered phagocytic activity and decreased activity of such enzymes as glucose-6-phosphate dehydrogenase, glutathione reductase, and glutathione peroxidase (Cohen and Wei, 1988). Mice exposed to vanadium for a period of 6 weeks and then infected with *Listeria monocytogenes* showed a higher level of these bacteria in the liver and spleen than did unintoxicated controls. Peritoneal macrophages isolated from these mice had depressed phagocytic activity. These results indicate a lowered resistance to *Listeria monocytogenes* in mice as a result of vanadium intoxication (Cohen et al., 1989). This depression in phagocytic activity was thought to result from alterations in microfilament function, resulting in impaired cytoplasmic rearrangements crucial to the process of membrane invagination (Wang and Choppin, 1981), and from disturbances in the cellular energy metabolism required for continued phagocytic activity (Cohen and Wei, 1988). Depressed phagocytic activity of peripheral blood granulocytes was also observed in rats intoxicated with vanadium (Zaporowska and Wasilewski, 1992; Zaporowska and Ścibior, 1996). Vaddi and Wei (1991) demonstrated a decrease in the activity of several key intraphagolysosomal enzymes, including alkaline phosphatase, *N*-acetylo-β-glucosaminidase and β-glucuronidase, whereas Zaporowska (1988) demonstrated that vanadate when administrated to rats at concentrations of 0.1 or 0.2 mg V/ml in their drinking water (which corresponded to about 12 and 22 mg V/kg b.w./day, respectively) caused an increase in granulocyte alkaline phosphatase activity. The alkaline phosphatase granulocytes (APG) coefficient in these animals was twice as large as that in controls.

In persons exposed to inhalation of a mixture of tungstate, titanium, vanadium, and cobalt dust and gaseous petrol Misiewicz (1980) found an increased percentage of formazane cells in the spontaneous nitroblue tetrazolium (NBT) reduction test. This was the result of the action of vanadium with other elements and petrol.

It is difficult to explain the mechanism of the action of vanadium on phagocytosis. As is known from the literature, L-ascorbic acid affects the immune response. Chemotaxis and phagocytosis are weakened in cases of L-ascorbic acid deficit (Padh, 1991; Sauberlich, 1994). A decrease in the level of the latter compound was observed in blood serum, liver, kidneys, and adrenals of rabbits and rats intoxicated with vanadium compounds (Roshchin, 1967; Chakraborty et al., 1977; Novakova et al., 1981; Zaporowska, 1994). A decrease in the L-ascorbic acid level was also noted in blood serum of subjects exposed to inhalation of vanadium (Gniot-Szulzycka et al., 1988).

Thus, by inducing a decrease in the L-ascorbic acid level in the organism, vanadium may indirectly affect the metabolic pathway requiring the presence of this acid. Therefore, the mechanism of the action of vanadium on phagocytosis seems to be complicated.

At 2×10^{-4} M concentration vanadium increases the production of the superoxide ion in cultures of peritoneal murine macrophages (Lison et al., 1988). This element increases oxygen uptake and phosphotyrosin accumulation in human neutrophils (Grinstein et al., 1990). Induction of tyrosine phosphorylation was also observed in HL60 granulocytes (Bianchini et al., 1993) and human lymphocytes (Schieven et al., 1993) treated with H_2O_2 and vanadate in vitro. In HL60 granulocytes, as in other cells, pervanadate induced activation of phospholipase C and then a massive increase in inositol-1,4,5-trisphosphate (IP_3) content, and release of Ca^{2+} from intracellular stores (Bianchini et al., 1993).

The results of many investigations also suggest that vanadium may be an immunoinhibitor. An addition of this element to murine thymocyte cultures inhibits the mitogenic response induced by concavalin A (Ramanadham and Kern, 1983). However, in cultures of human lymphocytes with V_2O_5 added, a reduction of the mitotic index and an increase in the mean time of generation and in the number of polyploid cells were observed (Roldan and Altamirano, 1990). Cohen et al. (1993) showed that mouse myelomonocytic macrophage-like WEHI-3 cells treated with NH_4VO_3 or V_2O_5 had reduced capacities for synthesis and release of monokines, such as tumor necrosis factor-α (TNFα) and interleukin-1, which are crucial for maintaining most immunocompetence.

A phosphorus deficiency in the diet of broiler chickens subjected to acute heat stress inhibited expression of heat shock proteins (HSPs). Leukocytes from these chicks were not able to respond to acute heat stress with the same level of expression of HSPs as the groups of birds that had normal or increased levels of phosphorus in the diet. However, addition of vanadium (5, 10, or 50 mg/kg) to the diet with phosphorus deficit allowed the chicks to respond with increased expression of HSPs. Thus, vanadium in trace amounts in the diet with deficiency of phosphorus improved chick tolerance of acute high temperature stress through vanadium-associated induction of HSPs (Edens and Hill, 1994).

5. INFLUENCE OF VANADIUM ON THROMBOCYTES

Human platelet activation is associated with phosphorylation of many proteins. Recently, attention has been focused on tyrosine phosphorylation of proteins and their function in platelet activation. Tyrosine phosphorylation has been indirectly implicated in several platelet functions, such as phospholipase C activation, aggregation, and secretion. Moreover, platelets have a very high protein tyrosine phosphatase (PTPase) activity. Pumiglia et al. (1992) concluded that PTPases are important regulators of signal transduction in platelets, whereas vanadate, as a known inhibitor of the PTPase, can stimulate protein tyrosine phosphorylation.

It was shown that vanadate (0.1 mM) or hydrogen peroxide (1 mM) alone had little or no effect on tyrosine phosphorylation in human platelets (Pumiglia et al., 1992). However, in many investigations a mixture of orthovanadate and hydrogen peroxide was used. In this mixture vanadyl hydroperoxide is formed, which was termed as "pervanadate" (Schieven et al., 1993). It is a better inhibitor of PTPase activity than orthovanadate, penetrates cells more readily, and is a much more potent stimulator of tyrosine phosphorylation than vanadate or H_2O_2 alone in intact cells (Pumiglia et al., 1992).

It was shown that pervanadate stimulated tyrosine phosphorylation 29 times more than thrombin in intact and saponin-permeabilized human platelets. Optimal stimulation of tyrosine phosphorylation was observed at final concentrations of 1 mM H_2O_2 and 0.1 mM vanadate. Increased tyrosine phosphorylation in platelets coincided with a decrease in $I(4,5)P_2$ levels, synthesis of $I(3,4)P_2$, mobilization of intracellular Ca^{2+}, and stimulation of protein kinase C-dependent protein phosphorylation (Pumiglia et al., 1992).

Vanadium can also affect aggregation of platelets. For example, vanadate inhibited type I collagen-induced platelet aggregation and ATP release in a dose-dependent manner. Lower vanadate concentrations of (0.5, 1, 2 mM) had no effect on aggregation but delayed the latency period. The ATP release was inhibited in a concentration-dependent manner (at 0.5, 1, 2 mM) (Chiang, 1992). It completely inhibited both aggregation and ATP release at a concentration of 3.5 mM. However, McNicol et al. (1993) demonstrated that sodium orthovanadate (7.5–100 μM) stimulated dose-dependent aggregation of saponin-permeabilized, but not intact, platelets. This effect was mediated by the liberation of arachidonic acid and its subsequent conversion to thromboxane A_2, which produces secondary effects. For example, the vanadate-induced platelet aggregation was also associated with stimulation of phospholipase C and phosphorylation of platelet proteins, notably pleckstrin and myosin light chain. Immunoblotting studies indicated that vanadate caused thyrosine phosphorylation of proteins of approximate molecular weights 26, 29, 32, 40, 42, 80, and 90 kDa.

Vanadate can also affect platelet ultrastructure. The addition of vanadate caused platelets to extend pseudopods, centralize, and evacuate their granules, and form small aggregates (McNicol et al., 1993).

6. SUMMARY

Vanadate at micromolar concentrations has been shown not to affect RBC shape, deformability, osmotic fragility, or metabolism, but it can affect these modalities at higher concentrations (Schrier et al., 1986).

Studies in recent years have demonstrated that vanadium can act as pro-oxidant (Ahmad, 1995). Thus, vanadium also produces peroxidative changes in the erythrocyte membrane that lead to reduction of RBC deformability, and increase in cell volume, and then to hemolysis. Therefore, the observed decrease in the erythrocyte count in animals intoxicated with vanadium may be mostly the result of the hemolytic activity of this element. However, in laboratory animals with chronic vanadium intoxication the decrease in the erythrocyte count need not always be associated with a significant decrease of hematocrit. This phenomenon resulted from the pro-oxidant activity of vanadium and increased RBC volume in these animals. It can lead to decreased overall respiratory surface of the erythrocytes as well as decreased oxygen-carrying capacity of the blood unit (Gill, 1989). The decreased values of the above-mentioned parameters in rats intoxicated with vanadium (Kaczanowska and Zaporowska, 1995) indicate decreased ability of the blood volume unit to transfer oxygen. But in vitro vanadium ions decreased 2,3-DPG levels in RBC, causing an increased affinity of hemoglobin to oxygen (Vives-Corrons et al., 1981; Ninfali et al., 1983). Vanadyl ion in vitro also enhanced oxygen affinity to hemoglobin and myoglobin (Sakurai et al., 1982).

The decreased erythrocyte count in vanadium-treated animals results from the hemolytic activity of the vanadium, whereas the increased erythrocyte count in vanadium-treated animals observed by other authors (Gulko, 1956; Blalock and Hill, 1987; Kaczanowska and Zaporowska, 1990) can explain dehydration of the organism and thickening of blood cells; the lack of hemato-logical effect in orally vanadium-treated animals (Platonow, and Abbey 1968; Hansard et al., 1978; Novakova et al., 1981; Kubena et al., 1985; Uthus and Nielsen, 1990; Dai and McNeill, 1994; Dai et al., 1995) may be considered the result of the cooperation of these two mechanisms.

The changes in the RBC system of animals intoxicated with vanadium were often accompanied by an increased percentage of reticulocytes and polychromatophilic erythrocytes in the peripheral blood (Zaporowska and Wasilewski, 1989a, 1992). This suggests a disorder in erythropoiesis and erythrocyte maturation.

Changes of the heme precursor level in blood serum and urine (Garlej et al., 1975; Missenard et al., 1989) observed in humans exposed occupationally to vanadium suggest a certain influence of this element on heme synthesis. This influence can depend, among other things, on inhibition of δ-aminolevulinic dehydratase activity during vanadium intoxication (Zaporowska, 1996). However, this problem requires further studies and observations.

Vanadium also affects the white blood cell system of animals, but the changes in this system seem to be less extensive. It has been shown that vanadium compounds decrease the phagocytic activity of macrophages (Waters et al., 1974; Cohen and Wei, 1988) and peripheral blood granulocytes (Zaporowska and Wasilewski, 1992; Zaporowska and Ścibior, 1996) both in vitro and in vivo. The mechanism by which vanadium causes decreased phagocytosis has not been fully known. On the one hand vanadium element decreases not only the L-ascorbic acid level in blood, which deficit can weaken phagocytosis, but also, the activity of some intraphagolysosomal enzymes such as alkaline phosphatase, N-acetylo-β-glucosaminidase, and β-glucuronidase. Moreover, it should be noted that protein tyrosine phosphorylation can also regulate phagocytosis. As mentioned above, vanadium is the inhibitor of protein tyrosine phosphatase (PTPase). For example, it was shown that inhibition of PTPase by pervanadate also markedly diminished Fcτ-receptor-mediated phagocytosis in vitro (Pyrzyńska et al., 1996). In the literature, the anticoagulant action of vanadium is described. It can inhibit aggregation of platelets (Chiang, 1992) and the activities of coagulation factors, including thrombin and Xa factor (Funakoshi et al., 1992).

The pro-oxidant action of vanadium suggests the necessity of supplementation of the diet with a natural antioxidant in situations of professional exposure to vanadium or high concentrations of this element in food. The hematological effects of excess vanadium in the animal organism under these conditions may thus be lessened.

REFERENCES

Ahmad, S. (1995). Oxidative stress from environmental pollutans, *Arch. Insect. Biochem. Physiol.* **29**, 135–157.

Barannik, P. I., and Michaliuk, I. A. (1970). Nekotorye dannye o roli chroma, svinca, vanadija i estestvennoj radioaktivnosti pishchevych produktov v etiologii endemitcheskovo zoba. *Gig. Sanit.* **35**, 103–105.

Barrabin, H., Garrahan, P. J., and Rega, A. F. (1980). Vanadate inhibition of the Ca^{2+}-ATPase from human red cell membranes. *Biochim. Biophys. Acta* **600**, 796–804.

Beaugé, L. A., Cavieres, J. J., Glynn, I. M., and Grantham, J. J. (1980). The effects of vanadate of the fluxes of sodium and potassium ions through the sodium pump. *J. Physiol.* **301**, 7–23.

Bello-Reuss, E. N., Grady, T. P., and Mazumdar, D. C. (1979). Serum vanadium levels in chronic renal disease. *Ann. Intern. Med.* **91**, 743.

Benabe, J. E., Echegoyen, L. A., and Martinez- Maldonado, M. (1985). Mechanism of inhibition of glycolisis by vanadate. *Adv. Exp. Med. Biol.* **208**, 517–528.

Beutler E. (1990). Red cell enzyme defects. *Hematol. Pathol.* **4**, 103–114.

Bianchini, L., Todderud, G., and Grinstein, S. (1993). Cytosolic $[Ca^{2+}]$homeostasis and tyrosine phosphorylation of phospholipase Cγ2 in HL60 granulocytes. *J. Biol. Chem.* **268**, 3357–3363.

Blalock, T. L., and Hill, C. H. (1987). Studies on the role of iron in the reversal of vanadium toxicity in chicks. *Biol. Trace Elem. Res.* **14**, 225–235.

Bond, G. H., and Hudgins, P. M. (1980). Inhibition of red cell Ca^{2+}-ATPase by vanadate. *Biochim. Biophys. Acta* **600**, 781–790.

Byrne, A. R., and Versieck, J. (1990). Vanadium determination at the ultra-trace level in biological reference materials and serum by radiochemical neutron activation analysis. *Biol. Trace Elem. Res.* **27**, 529–540.

Cantley, L. C., Resh, M. D., and Guidotti, G. (1978). Vanadate inhibits the red cell (Na^+,K^+)-ATPase from the cytoplasmic side. *Nature* **272**, 552–554.

Chakraborty, D., Bhattacharyya, A., Majumdar, K., and Chatterjee, G. C. (1977). Effects of chronic vanadium pentoxide administration on L-ascorbic acid metabolism in rats. Influence of L-ascorbic acid supplementation. *Int. J. Vitam. Nutr. Res.* **47,** 81–87.

Chasteen, N. D., Lord, E. M., Thompson, H. J., and Grady, J. K. (1986). Vanadium complexes of transferrin and ferritin in the rat. *Biochim. Biophys. Acta* **884,** 84–92.

Chatterjee, S. N., Agarwal, S., Jana, A. K., and Base, B. (1988). Membrane lipid peroxidation and its pathological consequences. *Ind. J. Biochem. Biophys.* **25,** 25–31.

Chiang, T. M. (1992). Okadaic acid and vanadate inhibit collagen-induced platelet aggregation; the functional relation of phosphatases on platelet aggregation. *Throm. Res.* **67,** 345–354.

Cohen, M. D., Chen, C. M., and Wei, C. I. (1989). Decreased resistence to *Listeria monocytogenes* in mice following vanadate exposure: Effect upon the function of macrophages. *Int. J. Immunopharmacol.* **11,** 285–292.

Cohen, M. D., Parsons, E., Schlesinger, R. B., and Zelikoff, J.T. (1993). Immunotoxicity of *in vitro* vanadium exposures: Effects on interleukin-1, tumor necrosis factor-α, and prostaglandin E_2 production by WEHI-3 macrophages. *Int. J. Immunopharmacol.* **15,** 437–446.

Cohen, M. D., Wei, C. I., Tan, H., and Kao, K. J. (1986). Effect of ammonium matavanadate on the murine immune response. *J. Toxicol. Environ. Health.* **19,** 279–298.

Cohen, M. D., and Wei, C. I. (1988). Effects of ammonium matavanadate treatment upon macrophage glutathione redox cycle activity, superoxide production, and intracellular glutathione status. *J. Leuk. Biol.* **44,** 122–129.

Conri, C., Simonoff, M., Fleury, B., and Moreau, F. (1986). Does vanadium play a role in depressive states? *Biol. Psychiatry* **21,** 546–548.

Corrigan, F. M., Coulter, F., Ijomah, G., Holliday, J., Macintyre, F., Skinner, E. R., Horrobin, D. F., and Ward, N. I. (1992). Plasma vanadium concentrations, lipoproteins and fatty acids in elderly people living near a smelter and in a control area. *Trace Elem. Med.* **9,** 93–96.

Crans, D. C., Williging, E. M., and Butler, S. R. (1990). Vanadate tetramer as the inhibiting species in enzyme reactions *in vitro* and *in vivo*. *J. Am. Chem. Soc.* **112,** 427–432.

Dai, C. H., and Krantz, S. (1992). Vanadate mimics the effect of stem cell factor on highly purified human erythroid burst-forming units in vitro, but not the effect of erythropoietin. *Exp. Hematol.* **20,** 1055–1060.

Dai, S., and McNeill, J. H. (1994). One-year treatment of non-diabetic and streptozotocin-diabetic rats with vanadyl sulphate did not alter blood pressure or hematological indices. *Pharmacol. Toxicol.* **74,** 110–115.

Dai, S., Vera, E., and McNeill, J. H. (1995). Lack of haematological effect of oral vanadium treatment in rats. *Pharmacol. Toxicol.* **76,** 263–268.

Domingo, J. L., Gómez, M., Llobet, J. M., Corbella, J., and Keen, C. L. (1991). Improvement of glucose homeostasis by oral vanadyl or vanadate treatment in diabetic rats is accompanied by negative side effects. *Pharmacol. Toxicol.* **68,** 249–253.

Edel, J., and Sabbioni, E. (1989). Vanadium transport across placenta and milk of rats to the fetus and newborn. *Biol. Trace Elem. Res.* **22,** 265–275.

Edens, F. W., and Hill, C. H. (1994). Influence of dietary phosphorus and trace amounts of vanadium on expression of heat shock proteins in peripheral blood leukocytes of broiler chickens. *Poult. Sci.* **73** (Suppl. 1), 79.

English, L. H., Macara, I. G., and Cantley, L. C. (1983). Vanadium stimulates the (Na^+,K^+) pump in Friend erythroleukemia cells and block erythropoiesis. *J. Cell Biol.* **97,** 1299–1302.

Fuhrmann, G. F., Hüttermann, J., and Knauf, P. A. (1984). The mechanism of vanadium action on selective K^+-permeability in human erythrocytes. *Biochim. Biophys. Acta* **769,** 130–140.

Funakoshi, T., Shimada, H., Kojima, S., Shoji, S., Kubota, Y., Morita, T., Nobuaki, T., and Ueki, H. (1992). Anticoagulant action of vanadate. *Chem. Pharm. Bull.* **40,** 174–176.

Garlej, T., Lis, E., and Mejran, S. (1975). Koproporfiryny i ich prekursory w moczu osób narażonych na działanie pyłów zawierających tlenki wanadu. *Bromat. Chem. Toksykol.* **8**, 213–217.

Gill, J. (1989). Seasonal changes in the red blood cell system in the european bison, *Bison Bonasus* L. *Comp. Biochem. Physiol.* **92A**, 291–298.

Gniot-Szulżycka, J., Berendt, T., Pokorska, K., and Rak, D. (1988). Zawartość kwasu askorbinowego, 2-wodorosiarczanu kwasu askorbinowego oraz cholesterolu i fosfolipidów w surowicy krwi ludzi nie narażonych i narażonych na toksyczne związki tytoniu i związki wanadu. *Bromat. Chem. Toksykol.* **21**, 85–91.

Grinstein, S., Furuya, W., Lu, D. J., and Mills, G. B. (1990). Vanadate stimulates oxygen consumption and tyrosine phosphorylation in electropermeabilized human neutrophils. *J. Biol. Chem.* **265**, 318–327.

Gulko, A. G. (1956). K charakteristike vanadija kak promyshlennogo jada. *Gig. Sanit.* **11**, 24–28.

Hamada, T. (1994). Vanadium induced hemolysis of vitamin E deficient erythrocytes in Hepes buffer. *Experentia* **50**, 49–53.

Hansard, S. L., Ammerman, C. B., Fick, K. R., and Millar, S. M. (1978). Performance and vanadium content of tissues in sheep as influenced by dietary vanadium. *J. Anim Sci.* **46**, 1091–1095.

Hansen, W. V., Aaseth J, and Skaug, V. (1985). Hemolytic activity of vanadylsulphate and sodium vanadate. *Acta Pharmacol. Toxicol.* **59**, 562–565.

Heinz, A., Rubinson, K. A., and Grantham, J. J. (1982). The transport and accumulation of oxyvanadium compounds in human erythrocytes *in vitro. J. Lab. Clin. Med.* **100**, 593–612.

Heller, K. B., Jahn, B., and Deuticke, B. (1987). Peroxidative membrane damage in human erythrocytes induced by a concerted action of iodoacetate, vanadate and ferricyanide. *Biochim. Biophys. Acta* **901**, 67–77.

Hill, C. H. (1990). Interaction of vanadate and chloride in chicks. *Biol. Trace Elem. Res.* **23**, 1–10.

Hogan, G. R. (1990). Peripheral erythrocyte levels, hemolysis and three vanadium compounds. *Experientia* **46**, 444–446.

Hosokawa, S., and Yoshida, O. (1990). Vanadium in chronic hemodialysis patients. *Int. J. Artif. Organs* **13**, 197–199.

Kaczanowska, E., and Zaporowska, H. (1990). Układ czerwonokrwinkowy przepiórek rasy Faraon (Coturnix coturnix Pharaon) w zatruciu wanadem. In *IV Zjazd Naukowy Pol. Tow. Toksykol., Jastrzębia Góra 24–26 IX 1990.* Streszczenia referatów, p. 61.

Kaczanowska, E., and Zaporowska, H. (1995). Parametry hematologiczne charakteryzujące funkcję oddechową krwi szczurów w zatruciu wanadem. *IV Sympozjum nt.: Molekularne i fizjologiczne aspekty regulacji ustrojowej.* Kraków 7–8 czerwca, Materiały, 138–140.

Kogawa, H., Sudo, K., and Imai, K. (1976). Splenic sequestration of reticulocytes in the rabbit, (II). Factors influencing splenic trapping capacity with special reference to osmotic fragility. *Memoirs of Osaka Kyoiku University Ser. III* **27**, 49–55.

Krawietz, W., Downs, R. W., Spiegel, A. M., and Aurbach, G. D. (1982). Vanadate stimulates adenylate cyclase via the guanine nucleotide regulatory protein by a mechanism differing from that of fluoride. *Biochem. Pharmacol.* **31**, 843–848.

Kretzschmar, M., and Bräunlich, H. (1990). Role of glutathione in vanadate reduction in young and mature rats: Evidence for direct participation of glutathione in vanadate inactivation. *J. Appl. Toxicol.* **10**, 295–300.

Kubena, L. F., Harvey, R. B., Fletcher, O. J., Phillips, T. D., Mollenhauer, H. H., Witzel, D. A., and Heidelbaugh, H. D. (1985). Toxicity of ochratoxin A and vanadium to growing chicks. *Poult. Sci.* **64**, 620–628.

Legrum, W. (1986). The mode of reduction of vanadate ($+$ V) to oxovanadium ($+$ V) by glutathione and cysteine. *Toxicology* **42**, 281–289.

Liochev, S. I., and Fridovich, I. (1989). Vanadate-stimulated oxidation of NAD/P/H. *Free Rad. Biol. Med.* **6**, 617–622.

Lison, D., Dubois, P., and Lauwerys R. (1988). *In vitro* effect of mercury and vanadium on superoxide anion production and plasminogen activator activity of mouse peritoneal macrophages. *Toxicol. Lett.* **40,** 29–36.

Lopez-Novoa, J. M., and Garrido, M. C. (1986). Hemodynamic effects of vanadate administration in rats with different levels of sodium intake. *Am. J. Med. Sci.* **291,** 152–156.

Macara, I. G., Kustin, K., and Cantley, L. C. (1980). Glutathione reduces cytoplasmic vanadate— Mechanism and physiological implications. *Biochim. Biophys. Acta* **629,** 95–106.

McNicol, A., Robertson, C., and Gerrard, J. M. (1993). Vanadate activates platelets by enhancing arachidonic acid release. *Blood* **81,** 2329–2338.

Mendz, G. L., Hyslop, S. J., and Kuchel, P. W. (1990). Stimulation of human erythrocyte 2,3-bisphosphoglycerate phosphatase by vanadate. *Arch. Biochem. Biophys.* **276,** 160–171.

Minden, H. (1968). Die Vanadiumvergiftung, klinisches Bildarbeitsmedizinische Problematik. *Zentralbl. Ges. Hyg.* **14,** 344–347.

Misiewicz, A. (1980). Wpływ powietrza zawierającego benzynę, wolfram, tytan, kobalt i wanad na czynność fagocytarną leukocytów. *Pol. Tyg. Lek.* **35,** 1965–1967.

Missenard, C., Hansen, G., Kutter, D., and Kremer, A. (1989). Vanadium induced impairment of haem synthesis. *Br. J. Ind. Med.* **46,** 744–747.

Naylor, G. J., Cirrigan, F. M., Smith, A. H. W., Connelly, P., and Ward, N. J. (1987). Further studies of vanadium in depressive psychosis. *Br. J. Psychiatry* **150,** 656–661.

Ninfali, P., Accorsi, A., Fazi, A., Palma, F., and Fornaini, G. (1983). Vanadate effects glucose metabolism of human erythrocytes. *Arch. Biochem. Biophys.* **226,** 441–447.

Novakova, S., Nikolthev, G., Angeleva, R., Dinoeva, S., and Mautner, G. (1981). Izutchenie vlijanija vanadija na eksperimentalnyj ateroskleroz. *Gig. Sanit.* **11,** 58–59.

Nozdrjuchina L. R., Korjakin, A. V., and Gribovskaja, N. F. (1974). Mikroelementy krovi rozlitchnych stadjach ischemitcheskoj bolezni serdca. In L. R. Nozdrjuchina (Ed.) *Biologitcheskaja Rol Mikroelementov i ich Primenenie v Selskom Chozjajstve i Medicine.* Izdatelstvo Nauka, Moskva, pp. 367–371.

O'Neal, S. G., Rhoads, D. B., and Racker, E. (1979). Vanadate inhibition of sarcoplasmic reticulum Ca^{2+}-ATPase and other ATPases. *Biochem. Biophys. Res. Commun.* **89,** 845–850.

Padh, H. (1991). Vitamin C: Newer insights into its biochemical functions. *Nutr. Rev.* **49,** 65–70.

Patel, V. P., and Fairbanks, G. (1986). Relationship of major phosphorylation reactions and Mg^{2+}-ATPase activities to ATP-dependent shape change of human erythrocyte membranes. *J. Biol. Chem.* **261,** 3170–3177.

Pfafferott, C., Meiselman, H. J., and Hochstein, P (1982). The effect of malonyldialdehyde on erythrocyte deformability. *Blood* **59,** 12–15.

Platonow, N., and Abbey, H. K. (1968). Toxicity of vanadium in calves. *Vet. Rec.* **82,** 292–293.

Pumiglia, K. M., Lau, L.-F., Huang, Chi-K., Burroughs, S, and Feinstein, M. B. (1992). Activation of signal transduction in platelets by the tyrosine phosphatase inhibitor pervanadate (vanadyl hydroperoxide). *Biochem. J.* **286,** 441–449.

Pyrzyńska, B., Strzelecka, A., Kwiatkowska, K., and Sobota, A. (1996). Tyrosine phosphorylation/ dephosphorylation controls onset of capping and Fcτ receptor-mediated phagocytosis. *Folia Histochem. Cytobiol.* **34,** 75.

Ramanadham, M., and Kern, M. (1983). Differential effect of vanadate on DNA synthesis induced by mitogens in T- and B-lymphocytes. *Mol. Cell. Biochem.* **51,** 67–71.

Roldan, R. E., and Altamirano, L. M. A. (1990). Chromosomal aberrations sister-chromatid exchanges, cell cycle kinetics and satellite association in human lymphocyte cultures exposed to vanadium pentoxide. *Mutat. Res.* **245,** 61–65.

Romero, R. A. (1994). Aluminum, vanadium, and lead intoxication of uremic patients undergoing hemodialysis in Venezuela. *Transplant. Proc.* **26,** 330–332.

156   Hematological Effects of Vanadium on Living Organisms

Roshchin, I. V. (1967). Toksikologija soedinenij vanadija primeniaemych v sovremennoj promyshlennosti. *Gig. Sanit.* **6**, 26–32.

Rossi, J. P. F. C., Garrahan, P. J., and Rega, A. F. (1981). Vanadate inhibition of active Ca^{2+} transport across human red cell membranes. *Biochim. Biophys. Acta* **648**, 145–150.

Sabbioni, E., Rade, J., and Bertolero, F. (1980). Relationships between iron and vanadium metabolism: The exchange of vanadium between transferrin and ferritin. *J. Inorg. Biochem.* **12**, 307–315.

Sakurai, H., Goda, T., and Shimomura, S. (1982). Vanadyl(IV) ion dependent enhancement of oxygen binding to hemoglobin and myglobin. *Biochem. Biophys. Res. Commun.* **107**, 1349–1354.

Sakurai, H., Shimomura, S., and Ishizu, K. (1981). Reduction of vanadate(V) to oxovanadium(IV) by cysteine and mechanism and structure of the oxovanadium(IV)–cysteine complex subsequently formed. *Inorg. Chim. Acta* **55**, L67–L69.

Sauberlich, H. E. (1994). Pharmacology of vitamin C. *Annu. Rev. Nutr.* **14**, 371–391.

Schieven, G. L., Kirihara, J. M., Myers, D. E., Ledbetter, J. A., and Uckun, F. M. (1993). Reactive oxygen intermediates activate NF-κB in a tyrosine kinase-dependent mechanism and in combination with vanadate activate the p56[lck] and p59[fyn] tyrosine kinases in human lymphocytes. *Blood* **82**, 1212–1220.

Schrier, S. L., Junga, I., and Ma, L. (1986). Studies on the effect of vanadate on endocytosis and shape changes in human red blood cells and ghosts. *Blood* **68**, 1008–1014.

Seljankina, K. P. (1961). Materialy k gigienitcheskomu normirovaniju soderzhanija vanadija v vode vodoemov. *Gig. Sanit.* **10**, 6–12.

Shi, X., and Dalal, N. S. (1990). Glutathione reductase functions as vanadate(V) reductase. *Arch. Biochem. Biophys.* **278**, 288–290.

Sitprija, V., Tangsanga, K., Eiam-Ong, S., Leelhaphunt, N., and Sriboonlue, P. (1990). Renal tubular acidosis, vanadium and buffaloes. *Nephron* **54**, 97–98.

Sokolov, S. M. (1983). Gigienitcheskaja ocenka nepreryvnogo i preryvistogo vozdejstvija pjatiokisi vanadija. *Gig. Sanit.* **9**, 77–79.

Staniszewska, K. (1993). 2,3-Dwufosfoglicerynian jako regulator powinowactwa hemoglobiny do tlenu. *Diagn. Lab.* **29**, 429–433.

Stankiewicz, P. J., (1989). Vanadium(IV)-stimulated hydrolysis of 2,3-diphosphoglycerate. *Arch. Biochem. Biophys.* **270**, 489–494.

Stankiewicz, P. J., Gresser, M. J., Tracey, A. S., and Hass, L. F. (1987). 2,3-Diphosphoglycerate phosphatase activity of phosphoglycerate mutase: Stimulation by vanadate and phosphate. *Biochemistry* **26**, 1264–1269.

Tanaka, K., Waxman, L., and Goldberg, A. L. (1984). Vanadate inhibits the ATP-dependent degradation of proteins in reticulocytes without affecting ubiquitin conjugation. *J. Biol. Chem.* **259**, 2803–2809.

Trummert, W., and Boehm, G. (1957). Über das Spurenelement Vanadium und seine hämopoetische Wirkung. *Blut* **3**, 211–216.

Turturro, A., Duffy, P. H., and Hart, R. W. (1993). Modulation of toxicity by diet and dietary macronutrient restriction. *Mutat. Res.* **295**, 151–164.

Uthus, E. O., and Nielsen, F. H. (1990). Effect of vanadium, iodine and their interaction om growth, blood variables, liver trace elements and thyroid status indices in rats. *Magnesium Trace Elem.* **9**, 219–226.

Vaddi, K., and Wei, C. I. (1991). Effect of ammonium metavanadate on the mouse peritoneal macrophage lysosomal enzymes. *J. Toxicol. Environ. Health* **33**, 65–78.

Varečka, Ĺ., and Carafoli, E. (1982). Vanadate-induced movements of Ca^{2+} and K^+ in human red blood cells. *J. Biol. Chem.* **257**, 7414–7421.

Varečka, Ĺ., and Peterajová, E., Pišova, E., and Ševčik, F. (1994). Vanadate and fluoride activate red cell Na$^+$ permeability by different mechanism. *Gen. Physiol. Biophys.* **13**, 127–135.

Varečka, Ĺ., Peterajová, E., and Pogády, J. (1986). Inhibition by divalent cations and sulphydryl reagents of the passive Ca^{2+} transport in human red blood cells observed in the presence of vanadate. *Biochim. Biophys. Acta* **856**, 585–594.

Vijaya, S., Crane, F. L., and Ramasarma, T. (1984). A vanadate-stimulated NADH oxidase in erythrocyte membrane generates hydrogen peroxide. *Mol. Cell. Biochem.* **62**, 175–185.

Vives-Corrons, J. L., Jou, J. M., Ester, A., Ibars, M., Carreras, J., Bartrôns, R., Climent, F., and Grisolia, S. (1981). Vanadate increases oxygen affinity and affects enzyme activities and membrane properties of erythrocytes. *Biochem. Biophys. Res. Commun.* **103**, 111–117.

Vogel, W. H. (1993). The effect of stress on toxicological investigations. *Hum. Exp. Toxicol.* **12**, 265–271.

Wang, E., and Choppin, P. W. (1981). Effect of vanadate on intracellular distribution and function of 10 nm filaments. *Proc. Nat. Acad. Sci. USA* **78**, 2363–2367.

Waters, M. D., Gardner, D. E., and Coffin, D. L. (1974). Cytotoxic effects of vanadium on rabbit alveolar macrophages *in vitro. Toxicol. Appl. Pharmacol.* **28**, 253–263.

Xu, Y.-H., Lu, Z.-Y., Conigrave, A. D., Auland, M. E., and Roufogalis, B. D. (1991). Association of vanadate-sensitive Mg^{2+}-ATPase and shape change in intact red blood cells. *J. Cell. Biochem.* **46**, 284–290.

Yoshino, S., Sullivan, S. G., and Stern, A. (1988). Inhibition of vanadate-induced NADH and NADPH oxidation in intact red blood cells. *Proc. Fourth Biennial General Meeting of the Society for Free Radical Research, Kyoto, Japan, April 9–13.* Elsevier Science Publishers, B. V., Amsterdam, pp. 197–200.

Zaporowska, H. (1988). Obraz krwi obwodowej u szczurów szczepu Wistar w przewlekłym zatruciu wanadem. *Bromat. Chem. Toksykol.* **21**, 234–240.

Zaporowska, H. (1994). Effect of vanadium on L-ascorbic acid concentration in rat tissues. *Gen. Pharmacol.* **25**, 467–470.

Zaporowska, H. (1996). Wybrane aspekty działania wanadu *in vivo* oraz *in vitro.* Wyd. UMCS, Lublin.

Zaporowska, H., and Ścibior, A. (1996). Phagocytic activity of rat granulocytes in vanadium intoxication. *Folia Histochem. Cytobiol.* **34** (Suppl. 2), 45.

Zaporowska, H., and Słotwińska M. (1996). Effect of vanadium on rat erythrocytes *in vitro. Folia Histochem. Cytobiol.* **34** (Suppl. 1), 99–100.

Zaporowska, H., and Wasilewski, W. (1989a). Some selected peripheral blood and haemopoietic system indices in Wistar rats with chronic vanadium intoxication. *Comp. Biochem. Physiol.* **93C**, 175–180.

Zaporowska, H., and Wasilewski, W. (1989b). Wpływ wanadu na układ krwiotwórczy i wybrane wskaźniki krwi obwodowej szczurów szczepu Wistar. *Bromat. Chem. Toksykol.* **22**, 121–125.

Zaporowska, H., and Wasilewski, W. (1992). Haematological results of vanadium intoxication in Wistar rats. *Comp. Biochem. Physiol.* **101C**, 57–61.

Zaporowska, H., Wasilewski, W., and Słotwińska, M. (1993). Effect of chronic vanadium administration to rats in drinking water. *BioMetals* **6**, 3–10.

9

GENETIC TOXICOLOGY OF VANADIUM COMPOUNDS

Mario A. Altamirano-Lozano
and M. E. Roldán-Reyes

Laboratorio de Citogenética, Mutagénesis y Toxicología
Reproductiva, UIBR, Facultad de Estudios Superiores-
Zaragoza,UNAM, A.P. 9-020, México 15000, D.F., México

E. Rojas

Laboratorio de Genética Toxicología Molecular,
Departamento de Toxicología Ambiental, Instituto de
Investigaciones Biomédicas, UNAM., A.P. 70228, Cuidad
Universitaria, México 04510, D.F. México

Vanadium in the Environment. Part 2: Health Effects, Edited by Jerome O. Nriagu.
ISBN 0-471-17776-8.   © 1998 John Wiley & Sons, Inc.

---

1. INTRODUCTION

Of 109 identified elements about 80 are usually considered metals. Thus, the chemistry of metals represents a major part of inorganic chemistry (Vouk, 1986). The properties of chemical elements depend on the electronic configuration of the atom and vary with the atomic number in a systematic way. Vanadium is a member of group VB of the periodic system and the first transition series. Vanadium is a greyish metal with an atomic number of 23, an atomic weight of 50.95, a melting point of 1,890 °C, a boiling point of 3,380 °C at 1 atm, and a specific gravity of 6.11 at 18.7°C. It forms oxidation states of -1, 0, $+2$, $+3$, $+4$, and $+5$ the oxidation states $+3$ (vanadic form), $+4$ (vanadyl form), and $+5$ (vanadate form) being the most common. the oxidation state $+4$ is the most stable (Table 1) (WHO, 1988).

Vanadium is widely distributed throughout the earth but in low abundance. It represents 0.07 wt % in the lithosphere, and a few deposits contain more than 1–2 wt %. Some 70 vanadium minerals are known, of which 40 are vanadates, (Chapter 1, Vol. 30). Vanadium occurs in uranium-bearing minerals and in copper, lead, and zinc vanadates and is a constituent of titaniferious magnetites (Baroch, 1983; Rosenbaum, 1983). Most of the vanadium reserves are in deposits in which the vanadium is a by-product or co-product with other minerals, including iron, titanium, and phosphate (Baroch, 1983; WHO, 1988). However, vanadium is the major trace element in fossil matter such as crude oil, coal, and carbonaceous fossils (Al-Swaidan, 1993; Baroch, 1983; Rehder, 1991; WHO, 1988). Vanadium is used in the steel industry for manufacture of alloys, and vanadium salts are also used in the synthesis of sulfuric acid, for coloring glass and ceramic glazes, for the production of dyes and inks, as well as paints and varnish dyers, in insecticides, in photographic materials, and for the oxidation of organic molecules such as benzene, aniline, and toluene (Lëonard and Gerber, 1994; Petrucci, 1989; WHO, 1988; Zhong et al., 1994). However, in recent years the industrial application of vanadium compounds has been expanding from its classic uses to new fields such as elements in superconductive materials (Kawal et al., 1989).

Research on the biological effects of vanadium has rapidly increased during the last 10 years, owing to its potential toxicological impact as environmental pollutant, either from natural sources or from industrial processes (Altamirano-Lozano et al., 1996; Cirani et al., 1995; Rojas et al., 1996a), and owing to its role as a possible pharmacological and essential element in mammals

Table 1 Physical and Chemical Properties of Vanadium and Some Vanadium Compounds

	At. Wt. or Mol. Wt	Melting Point (°C)	Boiling Point (°C)	Oxidation State	Solubility in Water (g/L)	
					Cold	Hot
Vanadium (V)	50.92	1,890 + 10	3,380	+5	Insol.	Insol.
Vanadium pentoxide (V_2O_5)	181.88	690	1,750	+4	8.0 (20°C)	No data
Vanadium tetraoxide (V_2O_4)	165.88	1,967	No data	+3	Insol.	Insol.
Vanadium trioxide (V_2O_3)	149.88	1,970	No data	+4	Slightly sol.	Soluble
Vanadyl sulfate ($VOSO_4$)	163.00	No data	No data	+5	Very sol.	—
Amonium metavanadate (NH_4VO_3)	116.98	1,200 decomposes	No data	+5	5.2	69.5
Sodium orthovanadate (Na_3VO_4)	183.91	No data	No data	+5	No data	No data
Sodium metavanadate ($NaVO_3$)	121.93	630	No data	+4	211	288
Vanadium tetrachloride (VCl_4)	192.75	−28	154		Decomp.	Decomp.

(Jandhyala and Hom, 1983; Rehder, 1991, 1992). Pharmacological uses of vanadium include lowering of cholesterol and glucose levels, diuretic and natriuretic effects, and contraction of blood vessels. It also enhances the oxygen affinity of hemoglobin and myoglobin. For these reasons, vanadate is considered useful for the symptomatic treatment of sickle-cell anemia (Meyerovitch et al., 1989; Rehder, 1991; Sakurai et al., 1982; Vives-Corrons et al.,1981). Vanadium has a direct cardiac effect, similar to the digitalis and insulin-mimetic effects; and several vanadium compounds have been studied with respect to their cytostatic activity (Djordjevic and Wampler, 1985; Dubyak and Kleinzeller, 1980; Rehder, 1991). The peroxovanadate complexes show antitumor activity in certain forms of leukemia, and vanadyl sulfate inhibits in rats the formation of mammary carcinoma induced by methyl nitrosourea (Thomson et al., 1984). Another interesting fact about VO^{2+} in vanadium compounds is that it forms a stable compound with an antibiotic used in cancer chemotherapy, bleomycin, and this compound is considered a supercancerostatic agent (Banci et l., 1982; Rheder, 1991).

Metals are an important class of environmental hazards and some of them are implicated as causing cancer in experimental animals and humans; one possible explanation for this effect is that particular metal ions can interact and induce DNA damage (Zakour and Glickman, 1984). Vanadium occurs naturally and is a good example of metals that constantly interplay with human life environmentally, industrially, occupationally, and biologically. Epidemiologic studies have indicated a correlation between exposure to airborne vanadium particles and the incidence of cancer in residents of metropolitan areas (Stock, 1960; Hickey et al., 1967). Hematological and biochemical alterations, renal toxicity, immunotoxicology, and reproductive effects were observed following exposure to vanadium compounds (Altamirano et al., 1991; Altamirano-Lozano et al., 1993; Domingo et al., 1986; Domingo et al., 1995; Jandhyala and Hom, 1983; Zaporowska and Wasilewski, 1992).

Vanadium can be highly toxic for humans and other animals because it is a potent inhibitor of many enzymes such as kinases, ribonucleases, phosphatases, and several ATPases, and it has been reported to accumulate in critical organs (Sabbioni et al., 1991; WHO, 1988). The most toxic form of vanadium to mammals is the pentavalent form (Sharma et al., 1987), and the toxicity of vanadium compounds appears to depend on valence (WHO, 1988). Among vanadium compounds, tetra- and pentavalent forms are the most studied because both are capable of reacting with genetic machinery; they can interact with the phosphate groups and the sugar alcohol groups of nucleotides to form complexes that inhibit or stimulate the activity of many DNA or RNA enzymes (WHO, 1988; Rehder, 1991), inducing several genotoxic and mutagenic effects.

2. BIOCHEMICAL ASSAYS

Vanadium is essential for some animals; however, vanadium has not yet achieved essential status for human beings (Domingo, 1996; Harland and

Harden-Williams, 1994). Vanadium has a profound effect on a broad variety of cellular processes including stimulation of cell differentiation (Rehder, 1991; Thomson et al., 1984), alterations in gene expression (Bosch et al., 1990), and several biochemical and metabolical systems (WHO, 1988). Cell growth appears to be controlled, to a large degree, by extracellular signals, and in some reports it has been shown that vanadium, both inhibits and enhances DNA synthesis in vitro, depending on the concentration in the media (Carpenter, 1981; Sabbioni et al., 1983; Smith, 1983). To evaluate the action of vanadium compounds on cell genetic functions and genome-related events many assays have been performed.

When rabbit alveolar macrophage culture were exposed to vanadium pentoxide (V_2O_5; 13.01 μg/ml), vanadium trioxide (V_2O_3; 21.35 μg/ml), and vanadium dioxide (VO_2; 32.75 μg/ml), cell viability after 20 h exposure was reduced by 50%. The results showed that cytotoxicity was related to the solubility of the vanadium compounds, in the order $V_2O_5 > V_2O_3 > VO_2$ (Waters et al., 1974).

Cande and Wolniak (1978), working with mitotic PtK1 cells, found that vanadate (sodium orthovanadate; Na_3VO_4) at 10–100 μM reversibly inhibits anaphase movement of chromosomes and spindle elongation by inhibiting dynein ATPase. After lysis in vanadate, spindles lose their characteristic form and become more barrel-shaped; however, in vitro microtubule polymerization is insensitive to vanadate.

Using cultures of quiescent human fibroblasts to evaluate changes in cell proliferation and assess the possible mitogen effect of Na_3VO_4 (1.0–40.0 μM), Carpenter (1981) found that control cultures after 48 h exhibited a 9% increase in cell numbers, while cultures containing 4.0 μM of sodium orthovanadate had a 61% increase in cell numbers; exposure to concentrations of vanadate higher than 40 μM resulted in a toxic effect on the cells. In this case the capacity of vanadate to induce DNA synthesis was measured by the incorporation of labeled thymidine. Maximal stimulation of DNA synthesis occurred in 4–10 μM vanadate treatments. On the other hand, Smith (1983), treating Swiss mouse 3T3 and 3T6 cells with vanadyl sulfate ($VOSO_4$) or Na_3VO_4 in concentrations between 5 and 50 μM, found that both vanadium compounds are mitogenic for quiescent cultures. In this case the maximal stimulation of thymidine incorporation occurred between 25 and 50 μM vanadium, while higher concentrations of these compounds caused cell toxicity.

The effects of vanadium compounds on DNA-metabolizing enzymes was reported by Sabbioni and colleagues in 1983. In this study, the authors found that vanadate ions (10^{-7} to 10^{-3} M) inhibited calf thymus terminal deoxynucleotidyl transferase and the catalytic activity of mammalian DNA polymerase α (at I_{50} of 60 μM), while the bacterial DNA polymerase 1 was inhibited by vanadate when concentration was raised to about 0.5 mM.

To assess the direct effects of sodium vanadate (0.01, 0.1, 1.0, 10, or 100 μM) on bone DNA, Canalis (1985) used cultures of 21-day-old fetal rat calvariaè. In this case after 24 h of culture vanadate (0.1–10 μM) stimulated the incorporation of [^{3}H]thymidine into DNA by 35–57%, and after 96 h at 1–10 μM by

29–38%. In all cultures treated with 100 μM sodium vanadate for 24 or 96 h an inhibitory effect on thymidine was observed. Vanadate also increased the bone DNA content from 7.0 $\pm$ 0.6 μg DNA per half calvaria in the control cultures to 10. 1 $\pm$ 1.1 μg DNA per half calvaria in the cultures treated with 1.0 μM sodium vanadate, and increased the mitotic index (MI), per histological section after 24 h of culture and 3 h of colcemid arrest (4.2 $\pm$ 0.3 in controls and 5.3 $\pm$ 0.4 in 1 μM sodium vanadate-treated calvariae). After 96 h of culture, controls had 5.3 $\pm$ 0.4 and sodium vanadate-treated calvariae (1.0 μM) had 8.7 $\pm$ 0.4 mitoses per section.

To determine the effect of sodium vanadate (1 μg was used as the highest test concentration) on proliferation and morphological cell changes, Sheu and Rodriguez (1991) used BALB/3T3 cells. After 48 h to 2 weeks of treatment with vanadate no effects were observed.

3. BACTERIAL SYSTEMS

Conflicting results in regard to the mutagenic activity of vanadium have been shown in bacterial assay systems using *Escherichia coli* and *Salmonella typhimurium* (Graedel et al., 1986; Hansen and Stern, 1984; Léonard and Gerber, 1994).

Early studies demonstrated that ammonium metavanadate (NH_4VO_3) and V_2O_5 were more genotoxic in recombination-repair-deficient (rec$^-$) strains of *Bacillus subtilis* than in the wild-type rec$^+$ (Kanematsu and Kada, 1978; Kanematsu et al., 1980). However, these compounds were not mutagenic in several strains of *Escherichia coli* and *Salmonella typhimurium* (Kada et al., 1980). On the other hand NH_4VO_3 was found to be mutagenic in *S. typhimurium* TA1535 in a modified plate incorporation assay and in the fluctuation test with TA100 (WHO, 1988). In other work, Sun (1987) demonstrated the induction of reverse mutations by V_2O_5 with *E. coli* WP2, WP2uvrA, and Cm-981, but no frame shift mutations with strains ND-160 and MR102. This compound showed negative results with *S. typhimurium* strains TA135, TA1537, TA98, and TA100. Thus, the results of mutagenicity studies of vanadium compounds with bacterial assays are conflicting, and no firm conclusions can be drawn.

4. YEAST SYSTEMS

D7 diploid strain of *Saccharomyces cerevisiae* was used to measure induced mitotic gene conversion and point reverse mutation, and the D61M diploid strain of *Saccharomyces cerevisiae* was used to evaluate mitotic chromosome loss by NH_4VO_3 (1.0, 2.5, 4.0, and 5.0 mM) and $VOSO_4$ (2.0, 5.0, 7.5, and 10.0 mM). NH_4VO_3 was shown to be able to induce genetic activity in a yeast system. Although $VOSO_4$ was not able to induce mitotic gene conversion and point reverse mutation in stationary phase cells of strain D7 of *S. cerevisiae* with or without S9 hepatic fraction, a significant increase in gene conversion and point reverse mutation was obtained in the same assay in metaphase-harvested cells that contained high levels of cytochrome P-450. In other experiments, similar results were obtained with strain D61M cells treated with metavanadate and vanadyl, indicating that these compounds are able to induce aneuploidy (Bronzetti et al., 1990; Galli et al., 1991).

5. DROSOPHILA TEST

Only two studies have examined the effects of vanadium compounds on this system. In 1994, Abundis utilizing chronic, subchronic, and acute treatments over different concentration ranges found that vanadium pentoxide could be a promoter in the wing smart test. In a more recent study the same group reported that vanadium pentoxide, vanadium chloride, and vanadyl sulfate are indirect mutagens and that vanadium(IV) is a weak mutagen that in low concentrations (6.5 and 8 ppm) can induce mitotic recombination (Abundis, 1996).

6. PLANT TEST

In a few studies using plant systems, it was found that when *Allium cepa* was exposed to higher doses of V_2O_5, this compound induced pycnosis and loss of chromatin matter, while lower doses poisoned the spindle (Singh, 1979, cited in Sharma and Talukder, 1987). In 1982, Jackson and Linskens, using an unscheduled DNA synthesis (UDS) assay in *Petunia hybrida* W166K pollen cultures treated with 2 mM V^{2+}, found that this metal ion gave a very strong UDS response.

Navas and associates (1986) evaluated the inhibitory action of Na_3VO_4 (0.01, 0.1, 1.0, 5.0, 10.0, and 100.0 mM) in plant cytokinesis, using root tips of *Allium cepa* L., flat violet variety. After 4 h of treatment with the vanadium compound, 1 h of recovery was given. The results showed that vanadate concentrations higher than 10 mM proved to be lethal. A 1 mM concentration of vanadate produced the highest frequency of binucleate cells, 1% at 4 h and 1.5% at 6 h; the percentage of binucleate cells arising per hour is thus stable for several hours.

7. IN VITRO MAMMALIAN CELL SYSTEMS

Information about the genotoxicity and mutagenicity in vitro of vanadium compounds is limited (Table 2).

Table 2 In Vitro Genotoxic and Mutagenic Effects of Vanadium Compounds in Mammalian Cells

Assay System[a]	V(V) Compound	Result	V(IV) Compound	Result	Reference
Human lymphocytes					Paton and Allison, 1972
Structural CA	$NaVO_3$	–			
	Na_3VO_4	–			
SCE	V_2O_5	–			
Human lymphocytes					Roldán and Altamirano, 1990
Structural CA	V_2O_5	–			
Numerical CA		–			
SCE		+			
MI		+			
SA		+			
Human lymphocytes					Migliore et al., 1993
Structural CA	NH_4VO_3	+	$VOSO_4$	+	
	Na_3VO_4	+			
SCE	NH_4VO_3	+	$VOSO_4$	–	
	Na_3VO_4	+			
MN	NH_4VO_3	+	$VOSO_4$	+	
	Na_3VO_4	+			
SA	NH_4VO_3	+	$VOSO_4$	+	
	Na_3VO_4	+			

Cell type / Endpoint	Compound		Compound		Reference
Human lymphocytes					
MN	Na_3VO_4	+			Migliore et al., 1995; Migliore and Scarpato, 1996
	NH_4VO_3	+			
	$NaVO_3$	+	$VOSO_4$	+	
Human lymphocytes					
DNA-SSB	V_2O_5	+			Rojas et al., 1996a, 1996b
V79 cells					
hprt locus mutations	NH_4VO_3	−	$VOSO_4$	−	Galli et al., 1991
	NH_4VO_3	+			Cohen et al., 1992
CHO and MOLT4 cells					
Protein–DNA crosslinks	NH_4VO_3	+			Cohen et al., 1992
V79 cells					
Numerical CA	V_2O_5	+			Zhong et al., 1994
MN		+			
SCE		−			
MI		+			
hprt locus mutations		−			
CHO cells					
Structural CA	NH_4VO_3	+	$VOSO_4$	+	Owusu-Yaw et al., 1990
SCE		−		−	

[a] CA, chromosomal aberrations; MI, mitotic index; MN, micronuclei; SA, satellite associations; SCE, sister chromatid exchanges; SSB, single-strand breaks.

7.1. Human Lymphocytes

In an early study, Paton and Allison (1972) found no increase in chromosome aberrations (CA) when human leukocytes were treated with sodium metavanadate (NaVO$_3$) or Na$_3$VO$_4$. In another in vitro study on human peripheral lymphocyte cultures with V$_2$O$_5$ (0.047, 0.47, and 4.7 moles) no increase in the frequency of sister chromatid exchange (SCE) was observed by Sun (1987).

In 1990 Roldán and Altamirano evaluated the CA, SCE, cell cycle kinetics (CCK), and satellite associations (SA) in human lymphocyte cultures exposed to V$_2$O$_5$ (2, 4, or 6 μg/ml). They found that the frequency of structural chromosomal aberrations (SCA) was not increased. However, the treatment with vanadium produced a significant increase in the frequency of polyploid cells (1.7% in the controls vs. 4.4% at 2 μg/ml, 4.0% at 4 μg/ml, and 4.6% at 6 μg/ml); the mitotic index (MI) was reduced, the average generation time was increased, and the frequency of cells with SA, the frequency of SA by cell, and the frequency of chromosome association was also increased in treated cultures. In another study using the same end points, Migliore et al., (1993) evaluated the effects of four vanadium compounds (2.5, 5, 10, 20, 40, 80, and 160 μM of NaVO$_3$, NH$_4$VO$_3$, Na$_3$VO$_4$, and SOVO$_4$). In this study the authors found that the frequency of SCA did not increase, but the yield of SCE was increased slightly for all compounds, although only at the higher concentrations. Data on the induction of SA showed a significant increase for all compounds in one or more doses compared to the controls. In lymphocytes treated with VOSO$_4$ a significant increase was found in almost the whole range of concentrations used (5–40 μM). Even at dose of 10 μM a significant increase in frequency of micronuclei (MN) was observed for all compounds, and a linear dose–effect relationship was observed. The results obtained through the fluorescence in situ hybridization (FISH) technique showed that the percentage of positive micronuclei varied from 76% to 84% for sodium metavanadate, from 68% to 83% for sodium orthovanadate, from 79% to 81% for vanadyl sulfate, and from 73% to 79% for ammonium metavanadate. In all cases and all doses the percentage of positive signals was significant in comparison with controls.

Recently Migliore et al. (1995) and Migliore and Scarpato (1996) assessed the formation of vanadium-induced micronuclei by applying the FISH technique to the human lymphocyte micronucleus assay. Cells were treated after 24 h with 10, 40, or 80 μM, Na$_3$VO$_4$, NH$_4$VO$_3$, NaVO$_3$, and SOVO$_4$. The slides obtained were then hybridized with biotin-labeled β-satellite DNA probes specific for human acrocentric chromosomes. These authors' data showed an increase in the MN frequency in all vanadium-treated cultures. Average MN levels ranged from 0.62% in control cultures to 5.32% in 80 μM SOVO$_4$ treatment, and 3.64% in 80 μM Na$_3$VO$_4$ cultures. Both vanadium salts produced a dose-dependent increase in MN frequency. The distribution of micronuclei determined as fluorescent β-satellite DNA signal showed that the proportion of β^+ micronuclei was statistically higher in treated cultures than in controls (32.8% in control cultures to 52.9–66.9% in treated cultures).

In a study performed by Rojas et al. (1996a, 1996b) the genotoxicity of vanadium pentoxide was evaluated directly in whole-blood leukocytes and in human lymphocyte cultures (obtained from 4 donors—2 males and 2 females) using the alkaline single-cell gel electrophoresis assay. Whole blood was treated 2 h with V_2O_5 (0.3, 30, and 3,000 μM), lymphocytes were cultured for 72 h, and V_2O_5 was added at 48 h, at the same concentrations. In this case V_2O_5 seems to be nontoxic in both cultured lymphocytes and blood leukocytes. The induction of DNA migration in the leukocytes was observed in all doses and donors; migration increased in a dose-response fashion and the distribution of DNA damage among cells was very similar for all doses. In contrast, the results were very different for cultured lymphocytes. In one donor no increase in DNA migration was observed; in two donors a very weak response was observed, and in one donor a significant dose response was found.

In a more recent experiment, Ramirez et al. (in press) found that V_2O_5 (0.001–0.1 μM) induced a hyperdiploid frequency significantly greater than controls, measured in interphase nuclei by FISH using α-satellite DNA probes for chromosomes 1 and 7. When the cells were analyzed by immunostaining of the spindle apparatus using anti-β-tubulin antibodies, monopolar configurations and disruptions of microtubules were frequently observed. Additional in vitro assays using purified tubulin indicated that this compound inhibited microtubule assembly and induced tubulin depolymerization. These results demonstrated that vanadium pentoxide disrupted spindle formation by interacting with β-tubuline.

7.2. Chinese Hamster Cells

The V79 Chinese hamster cell line was used by Galli and colleagues (1991) to evaluate the resistance to 6-thioguanine for measuring mutation induction by NH_4VO_3 (1.0, 2.5, 4.0, and 5.0 mM) and $VOSO_4$ (2.0, 5.0 and 7.5 mM) with and without S9 hepatic fraction. The results obtained by these authors showed that these compounds are very toxic. Vanadate showed a strong toxic effect when the cells were incubated in the absence of metabolic activation. At the same concentration of vanadate the survival increased up to six times in the presence of S9 hepatic fraction. On the other hand, vanadyl was less toxic than metavanadate when the cells were incubated without metabolic activation, and the toxicity was eliminated by the presence of S9. In these experiments, no mutagenic effect was observed in the presence or in the absence of S9 fraction.

In a study performed by Cohen et al. (1992) NH_4VO_3 (5.0, 10.0, 20.0, 25.0, and 50.0 μM) after 24 h exposure yielded a dose-dependent increase in mutation frequency at the V79 *hprt* locus. Ammonium metavanadate also enhanced the mutation frequency in a V79 variant containing a transfected bacterial *gtp* gene. When this compound was tested in both CHO and human MOLT4 cells, an increase in DNA-protein crosslinks was observed.

In another study, Zhong et al. (1994) used the Chinese hamster lung fibroblast cell line (V79) to determine the genotoxic effects of vanadium pentoxide. Vanadium treatments were performed at concentrations of 1, 3, 6, 9, and 12 μg/ml for 24 h, and treated with cytochalasin-B (Cyt-B). In each treatment they evaluated the numbers of mononucleated and binucleated cells by cell cycle kinetics analysis. The data indicated that with increases in concentration, the number of binucleated cells significantly decreased from 89% in the control to 23–32% in cultures treated with 3–12 μg/ml of V_2O_5.

To determine the MN frequency, they used immunofluorescence staining of kinetochore in binucleated cells with antikinetochore antibodies. The results showed a clear concentration-related increase in MN frequencies. In the concentration range from 1 to 3 μg/ml V_2O_5, the incidence of MN was significantly increased from 4.15% to 7.55%. The results obtained by these authors by immunofluorescent staining of kinetochores in MN with anticentromere antibody showed that almost one-half (49%) of micronuclei in untreated cells contained kinetochores (KC$^-$); however, more than 69% of micronuclei induced by 1 μg/ml treatment with V_2O_5 reacted positively with the antibody (Zhong et al., 1994). Testing for induction of 6-thioguanine-resistant (TGr) mutation, the same authors designed an experiment using V79 cells treated with 1, 3, 6, 9, and 12 μg/ml of V_2O_5 for 24 h. The data obtained showed that none of the three concentrations tested significantly increased the frequency of TGr mutants.

To evaluate the SCE frequency, the V79 cultures were treated with vanadium pentoxide at concentrations of 1, 2, 3, and 4 μg/ml for 24 h, and with bromodeoxyuridine. In this assay Zhong and colleagues found no significant increase in SCE frequency with all concentrations tested; however, a decrease in the replicative index was observed in the vanadium-treated cells. Finally, in these experiments, an increase in the endoreduplication of chromosomes frequently was observed by these authors: 6% at 1 μg/ml, 8% at 2 μg/ml, and 19% at 3 μg/ml of V_2O_5.

Owusu-Yaw et al. (1990) used cultures of Chinese hamster cells to evaluate the effect of three vanadium salts. The results obtained showed that vanadium compounds are cytotoxic, and caused a dose-response increase in SCA, with and without S9 fraction at concentrations of 12–18 μg/ml, 6–24 μg/ml, and 4–16 μg/ml of V_2O_3, $VOSO_4$, and NH_4VO_3, respectively. The analysis of the SCE frequency showed that the increase was small and not dose-related.

8. IN VIVO MAMMALIAN SYSTEMS

In contrast to in vitro systems data, the information on in vivo genotoxic and mutagenic activity of vanadium is very small (Table 3).

Table 3 In Vivo Genotoxic and Mutagenic Effects of Vanadium Compounds

Assay System[a]	V(V) Compound	Result	V(IV) Compound	Result	Reference
Rat					
Structural CA	V_2O_5	−			Giri et al., 1979
MI		+			
Mice					
MN	V_2O_5[b]	+			Sun, 1987
	V_2O_5[c]	+			
	V_2O_5[d]	+			
	V_2O_5[e]	−			
Dominant-lethal	V_2O_5	−			
CD-1 mice					
Structural CA	Na_3VO_4	−	$VOSO_4$	+	Ciranni et al., 1995
	NH_4VO_3	−			
Numerical CA	Na_3VO_4	+	$VOSO_4$	+	
	NH_4VO_3	+			
MN	Na_3VO_4	+	$VOSO_4$	+	
	NH_4VO_3	+			
CD-1 mice					
Structural CA	V_2O_5	−			Altamirano-Lozano et al., 1993; Altamirano-Lozano and Alvarez-Barrera, 1996
SCE		−			
MI		+			
Dominant-lethal		+			
DNA-SSB		+			Altamirano-Lozano et al., 1996

[a] CA, chromosomal aberrations; MI, mitotic index; MN, micronuclei; SCE, sister chromatid exchanges; SSB, single-strand breaks.
[b] ip.
[c] Subcutaneous.
[d] Inhalation.
[e] Oral.

171

8.1. Rat Test

When rats were treated for 21 days with 4 mg/kg V_2O_5, no increase in chromosomal aberrations in bone marrow cells was observed, but a decrease in the mitotic index was detected (Giri et al., 1979).

8.2. Mouse Test

In a micronucleus test, V_2O_5 was administered to mice of the 615 and Kunming albino strains by ip injection at doses of 0.17, 2.13, and 6.4 mg/kg for 5 consecutive days. In these experiments, a significant increase in MN frequency was observed in both strains (Sun, 1987). In addition, using both subcutaneous injection of V_2O_5 (0.25, 1.0, and 54.0 mg/kg) and inhalation of V_2O_5 dust (0.5, 2.0, and 8.0 mg/m^3) Sun (1987) induced micronuclei in mouse strain 615. However, negative results were obtained following oral administration of 1.44, 2.83, 5.65, and 11.3 mg/kg V_2O_5 suspension, daily for 6 weeks, to Kunming albino mice.

In a dominant-lethal mutation assay, V_2O_5 (0.2, 1.0, and 4.0 mg/kg) was administered daily by subcutaneous injection to male mice, and the results obtained by the analysis of fetuses and resorptions were considered negative for the induction of this end point (Sun, 1987).

Ciranni et al. (1995) evaluated three vanadium salts for induction of genotoxic effects in bone marrow of CD-1 mice following intragastric treatment. Micronucleus induction in polychromatic erythrocytes (PCEs), SCA, and numerical chromosomal aberrations (NCAs) in bone marrow cells were evaluated. Vanadium compounds were applied in a single dose via intragastric intubation. Doses used were sodium orthovanadate, 75 mg/kg; vanadyl sulfate, 100 mg/kg; and ammonium metavanadate, 50 mg/kg (doses of elemental vanadium were 0.60, 0.40, and 0.42 mM for $SOVO_4$, Na_3VO_4, and NH_4VO_3, respectively). For the MN test, bone marrow cells were sampled 6, 12, 18, 24, 30, 36, 42, 48, and 72 h after treatment, and for chromosomal aberrations, animals were killed 24 and 36 h after treatment. The authors reported that $SOVO_4$ produced a remarkable increase in MN in the majority of the sampling intervals. A delayed genotoxic effect was observed after Na_3VO_4 treatment; however, at three different sampling times (24, 30, and 48 h) MN frequency was found to be statistically significant. On the other hand, NH_4VO_3 induced a significant increase in micronuclei at three different sampling times after treatment (18, 24, and 30 h). The analysis of chromosomal aberrations showed that SVO_5 proved to be the only compound tested that was capable of affecting chromosome structure. The highest effect was observed at 24 h, and the chromosome damage observed was due mainly to breaks and fragments. In the case of NCA, an increase in the frequency of hypoploid cells was found at all tested sampling times for each compound, and NH_4VO_3 induced hyperploid cells at two sampling times. All compounds tested by these authors were shown to be effective in inducing polyploid cells, but the frequency among the groups did not vary with statistical significance.

When V_2O_5 was injected ip into male CD-1 mice (5.75, 11.5, and 23 mg/kg), no changes in the CCK, CA, or the SCE frequencies were seen in the bone marrow cells of the treated animals; however, the MI was reduced for the higher dose only in comparison with the control group (Altamirano-Lozano et al., 1993; Altamirano-Lozano and Alvarez-Barrera, 1996).

Altamirano-Lozano et al. (1996) evaluated the effects of V_2O_5 administered to CD-1 male mice. This compound (8.5 mg/kg) was injected ip every third day for 60 days and the animals were mated with normal females. A significant decrease in fertility rate and an increase in dominant lethal (DL) mutations (DL, measured by the frequency of live vs. dead and resorbed fetuses occurring after mating of chemically treated males with untreated females) were seen following this treatment.

In other experiments, the DNA damage in individual testis cells was analyzed by the single-cell gel electrophoresis technique (COMET assay). A single ip injection of vanadium pentoxide (5.75, 11.5, and 23 mg/kg) for 24 h induced DNA damage (single-strand breaks; SSB) in the testis cells that was expressed and detected as DNA migration (Altamirano-Lozano et al., 1996).

9. CONCLUSIONS

The effects of vanadium compounds on cells are very complex, because the chemistry of this metal is also complex. Vanadium has several oxidation states (-1 to $+5$) and frequently can form polymers (Nechay, 1984; Nechay et al., 1986; WHO, 1988); like molybdenum, vanadium is available in anionic and cationic forms, the most common ones being, under physiological conditions, vanadate (H_2VO_4) and vanadyl ion (VO^{2+}). However, there are other forms of vanadium that may be present in aqueous media at physiological pH. The cationic forms that can exist at pH about 7 are VO^{3+}, VO_2^+, V(IV), and V(III), and the anions forms are $H_2VO_4^{2-}$ and $V_4O_{12}^{4-}$ (Crans and Schelble, 1990; Nechay et al., 1986; Rehder, 1992).

A summary of existing data about all test systems used to evaluate the genetic effects of vanadium compounds are presented in Table 4. The effects of vanadium on cells involve interactions in several metabolic and enzymatic pathways. It has been shown to modify DNA synthesis and repair in low concentrations, but in high doses it is cytotoxic (Carpenter, 1981; Léonard and Gerber, 1994; Sabbioni et al., 1983; WHO, 1988).

The data on the mutagenic potential of vanadium in bacterial systems are inconclusive. It could produce mitotic gene conversion in yeast, but no mutagenic effect was found in mammalian cells in the presence or absence of microsomal S9 fraction and results in dominant-lethal mutations are limited and inconclusive. However some authors classify vanadium as a weak mutagen (Altamirano et al., 1996; Bronzetti et al., 1990; Galli et al., 1991; Kanematsu et al, 1980; Léonard and Gerber, 1994).

Table 4 Summary of Assays Used to Detect the Genotoxic and Mutagenic Effects of Vanadium Compounds

	Nonmammalian Systems													Mammalian Systems																											
														In Vitro																In Vivo											
	Prokaryotes		Lower Eukaryotes				Plants				Insects			Animal Cells								Human Cells								Animals							Humans[b]				
	D	G	D	R	G	A	D	G	C	R	G	C	A	D	G	S	M	C	A	T	I	D	G	S	M	C	A	T	I	D	G	S	M	C	DL	A	D	S	M	C	A
NaVO₃																			+	+						+	+	+													
Na₃VO₄																					+			+	+	+	+	+								+					
V₂O₅		+	+	+				+							+	+	+	+	+		+	+	+	+	+	+	+	+				+	+	+	+						
NH₄VO₃		+			+	+		+								+								+	+	+	+	+	+				+	+	+						
VCl₃										+	+																														
V₂O₃														+						+																					
VOSO₄			+			+				+				+	+									+	+	+	+						+			+					
VO₂														+																											

[a] A, aneuploidy; C, chromosomal aberrations; D, DNA damage; DL, dominant lethal mutation; G, gene mutation; I, alterations in cellular proliferation and DNA replication; M, micronuclei; R, mitotic recombination and gene conversion; S, sister chromatid exchanges; T, cell transformation.

[b] Humans not reviewed.

Results obtained in mammalian cells, both in vivo and in vitro, indicate that vanadium compounds produce a marginal response in the induction of sister chromatid exchanges and structural chromosomal aberrations, but increase the frequency of micronuclei and polyploid cells, decrease mitotic index, and induce DNA single-strand breaks and DNA–protein crosslinks (Altamirano et al., 1996; Cohen et al., 1992; Migliori et al., 1993; Rojas et al., 1996a,b; Roldán and Altamirano, 1990).

The most evident and studied effect exhibited by vanadium compounds is their ability to disrupt microtubule function. However, it is almost impossible to explain all the effects produced by these kinds of compounds by this mechanism. Thus, other mechanisms must be taken into account (Migliori et al., 1995; Ramirez et al., in press; Sakurai, 1994; Zhong et al., 1994).

The toxicity of metal compounds has traditionally been regarded as a function of dose and the potency of the metal itself. In recent years it has become clear that several metals and metalloids undergo transformations in mammalian tissues and that metabolism may have important implications in clinical pharmacology, toxicology, and genotoxicity (Manzo et al., 1992). Vanadate ($+5$ oxidation state) in body fluids and cells is reduced to tetravalent vanadium (VO^{2+}, vanadyl) by agents that are common in cells, for example, ascorbic acid, norepinephrine, pyridine, nucleotides, cysteine, and glutathione (Domingo et al., 1990; Legrum, 1986), and stabilized by various ligands. Free VO^{2+} is unstable at neutral pH and hydrolyzes mainly to $VOOH^+$; this form would be oxidized to VO_3^-. It has been proposed that the V(V) $\Leftrightarrow$ V(IV) redox reaction may regulate the (Na^+,K^+)-pump, since VO_3^- is a more potent inhibitor of (Na^+,K^+)-ATPase than VO^{2+}, and this reaction could be crucial in preventing the harmful effects of vanadium (Nechay et al., 1986; Sabbioni et al., 1991). Thus, this difference in cell metabolism could explain the differences observed when the genotoxic effects of vanadium compounds are studied.

Of the common oxidation states of vanadium, only V^{III}, V^{IV}, and V^V are involved in physiological systems; however, more recent investigations indicate that autoxidation and formation of hyperoxide also should be considered in neutral media (Rehder, 1991). Some reports indicate that vanadium(IV) possesses a pro-oxidant potential, and it has been shown to oxidize a variety of biochemical substrates (Domingo et al., 1990). This effect of vanadium also could generate free radicals and produce DNA damage (Sakurai, 1994; Shi et al., 1996); however, vanadium(V) fails to induce free radicals (Rojas., personal communication).

REFERENCES

Abundis, M. H. M. (1994). Valoración de la genotoxcidad del pentóxido de vanadio en células de las alas de *Drosophila melanogaster*. Comparación de tres protocolos. Bsc. thesis, Facultad de Ciencias, Universidad Nacional Autónoma de México.

Abundis, M. H. M. (1996). Determinación de la mutación y recombinación somáticas en la inducción de efectos genotóxicos por tres sales de vanadio en *Drosophila melanogaster*. Msc. thesis, Facultad de Ciencias, Universidad Nacional Autónoma de México.

Al-Swaidan, H. M. (1993). Determination of vanadium and nickel in oil products from Saudi Arabia by inductively coupled plasma mass spectometry (ICP/MS). *Anal. Lett.* **26,** 141–146.

Altamirano, M., Ayala, M. E., Flores, A., Morales, L., and Dominguez, R. (1991). Sex differences in the effect of vanadium pentoxide administration to prepubertal rats. *Med. Sci. Res.* **19,** 825–826.

Altamirano-Lozano, M., and Alvarez-Barrera, L. (1996). Genotoxic and reprotoxic effects of vanadium and lithium. In P. Collery, J. Corbella, J. L. Domingo, J. C. Elienne, and J. M. Llobet (Eds.), *Metal Ions in Biology and Medicine*. Jhon Libbey Eurotext, Paris, Vol. 4, pp. 423–425.

Altamirano-Lozano, M., Alvarez-Barrera, L., Basurto-Alcántara, F., Valverde, M., and Rojas, E. (1996). Reprotoxic and genotoxic studies of vanadium pentoxide in male mice. *Teratogenesis Carcinog. Mutagen.* **16,** 7–17.

Altamirano-Lozano, M., Alvarez-Barrera, L., and Roldán Reyes, E. (1993). Cytogenetic and teratogenic effects of vanadium pentoxide on mice. *Med. Sci. Res.* **21,** 711–713.

Banci, L., Dei, A., and Gatteschi, D. (1982). Vanadyl binding to bleomycin. *Inorg. Chim. Acta Bioinorg. Chem.* **67,** L53.

Baroch, E. F. (1983). Vanadium and vanadium alloys. In *Encyclopedia of Chemical Technology*. Jhon Wiley and Sons, New York, Vol. 23, pp. 673–687.

Bosch, F., Hatzoglou, M., Park, E. A., and Hanson, R. W. (1990). Vanadate inhibits expression of the gene for phosphoenolpyruvate carboxykinase (GTP) in rat hepatoma cells. *J. Biol. Chem.* **265,** 13677–13682.

Bronzetti, G., Morichetti, E., Della Croce, C., Del Carratore, R., Giromini, L., and Galli, A. (1990). Vanadium; genetical and biochemical investigations. *Mutagenesis* **5,** 293–295.

Canalis, E. (1985). Effect of sodium vanadate on deoxyribonucleic acid and protein syntheses in cultured rat calvariae. *Endocrinology* **116,** 855–862.

Cande, W. Z., and Wolniak, S. M. (1978). Chromosome movement in lysed mitotic cells is inhibited by vanadate. *J. Cell Biol.* **79,** 573–580.

Carpenter, G. (1981). Vanadate, epidermal growth factor and the stimulation of DNA synthesis. *Biochem. Biophys. Res. Commun.* **102,** 1115–1121.

Ciranni, R., Antonetti, M., and Migliore, L. (1995). Vanadium salts induce cytogenetic effects in *in vivo* treated mice. *Mutation Res.* **343,** 53–60.

Cohen, M. D., Klein, C. B., and Costa, M. (1992). Forward mutations and DNA–protein crosslinks induced by ammonium metavanadate in cultures mammalian cells. *Mutation Res.* **269,** 141–148.

Crans, D. C., and Schelble, S. M. (1990). Vanadate tetramer as the inhibiting species in enzyme reactions in vitro and in vivo. *J. Am. Chem. Soc.* **112,** 427–432.

Djordjevic, C., and Wampler, G. (1985). Antitumor activity and toxicity of peroxoheteroligand vanadate(V) in relation to biochemistry of vanadium. *J. Inorg. Biochem.* **25,** 51–56.

Domingo, J. L. (1996). Vanadium: A review of the reproductive and developmental toxicity. *Reprod. Toxicol.* **10,** 175–182.

Domingo, J. L., Gómez, M., Llobert, J. M., and Corbella, J. (1990). Chelating agents in the treatment of acute vanadyl sulphate intoxication in mice. *Toxicology* **62,** 203–211.

Domingo, J. L., Gómez, M., Sánchez, D. J., Llobert, J. M., and Keen, C. L. (1995). Toxicology of vanadium compunds in diabetic rats: The action of chelating agents on vanadium accumulation. *Mol. Cell. Biochem.* **153,** 233–240.

Domingo, J. L., Paternain, J. L., Llobert, J. M., and Corbella, J. (1986). Effects of vanadium on reproduction, gestation, parturition and lactation in rats upon oral administration. *Life. Sci.* **39,** 819–824.

Dubyak, G. R., and Kleinzeller, A. (1980). The insulin-mimetic effects of vanadate in isolated rat adipocytes. Dissociation from effects of vanadate as a (Na+, -K+)ATPase inhibitor. *J. Biol. Chem.* **255**, 5306–5312.

Galli, A., Vellosi, R., Fiorio, R., Della Croce, C., Del Carratore, R., Morichetti, E., Giromini, L., Rosellini, D., and Bronzetti, G. (1991). Genotoxicity of vanadium compounds in yeast and cultured mammalian cells. *Teratogenesis Carcinog. Mutagen.* **11**, 175–183.

Giri, A. K., Sanyal, R., Sharma, A., and Talukder, G. (1979). Cytological and cytochemical changes induced though certain heavy metals in mammalian systems. *Natl. Acad Sci. Lett.* **2**, 391–394.

Graedel, T. E., Hawkins, D. T., and Claxton, D. L. (1986). *Atmospheric Chemical Compounds, Sources, Occurrence and Bioassay.* Academic Press, New York.

Hansen, K. M. and Stern, R. (1984). A survey of metal-induced mutagenicity *in vitro* and *in vivo*. *Toxicol. Environ. Chem.* **9**, 87–91.

Harland, B. F., and Harden-Williams, B. A. (1994). Is a vanadium of human nutritional importance yet? *J. Am. Diet. Assoc.* **94**, 891–894.

Hickey, R. J., Schoff, E. P., and Clelland, R. C. (1967). Relationship between air pollution and certain chronic disease death rates. *Arch. Environ. Health* **15**, 728–738.

Jackson, J. F., and Linskens, H. F. (1982). Metal ion induced unscheduled DNA synthesis in petunia pollen. *Mol. Gen. Genet.* **187**, 112–115.

Jandhyala, B. S., and Hom, G. J. (1983). Physiological and pharmacological properties of vanadium. *Life Sci.* **33**, 1325–1340.

Kada, T., Hirano, K., and Shirasu, Y. (1980). Screening of environmental chemical mutagens by the rec-assay system with *Bacillus subtilis*. In F. J. de Serres and A. Hollaender (Eds.), *Chemical Mutagens: Principles and Methods for their Detection.* Plenum Press, New York, Vol. 5, pp. 149–173.

Kanematsu, N., Hare, M., and Kada, I. (1980). Rec assay and mutagenicity studies on metal compounds. *Mutat. Res.* **77**, 109–116.

Kanematsu, N., and Kada, I. (1978). Mutagenicity of metal compounds. *Mutat. Res.* **53**, 207–208.

Kawai, T., Seiji, K., Watanabe, T., Nakatsuka, H., and Ikeda, M. (1989). Urinary vanadium as a biological indicator of exposure to vanadium. *Int. Arch. Occup. Environ. Health* **61**, 283–287.

Legrum, W. (1986). The mode of reduction of vanadate (+V) to oxovanadium (+IV) by glutathione and cysteine. *Toxicology* **42**, 281–284.

Léonard, A., and Gerber, G. B. (1994). Mutagenicity, carcinogenicity and teratogenicity of vanadium compounds. *Mutat. Res.* **317**, 81–88.

Manzo, L., Costa, L. G., Tonini, M., Minoia, C., and Sabbioni, E. (1992). Metabolic studies as a for the interpretation of metal toxicity. *Toxicol. Lett.* **64/65**, 677–686.

Meyerovitch, J., Shechter, Y., and Amir, S. (1989). Vanadate stimulate-in vivo glucose uptake in brain and arrests food intake and body weight gain in rats. *Physiol. Behav.* **45**, 1113–1117.

Migliore, L., Bocciardi, R., Macri, C., and Lo Jacono, F. (1993). Cytogenetic damage induced in human lymphocytes by four vanadium compounds and micronucleus analysis by fluorescence *in situ* hybridization with a centromeric probe. *Mutat. Res.* **319**, 205–213.

Migliore, L., and Scarpato, R. (1996). Induction of aneuploidy in human lymphocytes by vanadium salts. In P. Collery, J. Corbella, J. L. Domingo, J. C. Elienne, and J. M. Llobet (Eds.), *Metal Ions in Biology and Medicine.* John Libbey Eurotext, Paris, Vol. 4, pp. 267–269.

Migliore, L., Scarpato, R., and Falco, P. (1995). The use of fluorescence *in situ* hybridization with a β-satellite DNA probe for the detection of accrocentric chromosomes in vanadium-induced micronuclei. *Cytogenet. Cell Genet.* **69**, 215–219.

Navas, P., Hidalgo, A., and Garcia-Herdugo, G. (1986). Cytokinesis in onion roots: Inhibition by vanadate and caffeine. *Experientia* **42**, 437–439.

Nechay, B. R. (1984). Mechanisms of action of vanadium. *Annu. Rev. Pharmacol. Toxicol.* **24,** 501–524.

Nechay, B. R., Nanninga, L. B., and Nechay, P. S. E. (1986). Vanadyl(IV) and vanadate(V) binding to selected endogonous phosphate, carboxyl, and amino ligands: Calculations of cellular vanadium species distribution. *Arch. Biochem. Biophys.* **251,** 128–138.

Owusu-Yaw, J., Cohen, M. D., Fernando, S. Y., and Wei, C. I. (1990). An assessment of genotoxicity of vanadium pentoxide. *Toxicol. Lett.* **50,** 327–336.

Paton G. R., and Allison, A. C. (1972). Chromosome damage in human cell cultures induced by metal salts. *Mutat. Res.* **16,** 332–336.

Petrucci, R. H. (1989). Vanadium. In *General Chemistry. Principles and Modern Applications.* 5th ed. Plenum Press, New York, 877 pp.

Ramírez, P., Eastmond, D. A., Laclette, J. P., and Ostrosky-Wegman, P. (in press). Disruption of microtubule assembly and spindle formations as a mechanism for the induction of aneuploid cells by sodium arsenite and vanadium pentoxide. *Mutat. Res.*

Rehder, D. (1991). The bioinorganic chemistry of vanadium. *Angew. Chem. Int. Ed. Engl.* **30,** 148–167.

Rehder, D. (1992). Structure and function of vanadium compounds in living organisms. *BioMetals* **5,** 3–12.

Rojas, E., Valverde, M., Herrera, L. A., Altamirano-Lozano, M., and Ostrosky-Wegman, P. (1996a). Genotoxicity of vanadium pentoxide evaluated by the single cell gel electrophoresis assay in human lymphocytes. *Mutat. Res.* **359,** 77–84.

Rojas, E., Valverde, M., Sordo, M., Altamirano-Lozano, M., and Ostrosky-Wegman, P. (1996b). Single cell gel electrophoresis assay in the evaluation of metal carcinogenicity. In P. Collery, J. Corbella, J. L. Domingo, J. C. Elienne, and J. M. Llobet (Eds.), *Metal Ions in Biology and Medicine.* John Libbey Eurotext, Paris, Vol. 4, pp. 375–377.

Roldán, R. E., and Altamirano, L. M. A. (1990). Chromosomal aberrations, sister chromatid exchanges, cell-cycle kinetics and satellite associations in human lymphocyte cultures exposed to vanadium pentoxide. *Mutat. Res.* **245,** 61–65.

Rosenbaum, J. B. (1983). Vanadium Compounds. In *Encyclopedia of Chemical Technology.* John Wiley and Sons, New York, Vol. 23, pp. 688–704.

Sabbioni, E., Clereci, L., and Brazzelli, A. (1983). Different effects of vanadium ions on some DNA-metabolizing enzymes. *J. Toxicol. Environ. Health* **12,** 737–748.

Sabbioni, E., Pozzi, G., Pintar, A., Casella, L., and Garattini, S. (1991). Cellular retention, cytotoxicity and morphological transformation by vanadium(IV) and vanadium(V) in BALB/ 3T3 cell lines. *Carcinogenesis* **12,** 47–52.

Sakurai, H. (1994). Vanadium distribution in rats and DNA cleavage by vanadyl complex: Implication for vanadium toxicity and biological effects. *Environ. Health Perspect.* **102**(Suppl 3), 35–36.

Sakurai, H., Tetsuko, G., and Shinomura, S. (1982). Vanadyl(IV) ion dependent enhancement of oxygen binding to hemoglobin and myoglobin. *Biochem. Biophys. Res. Commun.* **107,** 1349–1354.

Sharma, A., and Talukder, G. (1987). Effects of metals on chromosomes of higher organisms. *Environ. Mutagen.* **9,** 191–226.

Sharma, R. P., Flora, S. J. S., Drown, D. B., and Oberg, S. G. (1987). Persistence of vanadium compounds in lungs after intratracheal instillation in rats. *Toxicol. Ind. Health* **3,** 321–329.

Sheu, C. W., and Rodriguez, Y. (1991). Proliferation and morphological changes in BALB/3T3 cells treated with sodium vanadate. *Environ. Mol. Mutagen.* **17**(Suppl. 19), 68.

Shi, X., Jiang, H., Mao, Y., Ye, J., and Saffiotti, U. (1996). Vanadium(IV)-mediated free radical generation and related 2′-deoxyguanosine hydroxylation and DNA damage. *Toxicology* **106,** 27–38.

Singh, O. P. (1979). Effects of certain chemical pollutants on plant chromosomes. PhD thesis, University of Calcutta, India.

Smith, J. B. (1983). Vanadium ions stimulate DNA synthesis in Swiss mouse 3T3 and 3T6 cells. *Proc. Natl. Acad. Sci. USA* **80,** 6162–6166.

Stock, P. (1960). On the relation between atmospheric pollution in urban and rural localities and mortality from cancer, bronchitis, pneumonia, with particular reference to 3,4-benzopyrene, berylium, molybdenum, vanadium and arsenic. *Br. J. Cancer* **14,** 397–418.

Sun, M. (Ed.). (1987). Toxicity of vanadium and its environmental health standard. Changdu West China University of Medical Sciences, Report.

Thomson, H., Chasteen, N. D., and Meeker, L. D. (1984). Dietary vanadyl(IV) sulfate inhibits chemically-induced mammary carcinogenesis. *Carcinogenesis* **5,** 849–852.

Vives-Corrons, J. L., Jou, J. M., Ester, A., Ibars, M. Carreras, J., Bartrons, R., Climent, F., and Grisolia, S. (1981). Vanadate increase oxygen affinity and affects enzyme activities and membranes properties of erythrocytes. *Biochem. Biophys. Res. Commun.* **103,** 111–117.

Vouk, V. (1986). General chemistry of metals. In L. Frieberg, G. F. Nordberg, and B. V. Vouk (Eds.), *Handbook on the Toxicology of Metals.* Elsevier Science Publishers, New York, Ch. 2, pp. 14–35.

Waters, M. D., Gardner, D. E., and Coffin, D. L. (1974). Cytotoxic effects of vanadium on rabbit alveolar macrophages in vitro. *Toxicol. Appl. Pharmacol.* **28,** 253–263.

WHO (1988). Vanadium. *Environmental Health Criteria No. 80.* WHO, Geneva.

Zakour, R. A., and Glickman, B. W. (1984). Metal-induced mutagenesis in the *lacI* gene of *Escherichia coli. Mutat. Res.* **126,** 9–18.

Zaporowska, H., and Wasilewski, W. (1992). Hematological effects of vanadium on living organisms. *Comp. Biochem. Physiol.* **102C,** 223–231.

Zhong, B.-Z., Gu, Z.-W., Wallace, W. E., Whong, W.-Z., and Ong, T. (1994). Genotoxicity of vanadium pentoxide in Chinese hamster V79 cells. *Mutat. Res.* **321,** 35–42.

10

VANADIUM AND THE CARDIOVASCULAR SYSTEM: REGULATORY EFFECTS AND TOXICITY

Marco Carmignani, Anna Rita Volpe, Enrico Sabbioni, Mario Felaco, and Paolo Boscolo

Section of Pharmacology, Department of Basic and Applied Biology, University of L'Aquila, 67010 Coppito (AQ) (M. C., A. R. V.); Receptor Chemistry Centre, CNR, Catholic University School of Medicine, 00168 Rome (A. R. V.); European Commission, Environment Institute, Joint Research Centre, 21020 Ispra (VA) (E. S.); Institute of Biology and Genetics (M. F.) and Centre of Occupational Medicine and Ergophthalmology (P. B.), University of Chieti, 66100 Chieti; Italy

Vanadium in the Environment. Part 2: Health Effects, Edited by Jerome O. Nriagu.
ISBN 0-471-17776-8.   © 1998 John Wiley & Sons, Inc.

1.   INTRODUCTION

Degenerative cardiovascular diseases with associated renal pathologies affect a great part of the population in developed countries, and arterial hypertension is one of the most important public health problems. In about 90–95% of patients suffering from hypertension, specific alterations within the homeostatic mechanisms regulating blood pressure (BP) levels are not yet clearly understood. These patients are defined as suffering from "essential" hypertension, notwithstanding the fact that they may present cardiovascular functional abnormalities defined as secondary to the onset of hypertension. Secondary hypertension, recognized only when a primary cause or a specific alteration of an organ is demonstrated, includes renal and endocrine hypertension (e.g., hyperaldosteronism, pheochromocytoma, hyperparatiroidism) as well as hypertension depending on the intake of estrogen-containing contraceptives (Williams, 1994).

Although a genetic predisposition to suffer from essential hypertension was demonstrated (Kurtz and Spence, 1993; Williams, 1994), other factors such as obesity, diet, and psychosocial factors may be related to the increase of BP. Environmental toxic agents were also found to contribute to the development of hypertension. Large-scale studies on unselected population (males aged 40–59 years), based on the Second National Health and Nutrition Survey of the United States (1976–1980), indicated a significant relationship between blood lead and BP values (Pirkle et al., 1985). A significant correlation between these parameters was also found in middle-aged men from 24 British towns when the statistical analysis of the data took into consideration the variations related to the different towns (Pocock et al., 1984). It was supposed that modification of BP in the British men was due not only to lead exposure but also to environmental trace metals and others compounds, lead being a marker of environmental exposure to toxic agents. In this context, it was hypothesized that vanadium—which humans could be exposed to (Orvini and Lodola, 1979) and which is very active in inhibiting or stimulating a series of enzymatic activities (e.g., Na^+,K^+-ATPase and Ca^{++}-ATPase, and adenylate cyclase), that are involved in the mechanisms of cardiovascular regulation—would be a cause of essential hypertension (Nechay, 1984). Although the observation that vanadium increased in urine of patients suffering from essential hypertension (Schroeder et al., 1963) was not confirmed, its possible pathogenetic role in inducing hypertension cannot be excluded. Several studies on rats and rabbits, chronically exposed to different doses of vanadate, showed arterial hypertension and/or increase of the total peripheral vascular resistance (TPR) (Boscolo et al., 1994; Carmignani et al., 1989, 1991, 1992a, 1992b, 1996). This is in agreement with the observed increase of serum levels of vanadium in patients with renal disorders (Bello-Reuss et al., 1979).

Although the ethiopathogenesis of essential hypertension is far from being explained, the knowledge of the cardiovascular function has been improved in recent years. Thus, a review on possible effects of vanadium on the cardiovascular system appears well motivated. This chapter reviews current information available on the mechanisms of cardiovascular regulation that could be affected by vanadium exposure and on the levels of vanadium in animal and human tissues, particularly heart and kidney, under different conditions of exposure to vanadium compounds. In addition, results concerning recent experiments on the effects of chronic exposure to "pharmacological doses" of vanadium on the mechanisms of cardiovascular regulation of rats and rabbits are reported and discussed.

2. MECHANISMS OF CARDIOVASCULAR REGULATION

Hemodynamics may be defined through the fundamental relationship linking BP to cardiac output (CO) and TPR: $BP = CO \times TPR$. CO results from heart rate (HR) and stroke volume (SV): $CO = HR \times SV$; SV and TPR are mostly dependent on ventricular contractility (cardiac inotropism, CI) and arterial vascular tone, respectively (Antonaccio, 1990; Braunwald, 1994). If BP is measured in millimeters of mercury (mmHg), HR in beats/min, SV in milliliters (mL)/beat, and CO in mL/min, TPR will be calculated in dynes $\times$ sec/cm^{-5}. The maximum rate of rise of the left ventricular isovolumetric pressure (dP/dt), measured in mmHg/sec, represents, in experimental models, a suitable index of CI.

Cardiovascular function is regulated by homeostatic mechanisms involving heart, vascular bed, and circulating blood. Organs (such as kidney), endocrine glands (such as adrenals), neurogenic pathways (such as the sympathetic and parasympathetic ones), and humoral active substances (autacoids) play a primary role in regulating cardiovascular function.

Both the physiology and pharmacology of heart and vessels have been extensively reviewed (Antonaccio, 1990; Braunwald, 1994; Hardman et al., 1996; Marmé, 1987). In the last decade, particular emphasis was addressed to the endothelium because of its capacity of producing and releasing substances regulating vascular tone. Such substances induce vasoconstriction (e.g., endothelins and superoxide anion) or vasodilation (e.g., prostacyclin (PGI_2) and nitric oxide (NO); the former are also aggregating agents, whereas the latter have the opposite effect on platelets. In particular, NO is formed, as in other cells (e.g., neurons, macrophages), in equimolar amount with L-citrulline from L-arginine, in a complex reaction involving various isoforms (inducible or constitutive) of NO synthase (Moncada and Higgs, 1993). In this regard, recent data indicate that intracellular bradykinin is a physiological precursor for the synthesis of NO by supplying its carboxylic arginine by action of kininase I (Volpe et al., 1996).

The contractile processes of cardiac and vascular myocells are primarily regulated by cell surface receptors whose activation, by selective agonists, triggers transductional mechanisms such as ion fluxes and biochemical pathways (e.g., cyclic nucleotide, Ca^{++}-calmodulin, and inositol-1,4,5-trisphosphate systems). Neurotransmitters (e.g., noradrenaline (NA), acetylcholine (Ach), and peptides), hormones, local regulators of the peripheral circulation (autacoids such as bradykinin, prostaglandins, and angiotensins), and metabolic substances (e.g., aminoacids) interact with specific receptors located at both cardiac and vascular levels. In particular, catecholamines (adrenaline (A), NA, dopamine (DA), and L-DOPA) are released, as a consequence of the tonic nerve activity of the sympathetic system, from adrenal glands and postganglionic adrenergic nerve endings (Lefkowitz et al., 1996). Types, subtypes, related transductional mechanisms, and the physiological role of both presynaptic (autoreceptors) and postsynaptic adrenergic (catecholaminergic) receptors have been well determined as well as neuroanatomy, neurophysiology, and neuropharmacology of the central and peripheral parasympathetic (cholinergic) and sympathetic systems (Antonaccio, 1990; Braunwald, 1994; Lefkowitz et al., 1996; Marmé, 1987; Willebucher et al., 1992). Moreover, the tonic activity of these systems was found to be up- or down-regulated, within the central nervous system, in response to stimulations by baro- and chemoreceptors, hormones, autacoids such as angiotensin II, and modifications in plasma osmolality (Levinsky, 1994).

The kidney is involved in the regulation of cardiovascular function by producing, under homeostatic mechanisms, vasoactive substances such as kinins, prostaglandins, and angiotensins. Moreover, renal mechanisms that regulate cardiovascular function respond to stimulation of peptides produced in the brain (e.g., ADH) and atria (atrial natriuretic factor, ANF), as well as to sympathetic activity, hormones (such as aldosterone produced in the adrenal glands), and modifications of serum electrolytes and BP. Kidney determines both osmolality and plasma concentrations of electrolytes by directly regulating their excretion into urine (Bhoola et al., 1992; Levinsky, 1994; Scicli and Carrettero, 1986).

Renin is a proteolytic enzyme produced and stored in the renal cortex and juxtaglomerular cells surrounding the afferent arterioles of glomeruli. The juxtaglomerular apparatus consists of macula densa, a part of the distal tubule, and of modified granulated smooth myocells, located in the afferent arteriole of the glomerulus, with a minor contribution by those of the efferent arteriole and interstitial cells (Iannaccone et al., 1978). The regulation of synthesis and release of renin is complex and includes adrenergic mechanisms, hormone influences (e.g., by ADH and ANF), and changes in the transmembrane ion gradients (e.g., of Na^+, K^+, Ca^{++}) in juxtaglomerular cells (Churchill, 1985; Levinsky, 1994).

The kallikrein–kinin system is implicated in the regulation of renal blood flow and natriuresis; the latter determines water and electrolyte excretion and participates in the control of BP (Bhoola et al., 1992; Scicli and Carrettero, 1986). Kallikrein is present in the kidney in connecting tubule cells of the distal nephron. Renal kallikrein, present also in other organs, is called glandular kallikrein, being different from plasma kallikrein and similar to urine kallikrein. Renal kallikrein releases the kinin kallidin from two α_2-globulins (kininogens) synthesized in the liver. Kallidin (lys-bradykinin) is converted to bradykinin by an aminopeptidase present in plasma and urine. Renal kallikrein has a potent vasodilator and natriuretic effect; it is possible that a part of renal kallikrein produced in the kidney may act not only on that organ, but also on the whole vascular bed. Kinins are catabolized rapidly to inactive peptides in kidney, blood, and urine by the action of kininases. The renin–angiotensin and kallikrein–kinin systems are linked by an enzyme that acts as angiotensin I-converting enzyme, transforming angiotensin I to angiotensin II, and as kininase II, determining the breakdown of kinins; the kallikrein–kinin system is also linked to the prostaglandin–thromboxane–leukotriene system, since, for instance, bradykinin stimulates the synthesis of renal prostaglandins (Bhoola et al., 1992; Churchill, 1985; Vio and Figueroa, 1985).

Central and peripheral catecholaminergic mechanisms as well as renin–angiotensin–aldosterone, kallikrein–kinin, enkephalin, and nitric oxide systems will appear, along with Na^+, K^+-ATPase (sodium pump), the main targets of vanadium in altering cardiovascular function.

3. VANADIUM IN ANIMAL AND HUMAN TISSUES

Information on the vanadium content in animals is limited and almost always referred to tissues such as liver and kidney, with no attention to the cardiovascular system.

Cattle and heifers were poisoned in Sweden in 1990 by vanadium-contaminated fertilizers. A part of these animals died by acute intoxication, while the rest of the herd was slaughtered 3 months and 5 months later. In the heifers that died by acute poisoning, high vanadium concentrations were found in kidneys, liver, and spleen (5.5, 5.9, and 1.9 mg/kg wet weight, respectively), while the highest value in bone tissue was 0.68 mg/kg. After 5 months, the corresponding mean values were 0.058, 0.224, and 0.213 mg/kg wet weight, respectively. In cows, which were less exposed, the corresponding values were 0.095, 0.012, and 0.013 mg/kg. Unfortunately, the vanadium content of the cardiovascular tissues was not determined in this investigation (Frank et al., 1996).

In rats (3 weeks of age) without exposure to vanadium and simply grown with normal diet (mineral water and chow food at libitum; vanadium content 2.8 μg/L and 78 ng/g, respectively), the concentration of vanadium in heart tissue was 29.7 ng/g wet weight. This value decreased to 8.9 and 6.8 ng/g 77 and 115 days after exposure. This suggests a high retention capacity of undeveloped cardiac tissue for vanadium (Edel et al., 1984).

In an experimental study, rats received 150 μg/mL of sodium metavanadate, 230 μg/mL of sodium orthovanadate, and 230 μg/mL of vanadyl sulfate pentahydrate in drinking water for 28 days. Rats exposed to sodium orthovanadate showed lower levels of the element in tissues than those treated with sodium metavanadate. In these animals, at the end of the experiment, mean levels of vanadium in kidney, heart, liver, and spleen were 8.21, 0.54, 1.51, and 4.85 mg/kg wet weight, respectively. A similar disposition was found in the same tissues of animals treated with vanadyl sulphate (mean vanadium contents 9.30, 0.6, 2.26, and 5.07 mg/kg wet weight, respectively). The similarities in the distribution of the two chemical species of vanadium in rat tissues is not surprising, because in absence of reducing agents or organic ligands the tetravalent vanadium is oxidized in few minutes to vanadium by the air. So, the study (Domingo et al., 1991) probably involved exposure only to vanadate.

The data of these studies show that, after acute exposure, most of the vanadium was stored in spleen and liver while, after prolonged exposure, the content of the metal was greater in the kidneys of cows and rats but not of heifers. It is known that both vanadyl and vanadate bind to rat and human transferrin and ferritin, the transport and depot proteins of iron (Sabbioni and Marafante, 1981). Most transferrin is located in spleen and bone marrow as well as liver, kidneys, and testicles. Vanadium, after prolonged exposure, is excreted free or bound to proteins through the kidney, where it may accumulate in tubular cells. However, the uptake of vanadium in tissues may depend on animal species. This may explain the different distribution of vanadium in kidneys of heifers, cows, and rats. Different bindings of vanadium were demonstrated in rats exposed for 28 days to 150 μg/mL of sodium metavanadate in drinking water and treated with the chelators Tiron and desferoxamine for 2 weeks. Vanadium appeared to be bound in organs in a form "accessible for chelation" and another form, bound to the residual fraction, which is difficult to mobilize (Gomez et al., 1991).

Retention, subcellular distribution, and binding of vanadium in the rat after intratracheal instillation of 200 ng/kg body weight of ^{48}V-labeled tetravalent vanadyl or pentavalent vanadate were reported by Edel and Sabbioni (1988). The tissue vanadium was determined at 3 h, 1 day, and 12 days following intratracheal treatment. The levels of vanadium in the lung decreased from 12 ng/g wet weight after 3 h to 1.35 ng/g at 12 days. At 3 h after instillation the concentration of vanadium in heart was 0.17 ng/g, while in kidney and liver was 1.01 and 0.24 ng/g, respectively. At 12 days after installation the level in heart decreased to 0.04 ng/g, while 0.61 and 0.15 ng/g wet weight were the hepatic and renal concentrations. The major part of the vanadium in kidney, liver, and lung was associated to the nuclear fraction (30–40% of the homogenate), followed by the cytosol and mitochondrial fractions. Gel filtration chromatography of the lung cytosol showed two biochemical pools of vanadium. The first was protein-bound vanadium, which may be involved in the long-term accumulation of this element; the second pool was interpreted as a diffusible form of vanadium. Edel and Sabbioni (1988) also studied retention of vanadium 1 day after oral administration of 277 ng/kg as vanadyl or vanadate ions. At 1 day following administration mean levels of vanadium in cardiac tissue were 0.1 and 0.14 pg/g for vanadium(V) and vanadium(IV), respectively. Interestingly, the qualitative distribution of vanadium after intratracheal instillation and oral administration of pentavalent or tetravalent vanadium was similar, which suggests a similar metabolic pattern for both chemical forms of vanadium independently of the route of exposure.

Carmignani et al. (1989, 1992a) exposed two groups of 12 male weaning Sprague-Dawley rats to 100 μg/mL of vanadium as sodium metavanadate (NaVO$_3$) in drinking water for 7 and for 10 months. Ten animals (controls) were kept in the same animal house conditions and received drinking water not supplemented with vanadium. At 7 months after treatment, six vanadium-exposed rats and six control animals were used for determining cardiovascular parameters, and for histopathological examination by light and transmission electron microscopy and determination of the tissue content of vanadium as carried out by neutron activation analysis (Edel et al., 1984). Tissue vanadium levels were also determined in six other vanadium-exposed rats after 10 months of treatment.

The rats exposed for 7 months to vanadium showed a significant increase of BP and HR (but not of CI), compared with control animals. Histopathological examination did not show cardiac or vascular alterations. In the kidney, the lumen of proximal tubule cells was narrowed and contained amorphous protein material. The highest content of vanadium was found in renal cortex, followed by spleen and femur. The vanadium content of renal medulla was approximately half that of renal cortex; high levels of vanadium were also found in liver, testicles, and cardiac auricles. In the heart, vanadium was five times higher in atrium than in ventricles, while aorta contained higher amounts of the element than cardiac ventricles. Low levels of vanadium were found in brain and cerebellum. At 10 months after treatment liver, testicles, femur, brain, and cerebellum exhibited increased vanadium concentration compared with the corresponding tissues of rats treated for 7 months, while the vanadium content of spleen, renal cortex, and medulla was unchanged. Interestingly, the concentrations of vanadium of cardiac auricles and ventricles as well as of aorta were also increased in rats exposed for 10 months, as compared with those exposed for 7 months (Table 1) (Carmignani et al., 1989, 1992a). High levels of vanadium were also found in atria of the exposed rats, where the sinus node paces HR and ANF is produced to control the diuresis (Levinsky, 1994).

The lack of an increase in kidney vanadium at 8, 9, and 10 months of exposure could be explained by an impairment of this organ in reabsorption of vanadium or vanadium-binding proteins from urine. However, the observation that vanadium was higher in the renal cortex than in medulla is important, because renin and kallikrein are localized in the cortex. Renin and kallikrein are enzymes that exert a major role in the BP-regulating mechanisms (Bhoola et al., 1992; Churchill, 1985; Iannaccone et al., 1978). The high levels of vanadium found in aorta suggest that the vascular bed may be a target for this element (Carmignani et al., 1989, 1992b).

Boscolo et al. (1994) and Carmignani et al. (1992b) exposed male Sprague-Dawley rats to 10 and 40 μg/mL of vanadium (as sodium metavanadate,

Table 1   Vanadium Content in Tissues of Rats Exposed to 100 ppm of Vanadate in Drinking Water for 7 or 10 Months[a,b]

Tissue	Vanadium Content (ng/g Wet Weight)	
	7 Months	10 Months
Liver	1,553 ± 160	1,996 ± 471
Spleen	7,007 ± 1,225	7,768 ± 1,350
Testicles	1,881 ± 504	2,815 ± 622
Femur	7,857 ± 973	9,763 ± 2,035
Brain	137 ± 42	180 ± 26
Cerebellum	193 ± 55	296 ± 108
Kidney (medulla)	3,342 ± 778	3,419 ± 1,134
(cortex)	6,770 ± 515	6,280 ± 910
Aorta	870 ± 115	1,349 ± 363
Heart (auricle)	2,710 ± 890	3,197 ± 1,448
(ventricle)	554 ± 110	657 ± 137

[a] Values are means ± *SD* (n = 5 in both groups).
[b] Controls were six rats. All corresponding tissues analyzed had a vanadium content <25 ng/g wet weight.

NaVO$_3$) in drinking water for 7 months. At the end of the treatment with vanadate, the oxidation state of vanadium in heart and kidney was determined by electron paramagnetic resonance (EPR) after preparation of samples in a nitrogen atmosphere (Sabbioni et al., 1993). Heart and kidney vanadium was present almost completely in the form of tetravalent vanadium (Fig. 1).

Vanadium is able to cross the placental barrier and to be metabolized in the fetus and incorporated into the cardiac tissue. A mean value of 4.3 pg/g wet weight was found in the heart of rat fetuses 1 day before birth and 7 days after iv injection of 100 pg of [48]V vanadate to pregnant rats. The corresponding value in the mother was 100 pg/g (Edel and Sabbioni, 1989).

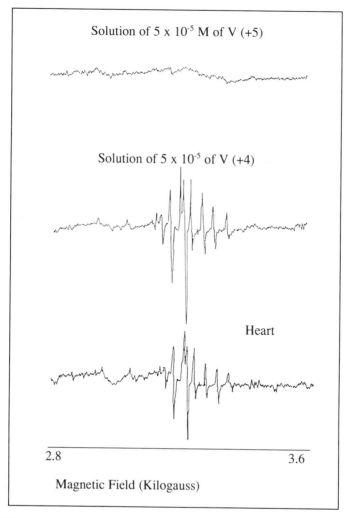

Figure 1. Electron paramagnetic resonance spectra showing that, in the heart of a rat exposed to 40 ppm of vanadate in drinking water for 7 months, vanadium is present as vanadyl (+4 oxidation state).

The total body pool of vanadium estimated for adult humans is reported to be about 100 μg, with higher values in city dwellers. Cardiac tissue contains 1–2 ng/g wet weight. The highest levels of vanadium are those of liver, kidney, bone, spleen, and thyroid (13, 5, 3, 3, 3 ng/g wet weight, respectively; Nechay et al., 1985). According to the TRACY project (Vesterberg et al., 1993) normal values of vanadium in biological fluids of the general population appear to be around 1 nmol/L (blood, serum) and around 10 nmol/L (urine) (Cornelis et al., 1981; Simonoff et al., 1984; Sabbioni et al., 1996). The above reported values suggest that vanadium in the human body is localized mainly inside cells.

Daily dietary intake of vanadium was estimated to be of the order of 15 μg; it is possible that less than 5% of this vanadium is absorbed, while airborne vanadium seems to be absorbed efficiently (French and Jones, 1993). Vanadium levels in environmental air are 1–2 pg/m^3 in continental atmosphere and 2 $\mu g/m^3$ in industrialized areas (Sabbioni et al., 1984). In freshwater vanadium can be present in tetravalent and pentavalent form (Orvini and Lodola, 1979). High levels of occupational vanadium exposure can occur during cleaning and maintenance operations of boilers at oil-fired power plants. No cardiovascular alterations have been reported in occupational exposure to vanadium ash. A group of ten workers in one of these plants was acutely intoxicated, showing irritative symptoms of upper airways but no evident cardiovascular alterations. On the other hand, five workers showed proteinuria during acute intoxication; their mean urinary vanadium level, determined 2 weeks and 6 months after acute exposure, were 92 and 38 mg/L, respectively (Todaro et al., 1991).

Sabbioni and Maroni (1983) measured blood and urinary vanadium in workers not engaged in maintainance operations at oil-fired power plants as well as in scaffelders, burner operators, welders, and cleaners protected by masks. The excretion of vanadium in these workers was rather slow, which is consistent with long-term accumulation of vanadium in the organism during chronic exposure. Thus, the absence of symptoms of illness observed in these workers during or after exposure cannot exclude possible long-term health effects, including those on the cardiovascular system.

We can hypothesize that environmental or occupational vanadium exposure may affect the cardiovascular system mainly by actions on renal mechanisms of regulation of the cardiovascular function.

4. CARDIOVASCULAR EFFECTS OF VANADIUM

4.1. Humans

The predominant form of vanadium in biological fluids at pH 4–8 is vanadate–metavanadate (+5 oxidation state), which may enter cells, where it is reduced almost completely to vanadyl (+4 oxidation state) (Nechay et al., 1985). Both vanadate and vanadyl were found to be very active in modifying enzymatic activities of humans and animals, including those of particular importance on the cardiovascular system (Boscolo et al., 1994; Carmignani et al., 1996), which suggests that exposure to vanadium compounds may also induce essential hypertension. Physiological levels of vanadium have been proposed to play a role in the pathogenesis of hypertension through an increase in the affinity of ouabain-like substances to Na^+,K^+-ATPase (Hansen, 1979). Vanadate increased cardiovascular reactivity to NA and angiotensin II in subjects not uniformly hypertensive (Jandhyala and Hom, 1983).

4.2. Isolated Biological Systems (In Vitro and Ex Vivo Studies)

Both vanadate and vanadyl are able to stimulate or inhibit in vitro several enzymatic activities and to interact with biomolecules, influencing cardiac and vascular myocell contractility (Nechay et al., 1985). Vanadate was able to inhibit sarcolemma Ca^{++} pump in embryonic chick heart cells (Sauvadet et al., 1995), and vanadate peroxide (pervanadate) induced contraction in rat gastric longitudinal muscle strips and in aortic rings, depending on its concentrations (Laniyonu et al., 1994). Moreover, vanadate was found to produce an increase of intracellular Ca^{++} in cultured aortic smooth muscle cells, suggesting that it affects vascular tone (Sandirasegarane and Gopalakrishnan, 1995).

Vanadate was found to activate the adenylate cyclase activity in many cell types. In the heart, the increased concentrations of 3′-5′-cyclic adenosine monophosphate (cAMP) causes release of Ca^{++} from intracellular stores, leading to an increase of HR and CI and to vasodilatation (Braunwald, 1994; Carmignani et al., 1996; Nechay, 1984; Willebucher et al., 1992). cAMP is a second messenger for the cardiovascular effects following activation of the peripheral β-adrenoreceptors (Braunwald, 1994; Lefkowitz et al., 1996).

Both vanadate and vanadyl inhibit, with different mechanisms, Na^+,K^+-ATPase activity. Inhibition by vanadate requires a divalent cation, such as Mg^{++}, and is enhanced by K^+ and antagonized by Na^+. Other enzymes are inhibited by vanadate including Ca^{++}-ATPase of the sarcoplasmic reticulum and plasma membrane (Jandhyala and Hom, 1983; Nechay, 1984).

The inhibition of Na^+,K^+-ATPase and Ca^{++}-ATPase by vanadium influences not only CI and vascular tone but also other mechanisms of cardiovascular regulation (e.g., secretion of renin). Vanadate was found to inhibit renin secretion from renal cortical slices. It was suggested that an increase of intracellular Ca^{++} (due to inhibition of both Na^+,K^+-and Ca^{++}-ATPase activities) was responsible for this inhibitory effect of the element (Churchill et al., 1990). However, these effects of vanadium differ from those of ouabain, a specific inhibitor of Na^+,K^+-ATPase activity, or K^+-free fluid in the presence of methoxyverapamil, a blocker of the "slow" (receptor-operated) Ca^{++} channels. In this case, the increase of intracellular Ca^{++} could be ascribed to a depolarization-induced influx of Ca^{++} through voltage-operated Ca^{++} channels or to inhibition of Na^+,K^+-ATPase activity, followed by both an increase in intracellular Na^+ and a decrease in Na^+/Ca^{++} exchange (Churchill et al., 1990; Hansen, 1979; Nechay, 1984).

Vanadate was also found to act on myocardial troponin: Ca^{++}-dependent regulation of tension and Na^+,K^+-ATPase activity in permeabilized porcine ventricular muscle were lost following incubation with vanadate; myocell contractility was in part restored after transfer of the muscle to a vanadate-free solution and was completely restored after incubation with troponin or troponin subunits (Strauss et al., 1992).

A series of metabolic effects of vanadate may be related to its ability to stimulate the oxidation of NADPH by glutathione reductase and other flavoprotein dehydrogenases with consumption of O_2 and formation of free radicals (Kalyani et al., 1992).

In vitro studies on cultured mouse embryo fibroblasts (BALB/3T3 cells) showed that vanadium(V) is cytotoxic and induces morphological neoplastic transformation, while vanadium(IV) has no effect (Sabbioni et al., 1991). Thus, the reduction of vanadium(V) to vanadium(IV) by glutathione observed in these cells would represent a physiological mechanism of detoxication of the toxic pentavalent species (Sabbioni et al., 1993). The key role of a glutathione-dependent mechanism in the reduction of the toxic vanadium(V) was also supported by studies in rats, in which the intensity and duration of renal vanadium(V) effects were found dependent on the glutathione reduction to vanadium(IV) (Kretzschmar and Braunlich, 1990).

4.3. Laboratory Animals (Acute Exposure)

The effects of the administration of vanadate on CI of living animals were dependent on the dose, experimental conditions, and animal species (Akera et al., 1983; Jandhyala and Hom 1983; Lopez-Novoa and Garrido, 1986).

In anesthetized cats and dogs, vanadate, infused either intravenously or by direct injection into the renal artery, induced vasoconstriction, increase of both BP and arterial vascular resistance, and reduction of renal blood flow and diuresis (Benabe et al., 1984; Inciarte et al., 1980; Larsen and Thomsen, 1980). The effects of vasoconstriction were reduced by decreasing the concentration of Ca^{++} in the blood perfusing the kidney. By contrast, the effects of vanadate were less intense in the rats, in which renal blood flow and diuresis were not reduced. The effects of vanadate were additive to those of other diuretics such as furosemide and chlorothiazide (Benabe et al., 1984; Day et al., 1980; Nechay et al., 1985; Dafnis et al., 1992).

In another study on conscious dogs sodium metavanadate, given by intravenous infusion, increased systemic BP, pulmonary arterial pressure, TPR, and CO, while it suppressed plasma renin activity (PRA). The effects of vanadate on systemic and pulmonary BP were attenuated after administration of the Ca^{++} antagonist verapamil; however, CO and PRA were unchanged, demonstrating that the cardiovascular modifications induced by vanadate were in part related to an increase of intracellular Ca^{++} (Sundet et al., 1984). Moreover, in another study on dogs the concomitant infusion of vanadate and ouabain in the renal artery induced marked natriuresis (Cruz-Soto et al., 1984).

4.4. Laboratory Animals (Chronic Exposure; New Studies)

A study on exposure via drinking water of male Sprague-Dawley rats to 100 μg/mL of vanadium, as $NaVO_3$, for several months showed increased systolic BP (Steffen et al., 1981). On this basis, we carried out new experiments on the same strain of rats as well as rabbits exposed to different doses of vanadate for a period ranging from 7 to 12 months.

4.4.1. Experimental Procedures

Rats received 0 or 100 ppm and 10, or 40 ppm of vanadium (as $NaVO_3$) via drinking water ad libitum for 7 or 10 months starting from weaning; other rats of the same sex and strain were given 0 or 1 ppm of $NaVO_3$ in drinking water for 7 months. Twelve 3-month-old grey-brown male rabbits, randomly divided in two groups of six animals, were given 0 or 1 ppm of $NaVO_3$ via drinking water for 12 months. The animals were housed in stainless steel cages and kept on a standard laboratory diet and in controlled environmental conditions.

At the end of the experiment, 24-h urine was collected in metabolic cages, while blood samples were obtained under thiopental anesthesia. Urine creatinine, total nitrogen, proteins, and electrolytes as well as the urinary activity of kallikrein, kininase I and II, and enkephalinase were determined (Boscolo et al., 1994). PRA and aldosterone level were determined by standard techniques (Boscolo et al., 1994), while catecholamines (NA, A, DA, and L-DOPA) (Carmignani et al., 1996) and NO (as L-citrulline) (Volpe et al., 1996) were determined by original HPLC methods able to detect femtomolar concentrations.

After cardiovascular measurements, rats were killed by decapitation. Tissues were then excised for histopathological and histochemical examination (Pearse, 1972), and for determining the content of vanadium by neutron activation analysis and/or atomic absorption spectrophotometry (Sabbioni et al., 1992). Brains of rats were also frozen for determining catecholamine levels in the hypothalamus, while heart and kidney from the rats treated with 40 and 10 ppm of vanadium were processed, following preparation of samples in a hydrogen atmosphere, by electron paramagnetic resonance spectroscopy for determining the oxidation status of vanadium (Carmignani et al., 1992a).

4.4.2. Cardiovascular Measurements

Systolic, diastolic, and mean aortic BP, HR, and dP/dt were monitored polygraphically in the rats under thiopental anesthesia; in the rabbits, aortic blood flow, SV, CO, and aortic vascular resistance were also measured, as previously described (Boscolo and Carmignani, 1986; Carmignani et al., 1996).

Bilateral carotid occlusion (BCO) and vagotomy at the neck below the nodose ganglion were performed before injecting the following substances

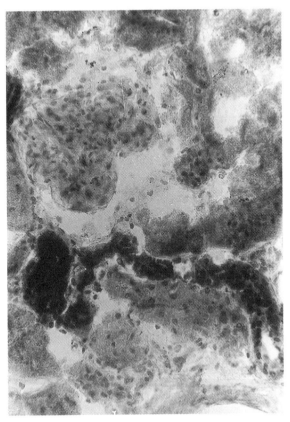

A

Figure 2. Na^+, K^+-ATPase in the kidney of a control rat (**A**) and of a rat exposed to 40 ppm of vanadate in drinking water for 7 months (**B**). In the control rat, Na^+/K^+-ATPase activity was higher in distal tubules than in proximal tubules and glomerulus; in the exposed rat, the enzymatic activity was reduced in all the examined structures. Light microscopy, original $\times$ 400.

(μg/kg body weight) by intravenous route: bradykinin (1, 11.5), ile-ser-bradykinin (T-kinin; 11.5), leu[5]- and met[5]-enkephalins (100), angiotensin I and II (0.50), phenylephrine (an α_1-adrenoreceptor agonist; 10), histamine (10), serotonin (10), isoprenaline (a $\beta_{1,2}$-adrenoreceptor agonist; 0.25), dibu-tyryl cAMP (dcAMP, able to cross plasma membranes; 5,000), veratrine (acti-vating baroreceptors; 10), verapamil (a C^{++} antagonist; 100), hexamethonium (a ganglioplegic drug; 625, 1,250), NA (1), DA (0.125, 1) and Ach (2.50). Tyramine (releasing the labile pool of NA from postganglionic sympathetic endings; 250), ouabain (a specific inhibitor of Na^+,K^+-ATPase; 20) and cloni-dine (an α_2-adrenoreceptor agonist; 10) were also administered by intravenous injection. The doses were expressed in terms of free bases, and peak values were considered. Each consecutive test was not performed until all parameters had spontaneously returned to the basal values and remained stable. Data

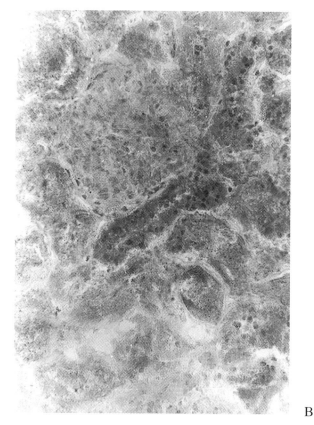

B

Figure 2. (*Continued*)

were compared by analysis of variance and/or Dunnet *t* test for multiple comparison. Only a *p* value less than 0.05 was considered to be significant.

4.4.3.   Histopathological Study

In the rats exposed to 100 ppm of vanadium, examination by light microscopy did not evidence cardiac or vascular alterations. In the kidney, the lumen of proximal tubule cells was narrowed and contained amorphous protein material; electron microscopy (EM) examination of these cells evidenced many swollen mitochondria and spheroidal electron-dense structures in the cytoplasm (Carmignani et al., 1991). Alterations were not evident, in the heart, by EM examination or, in the other organs, by light microscopy observation.

In the rats exposed to 40 ppm of vanadium the lumen of proximal tubules was narrowed and contained amorphous material. Such abnormalities were less evident in the rats treated with 10 ppm of vanadium and not significant in rats and rabbits exposed to 1 ppm of vanadium (Boscolo et al., 1994).

In the control rats, Na^+,K^+-ATPase activity was higher in distal tubules of the renal cortex (Fig. 2) and in straight tubules of the renal medulla; in the rats exposed to 40 ppm of vanadium, Na^+,K^+-ATPase activity was reduced. The decrease in enzymatic activity was more evident in nephrons with morphological alterations.

In the rabbits exposed to 1 ppm of vanadium, renal and liver monoaminooxidase (MAO) and NADH-diaphorase activities were augmented, while those of glucose-6-phosphate dehydrogenase (G6PDH) and NADPH-diaphorase were reduced in relation to the control animals (Carmignani et al., 1995, 1996).

4.4.4.   *Cardiovascular Parameters (BP, HR, and CI)*

The rats chronically exposed to 100 ppm of vanadium showed, as compared with the control group, increased systolic and diastolic BP and HR, while CI (dP/dt) was unchanged (Table 2). The rats chronically exposed to 40, 10, or 1 ppm of vanadium showed increased BP without a dose-dependent effect and normal HR and CI (Table 2).

The rabbits exposed for 1 year to 1 ppm of vanadium did not present significant changes of BP (in relation to the controls); their CI was reduced and HR unchanged (Table 2). Systolic and diastolic BP of the exposed rabbits were not modified since, in these animals, the increase in vascular resistance was counteracted by reduction of CI. CI is negatively related to cardiac after-

Table 2   Blood Pressure, Heart Rate, and Maximum Rate of Rise of Left Ventricular Isovolumetric Pressure (dP/dt) in Rats and Rabbits Chronically Exposed to Different Doses of Vanadium in Drinking Water[a]

	Blood Pressure (mmHg)		Heart Rate (beats/min)	dP/dt (mmHg/sec)
	Systolic	Diastolic		
Rats exposed to 0 or 100 ppm of V (for 7 months)				
Control	122 ± 3	95 ± 3	239 ± 4	$5,063 \pm 431$
100 ppm	$144 \pm 4*$	$115 \pm 5*$	$288 \pm 6*$	$4,588 \pm 426$
Rats exposed to 0, 40, or 10 ppm of V (for 7 months)				
Control	106 ± 7	85 ± 5	356 ± 12	$4,208 \pm 364$
40 ppm	$132 \pm 4*$	$114 \pm 7*$	365 ± 21	$4,648 \pm 323$
10 ppm	$137 \pm 5*$	$112 \pm 5*$	341 ± 18	$4,601 \pm 239$
Rats exposed to 0 or 1 ppm of V (for 7 months)				
Control	108 ± 5	84 ± 4	353 ± 8	$4,550 \pm 624$
1 ppm	$130 \pm 4*$	$106 \pm 3*$	348 ± 9	$4,732 \pm 521$
Rabbits exposed to 0 or 1 ppm of V (for 12 months)				
Control	115 ± 6	68 ± 4	211 ± 8	$2,226 \pm 74$
1 ppm	111 ± 5	69 ± 4	201 ± 6	$1,645 \pm 43*$

[a] Values are means $\pm$ *SEM* ($n = 6$ in each group).
* $p < 0.05$.

load, which is mainly determined by the vascular resistance. Moreover, in the exposed rabbits there was reduction of CO, SV, and aortic blood flow (Carmignani et al., 1996).

4.4.5. *Renin–Angiotensin–Aldosterone System*

PRA was augmented in the rats chronically treated with 40 or 10 ppm of vanadium, while it was unchanged in rats that received 1 ppm of vanadium (Table 3) (Boscolo et al., 1994). This datum is not in agreement with the findings that vanadate inhibits renin secretion from renal cortical slices (Churchill et al., 1990). Moreover, in another study on conscious dogs $NaVO_3$, given by intravenous infusion, suppressed PRA (Sundet et al., 1984). The increase of PRA in the rats chronically treated with 40 or 10 ppm of vanadate was referred to both slight morphological alterations and inhibition of Na^+,K^+-ATPase activity (in macula densa and distal tubule cells), which are able to modify metabolism and excretion of electrolytes. In fact, in the rats exposed to 40 or 10 ppm of vanadium, potassium excretion was augmented (Boscolo et al., 1994).

The cardiovascular responses to angiotensin I and II were unchanged in the rats exposed to 40, 10, or 1 ppm of vanadium, demonstrating that the reactivity of cardiovascular angiotensin receptors was not affected by vanadium exposure (Boscolo et al., 1994; Carmignani et al., 1992b).

Plasma aldosterone was augmented in the plasma of rats treated with 10 ppm of vanadium, unchanged in those exposed to 40 ppm, and reduced in those treated with 1 ppm of vanadium; plasma aldosterone was unchanged in vanadium-exposed rabbits (Table 3). The effect of vanadium on plasma aldosterone may in part depend on alterations of electrolyte metabolism in tubular cells and in part on modifications of PRA and kallikrein activity, which, by counteracting the reabsorption of Na^+ (potentiated by aldosterone) in the distal tubules, are linked to this hormone by a homeostatic mechanism (Scicli and Carrettero, 1986).

4.4.6. *Urinary Kallikrein, Kininase I and II, and Enkephalinase Activities*

In the rats exposed to 40 or 10 ppm of vanadium, urinary kallikrein activity was augmented, while it was reduced in both rats and rabbits exposed to 1 ppm of vanadium (Table 3). As reported above, kallikrein activity may be in part related to that of aldosterone, which counteracts the natriuretic effect of kallikrein (Scicli and Carrettero, 1986).

Urinary kininase I and kininase II activities were augmented in the rats exposed to 40 or 10 ppm of vanadium, with a dose-response effect, and in the rabbits treated with 1 ppm of vanadium, while they were unchanged in the rats treated with 1 ppm of vanadium. In this respect, the effects of vanadium on the kallikrein and kininase activities of rats and rabbits may be considered equivalent, also considering that the rabbits received the dose of 1 ppm of vanadium for a longer period (12 months instead of 7 months) (Table 3). The

Table 3 Plasma Renin Activity (PRA) and Aldosterone Levels and Urinary Kallikrein, Kininase I and II, and Enkephalinase Activities in Rats and Rabbits Chronically Exposed to Different Doses of Vanadate in Drinking Water[a]

	PRA	Plasma Aldosterone	Urine Kallikrein	Urine Kininase I	Urine Kininase II	Urine Enkephalinase
Rats exposed to 0, 40, or 10 ppm of V (for 7 months)						
40 ppm	Increased	Unchanged	Increased	Increased	Increased	Increased
10 ppm	Increased	Increased	Increased	Increased	Increased	Increased
Rats exposed to 0 or 1 ppm of V (for 7 months)						
1 ppm	Unchanged	Reduced	Reduced	Unchanged	Unchanged	Increased
Rabbits exposed to 0 or 1 ppm of V (for 12 months)						
1 ppm	—	Unchanged	Reduced	Increased	Increased	Increased

[a] Increase or reduction were significant at p values <0.05 with respect to controls ($n = 6$ in each group).

198

alteration of the kallikrein–kinin system in vanadium-exposed rats was also demonstrated by the increased cardiovascular responsiveness to intravenous bradykinin (reported in Section 4.4.10). Urinary enkephalinase activity was also augmented in both rats treated with 40, 10, or 1 ppm of vanadium, without a dose-response effect, and in rabbits exposed to 1 ppm of vanadium (Table 3). The increased cardiovascular responsiveness to intravenous enkephalins in vanadium-exposed rats (also reported in Section 4.4.10) may then be explained through an up-regulation of the peripheral opioid receptors, depending on increased destruction of circulating and tissue enkephalins by enkephalinase.

4.4.7. Urinary Electrolytes

In the rats treated with 40, 10, or 1 ppm (Boscolo et al., 1994) of vanadium as well as in the rabbits treated with 1 ppm of vanadium (Carmignani et al., 1996) urinary excretion of creatinine, total nitrogen, proteins, and sodium were unchanged in relation to the controls. In contrast, urine potassium was augmented in the rats exposed to 40 or 10 ppm of vanadium with a dose-response effect (Boscolo et al., 1994). Urinary calcium was unchanged in the rats treated with 40 or 10 ppm of vanadium and reduced in those treated with 1 ppm. Urinary Na^+ and K^+ were in the normal range in the vanadium-exposed rabbits, in which, however, the urinary sodium/potassium ratio was increased. The modifications of K^+ excretion in the rats exposed to 10 or 40 ppm of vanadium may be related to the slight histopathological and histochemical alterations of renal tubules, while the reduction of urinary Ca^{++} in the rats exposed to 1 ppm of vanadium may be related not only to an alteration of renal Ca^{++} metabolism but also to other metabolic disorders (including a possible reduced absorption of Ca^{++} in the gastrointestinal tract) (Boscolo et al., 1994; Carmignani et al., 1996).

4.4.8. Barosensitivity and Vagal Activity

Both hypertensive and inotropic responses to bilateral carotid occlusion (BCO) were reduced in all the groups of rats exposed to vanadium (Tables 4 and 5).

The hypotensive response to veratrine, a drug that activates baroreceptors, was reduced in the animals treated with 100, 40, or 10 ppm of vanadium (Table 4) and unchanged in those treated with 1 ppm of vanadium (Carmignani et al., 1992b).

The pressor responses to vagotomy were also reduced in the rats treated with 40 and 10 ppm of vanadium and unchanged in those treated with 1 ppm of vanadium (Tables 4 and 5). In all the groups of vanadium-exposed rats, cardiovascular responses to Ach were unchanged.

It seems evident that vanadium exposure reduces the baroreactivity and vagal tone in rats with a dose-dependent effect.

Table 4 Cardiovascular Responses to Bilateral Carotid Occlusion (BCO), Vagotomy, and Drugs (iv Route; $\mu g/kg$) in Rats Chronically Exposed to 0 or 100 ppm or 0, 10, and 40 ppm of Vanadate in Drinking Water for 7 Months[a]

Drugs or Experimental Procedures	Mean Blood Pressure Response		Inotropic Response	
	Control	Exposed	Control	Exposed
Rats exposed to 0 or 100 ppm of V				
BCO	Increased	Less than control	Increased	Less than control
Vagotomy	Increased	Less than control	Increased	Same as control
Veratrine (10)	Reduced	Less than control	Reduced	Less than control
Bradykinin (1)	Reduced	Reversed	Increased	More than control
Isoprenaline (0.25)	Reduced	More than control	Increased	More than control
Tyramine (250)	Increased	More than control	Increased	Same as control
Hexamethonium (625)	Reduced	More than control	Reduced	More than control
Clonidine (10)	Increased[b]	More than control	Reduced	Less than control
Ouabain (10)	Increased	Same as control	Increased	Less than control
Phenylephrine (10)	Increased	Same as control	Increased	Same as control
Angiotensin II (0.5)	Increased	Same as control	Increased	Same as control
Verapamil (100)	Reduced	More than control	Reduced	Same as control
Rats exposed to 0, 10, or 40 ppm of V				
BCO	Increased	Less than control	Increased	Less than control
Vagotomy	Increased	Less than control	Increased	Same as control
Veratrine (10)	Reduced	Less than control	Reduced	Same as control
Bradykinin (1)	Reduced	More than control	Increased	More than control
Isoprenaline (0.25)	Reduced	More than control	Increased	More than control
Tyramine (250)	Increased	More than control	Increased	Same as control
Hexamethonium (625)	Reduced	More than control	Reduced	More than control
Clonidine (10)	Increased[b]	More than control	Reduced	Same as control
Ouabain (10)	Increased	Same as control	Increased	Same as control
Phenylephrine (10)	Increased	Same as control	Increased	Same as control
Angiotensin II (0.5)	Increased	Same as control	Increased	Same as control
Verapamil (100)	Reduced	Same as control	Reduced	Same as control

[a] Less or more than control was significant at p values < 0.05 ($n = 6$ in each group).

[b] Initial pressor response depending on activation of the peripheral α_2-adrenoreceptors.

Table 5   Cardiovascular Responses to Bilateral Carotid Occlusion (BCO), Vagotomy, and Drugs (Given by iv Route; μg/kg) of Rats Exposed to 0 or 1 ppm of Vanadate in Drinking Water for 7 Months[a]

Drugs or Experimental Procedures	Mean Blood Pressure (ΔmmHg)		dP/dt (ΔmmHg/sec × 10⁻¹)	
	Control	Exposed	Control	Exposed
BCO	+32 ± 3	+23 ± 1*	+264 ± 38	+160 ± 16*
Vagotomy	+14 ± 2	+17 ± 2	+167 ± 24	+147 ± 19
Bradykinin (11.5)	−28 ± 4	−44 ± 1*	+652 ± 74	+703 ± 60
Isoprenaline (0.25)	−18 ± 3	−31 ± 2*	+749 ± 37	+809 ± 42
Leu⁵-enkephalin (100)	−7 ± 1	−16 ± 2*	−126 ± 41	−89 ± 19
Met⁵-enkephalin (100)	−13 ± 2	−27 ± 3*	−135 ± 31	−86 ± 27
Tyramine (250)	+22 ± 1	+37 ± 4*	−173 ± 18	−201 ± 21
Hexamethonium (625)	−34 ± 4	−19 ± 2*	−294 ± 30	−193 ± 21*
Clonidine[b] (10)	+12 ± 3	+24 ± 1*	−193 ± 23	−171 ± 12
Ouabain (20)	+11 ± 1	+13 ± 2	+91 ± 8	+195 ± 16*
Angiotensin II (0.5)	+20 ± 3	+27 ± 5	−272 ± 27	−240 ± 24

[a] Values are means ± SEM ($n = 10$ in each group).
[b] The initial hypertensive response (depending on activation of the peripheral α_2-adrenoreceptors) was followed by long-lasting hypotension.
* $p = 0.05$.

4.4.9. Sympathetic Tone and Catecholaminergic Mechanisms

The central sympathetic tone, studied by the cardiovascular responses to the ganglioplegic drug hexamethonium (blocking neurotransmission in the sympathetic ganglions through a specific antagonistic effect on the "nicotinic" receptors to Ach), was augmented in the rats exposed to 100, 40, and 10 ppm of vanadium and reduced in those treated with 1 ppm of vanadium (Tables 4 and 5). This apparent contrast may be explained by the complex neurogenic effects of vanadium within the central and peripheral nervous systems, including those on the baroreflex and vagal mechanisms. In this regard, higher concentrations (ng/mg protein) of adrenaline (exposed, 3.94 ± 0.63; control, 2.22 ± 0.18; $p < 0.05$) were found in the hypothalamus of the rats exposed to 40 ppm of vanadium but not in those treated with 10 ppm of vanadium; in the animals treated with 40 ppm of vanadium, the hypothalamic levels of DA were reduced (exposed, 0.87 ± 0.10; control, 1.57 ± 00.07; $p < 0.05$) and those of NA unchanged (means ± SEM; $n = 6$ in each group).

Plasma A, NA, DA, and L-DOPA were greatly augmented in the rats exposed to 1 ppm of vanadium (Fig. 3) as well as in the vanadium-exposed rabbits (Carmignani et al., 1996). These data were explained by a direct stimulation by vanadium of the catecholamine release from postganglionic sympathetic nerve endings and/or pheochromocytes. This was confirmed by

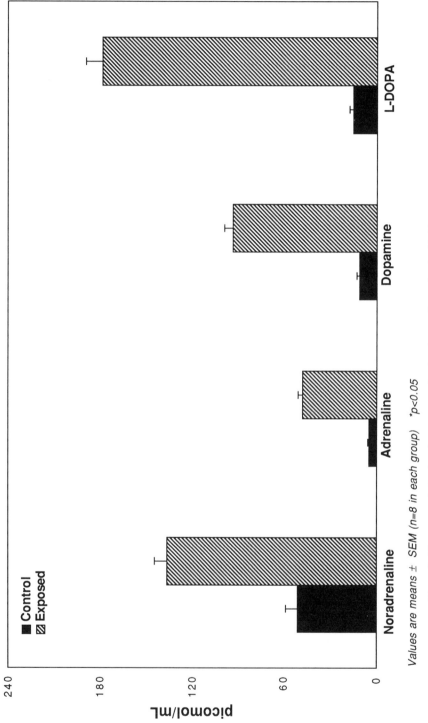

Figure 3. Plasma catecholamines in rats exposed to 0 or 1 ppm of vanadate in drinking water for 7 months.

Values are means ± SEM (n=8 in each group) *p<0.05*

the increased pressor responses to tyramine in all the groups of vanadium-exposed rats (Tables 4 and 5) and in vanadium-exposed rabbits (Carmignani et al., 1996). Such effects of tyramine were not dependent on nerve activity, since they were observed in the presence of reduced (rats exposed to 1 ppm of vanadium) or increased (rats exposed to 100, 40, or 10 ppm of vanadium) central sympathetic tone. In the rabbits exposed to 1 ppm of vanadium, MAO activity was augmented both in kidney (Fig. 4) and liver. The enhanced MAO activity, observed by histochemical methods, may be secondary to the observed increase of catecholamines in plasma of the same animals.

Rats exposed to 100, 40, 10, or 1 ppm of vanadium showed increased cardiovascular responses to isoprenaline and to clonidine (a central and peripheral α_2-adrenoreceptor agonist) but not to phenylephrine; however, the inotropic response to clonidine was reduced in the rats treated with 100 ppm of vanadium (Table 4 and 5). The cardiovascular effects following activation of the β-adrenoreceptors were related to an augmented production, in cardiac and vascular myocells, of cAMP by the vanadium-stimulated adenylate cyclase activity; cAMP regulates the availability of Ca^{++} by different mechanisms in myocardiocytes, the cardiac conduction system, and vascular myocells, thus causing increase in CI and HR, and arterial vasodilation (Willebucher et al., 1992). The effect of vanadium on cAMP was also confirmed by the increased cardiovascular responses observed following intravenous administration of dcAMP in the rats exposed to 40 or 10 ppm of vanadium (Carmignani et al., 1992b).

4.4.10 Cardiovascular Responses to Bradykinin and Other Physiological Agonists

In the rats exposed to 100 ppm of vanadium, the hypotensive response to bradykinin was reversed into a hypertensive response; this fact may be explained by the higher activation by this kinin of its cardiac BK_2 receptors, with the consequence of an increase of CI to levels overcoming the vasodilatation effects due to activation of the vascular BK_1 receptors (Table 4). Moreover, both hypotensive and inotropic responses to bradykinin were potentiated by vanadium in the rats treated with 40 or 10 ppm of vanadium, while only the hypotensive response was increased in rats treated with 1 ppm of vanadium (Table 4 and 5). These data are important considering that bradykinin is an autacoidal substance that plays a key role in the kallikrein–kinin (Scicli and Carretero, 1986) and NO systems (Volpe et al., 1996).

Both hypotensive and negative inotropic responses to leu[5]- and met[5]-enkephalins were potentiated by vanadium in the rats exposed to 40 or 10 ppm of vanadium (Carmignani et al., 1991, 1992b), while only the hypotensive response was increased in the rats treated with 1 ppm of vanadium (Table 5). It must be taken into account that the metabolism of enkephalins, as that of kinins, is greatly influenced by vanadium exposure, since the activities

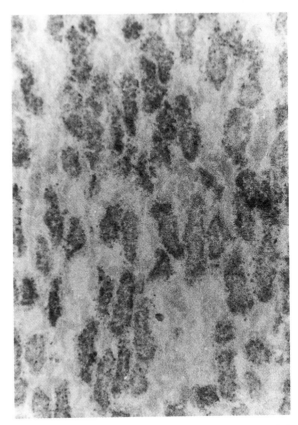

A

Figure 4. Monoamine oxidase activity in the renal medulla of a control rabbit (**A**) rabbit exposed to 1 ppm of vanadate in drinking water for 12 months (**B**). The enzymatic activity is increased in the tubules of the exposed animals. Light microscopy, original × 100.

of enkephalinases and kininases are increased in vanadium-exposed animals (Carmignani et al., 1995, 1996).

The cardiovascular responses to serotonin, histamine, Ach, and DA were not modified in the rats exposed to vanadium.

4.4.11. Cardiovascular Responses to Verapamil and Ouabain

The hypotensive response to verapamil was increased only in the rats exposed to 100 ppm of vanadium.

The positive inotropic response to ouabain was reduced in the rats treated with 100 ppm of vanadium and increased in those treated with 1 ppm of vanadium, without modifications of the pressor response (Tables 4 and 5). The rabbits exposed to 1 ppm of vanadium also presented an increase of the inotropic response to ouabain in relation to controls (Carmignani et al., 1995).

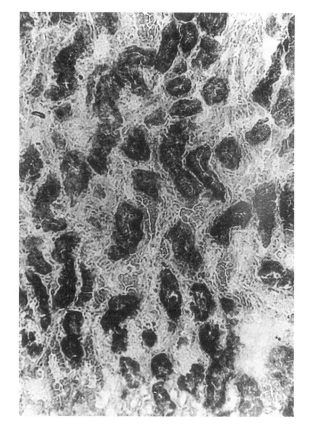

B

Figure 4. (*Continued*)

4.4.12. Plasma NO and Tissue NADH- and NADPH-Diaphorases and G6PDH Activities

Plasma levels of NO were reduced in the rabbits exposed to 1 ppm of vanadium (Carmignani et al., 1996). NADPH diaphorase (this enzyme is colocalized with NO synthase, of which it represents an indirect index of activity) was reduced in tubules and glomeruli of the vanadium-exposed rabbits, while NADH diaphorase activity was augmented at the same levels; also renal G6PDH activity was reduced in vanadium-exposed rabbits (Carmignani et al., 1995, 1996). The activity of this enzyme in macula densa and distal tubule cells was already related to the synthesis of renin in juxtaglomerular cells (Bhoola et al., 1992).

4.4.13. Discussion of New Data

The results obtained in rats and rabbits show that chronic exposure to vanadium is able to increase TPR and, although only at high doses (in our experiments, 100 ppm in rats), also HR, thus leading to arterial hypertension. Species differences (the rabbit, in our study) may explain the lack of systemic hypertension owing to a contemporary negative inotropic effect of vanadium, with consequent reduction of SV and CO. In this respect, it was found that vanadium (as vanadyl) may have opposite or biphasic effects on CI by complex interactions with a lot of cellular mechanisms that regulate the availability of free calcium for contractile processes (Akera et al., 1983; Carmignani et al., 1991, 1992a, 1992b; Jandhyala and Hom, 1983; Nechay, 1984). Among these mechanisms, inhibition of Na^+,K^+-ATPase (sodium pump) and stimulation of adenylate cyclase as well as activation of a series of receptors to physiological agonists by vanadium were shown to play a secondary role, whereas type of cardiac muscle (atrial, ventricular, papillar), experimental conditions (in vivo, ex vivo), doses, or concentrations of vanadium and/or animal species represented primary variables (Akera et al., 1983; Nechay, 1984; Vollkel and Czartolomna, 1991; Willebucher et al., 1992). Therefore, it was not surprising that the inotropic responses to ouabain were increased in both rats and rabbits exposed to 1 ppm of vanadium and reduced in the rats treated with 100 ppm of the metal. Analogously, considering that the isoprenaline-induced increase of cAMP in vascular (β_2-effect) and cardiac myocells (β_1-effect) leads to opposite effects (relaxation and contraction, respectively) (Marmé, 1987), this study has shown that vanadium potentiates the inotropic effect of that agonist in rabbits, the vascular effect in the rats exposed to the lowest dose of vanadium, and both effects in the rats treated with higher doses of the element (similar results were also obtained with dcAMP).

In all the exposed animals, vanadium interacted with central and peripheral catecholaminergic mechanisms. Vanadium stimulated a tyramine-like or -sensitive release of catecholamines from postganglionic sympathetic endings and, probably, pheochromocytes. This effect was independent of nerve activity, since it was observed in the presence of unchanged (rabbits), reduced (rats exposed to 1–40 ppm of vanadium), or increased sympathetic tone (rats exposed to 100 ppm of vanadium), as indicated by the cardiovascular responses to hexamethonium (Carmignani et al., 1989, 1991, 1992a). In this regard, it is likely that vanadium reduces or enhances sympathetic tone by changing the levels of some catecholamines (A and DA, in this study) in central areas (e.g., the hypothalamus); these levels might be altered differently depending on doses of exposure to vanadium, thus causing sympathetic hypotone or hypertone (Sharma et al., 1985). Conversely, vanadium reduced baroreflex reactivity and vagal tone in all the vanadium-treated animals and, already at the 1 ppm level of exposure, raised plasma catecholamines (through a tyramine-like effect) (Carmignani et al., 1989, 1992a; Marmé, 1987). The enhanced MAO activity, verified in the kidney, may be secondary to the increased catecholamine levels; however, a direct effect of vanadium cannot be excluded. Moreover, besides the capacity to potentiate cardiovascular responses following activation of the peripheral β_1- and β_2-adrenoreceptors, vanadium was found to constantly potentiate the vascular α_2-adrenoreceptor-dependent responses to clonidine. Since activation of the α_2-adrenoreceptors promotes contraction in vascular myocells by opening the "slow" (receptor-operated) Ca^{++} channels in plasma membrane (Akera et al., 1983; Marmé, 1987; Willebucher, 1992), the higher vasodilating effects of verapamil (blocking these channels on the external side) in vanadium-treated rats seemed to confirm that this element enhances the influx of Ca^{++} into vascular smooth myocells (Lefkowitz et al., 1996; Pirkle et al., 1985). On the other hand, the reflex reduction of the negative inotropic response to clonidine in the 100 ppm vanadium-treated rats may be related to the lower vagal activity and/or to higher sympathetic outflow (Carmignani et al., 1989, 1992b).

This study suggests that chronic exposure to vanadium affects the contractility of vascular and cardiac myocells also by acting on the kallikrein–kinin, renin–angiotensin–aldosterone, and enkephalin systems (Carmignani et al., 1991, 1992a, 1993). At the 1-ppm level, vanadium reduced the activity of kallikrein, while increasing that of enkephalinase and, only in the rabbits, also those of kininase I and II (in lower degree); at higher doses (10–100 ppm), vanadium augmented all these enzymatic activities, mostly those of kininase I and II (K1 > K2). Such effects may explain the vanadium-induced potentiation of the cardiovascular responses to enkephalins and bradykinin in the animals exposed to 1, 10, or 40 ppm of the metal; the reversal of the hypotensive effect of bradykinin, in the 100 ppm vanadium-exposed rats, might be related to the strong increase (more than 30 times) of the kininase I activity, causing higher activation of the cardiac BK_2 receptors (with positive inotropic effects able to increase CO to levels overcoming those of the reduced vascular resistance due to the vasodilating BK_1 effect of bradykinin) (Carmignani et al., 1992a, 1993). Moreover, the above results raise the possibility that there is a threshold effect (2–10 ppm) such that vanadium may increase PRA and the activities of kininase I and II and kallikrein, as well as K^+ excretion (Boscolo et al., 1994). In the range of 1–40 ppm of vanadium, there was good correlation between the changes in activities of kininase I and II, and of PRA. On the other hand, a partial relation existed between the activity of these enzymes and that of kallikrein and between urinary kallikrein activity and plasma aldosterone (Boscolo et al., 1994; Carmignani et al., 1992a). The results of these studies do not show whether all these effects of vanadium are the result of a direct action of the metal or whether they are secondary, although the biochemical and functional relationships among the kallikrein–kinin, renin–angiotensin–aldosterone, enkephalin, and prostaglandin systems are well known (for review, see Bhoola et al., 1992). However, by contrast with the inhibition of synthesis or release of renin "in vitro" (Churchill et al., 1990) and in dogs (Sundet et al., 1984), chronic exposure to 10 or 40 ppm of vanadium augmented PRA in rats, likely through an increased renin release related to the observed inhibition by vanadium of Na^+,K^+-ATPase in the cells of the macula densa and of the distal tubule (Boscolo et al., 1994). Also the observed reduction of G6PDH activity in the kidney of vanadium-exposed rabbits may be related to a decreased activity of the local renin–angiotensin–aldosterone system and/or to a reduced synthesis and/or release of renal kallikrein, as was shown in the same rabbits. In general, the activity of G6PDH was found to be related, in the kidney, to the local homeostasic equilibrium between the renin–angiotensin–aldosterone and kallikrein–kinin systems through a mechanism previously described (Bhoola et al., 1992; Iannaccone et al., 1978; Scicli and Carretero, 1986).

Even if higher levels of vanadium (as vanadyl) were found in the kidney of control and vanadium-exposed animals, relatively high levels of the metal were found also in arterial vessels (aorta), heart, and central nervous system. The accumulation in such tissues is likely to be in part responsible, through local mechanisms, of the cardiovascular effects of vanadium. In this regard, vanadium reduced synthesis and/or release of NO, the endothelium-derived vasodilating factor generated from L-arginine (transformed in L-citrulline) by NO synthase (Moncada and Higgs, 1993). A lot of vasodilating and vasoconstrictor agonists (including bradykinin, Ach, and catecholamines) act by stimulating or inhibiting, respectively, synthesis and/or release of NO (Moncada et al., 1991), and our data indicate that intracellular arginine-containing peptides (including bradykinin) are physiological precursors of NO (Carmignani and Volpe, 1996; Volpe et al., 1996). Therefore, from the evidence obtained in the rabbits exposed to vanadium it may be suggested that this element increases arterial resistance by reducing production of NO from bradykinin. However, the vanadium-induced reduction of plasma NO is also a result of positive and negative modulations brought on by agonists (including enkephalins and catecholamines) whose blood concentrations are changed by vanadium (Moncada et al., 1991). In this respect, the observed effects of vanadium on the activities of renal NADPH and NADH diaphorases add further evidence. NADPH is a cofactor of NO synthase in the reaction transforming L-arginine in L-citrulline through formation of the intermediate N^{ω}-hydroxyarginine. Moreover, NADPH diaphorase is colocalized and copurified with NO synthase, of which it represents an indirect index of both activity and release (Moncada et al., 1991; Moncada and Higgs, 1993). It may then be deduced that vanadium reduces formation of L-citrulline and NO (produced physiologically in equimolar amounts) because of reduced availability of NADPH, with the consequent decrease of its diaphorasic oxidation. The concomitant decreased activity of G6PDH in the vanadium-exposed rabbits might depend on a vanadium-induced minor formation of NADPH in the pentose cycle (Carmignani et al., 1995). On the other hand, the reduced activity of NADH diaphorase in the same rabbits may be explained in terms of biochemical equilibrium and/or reactions of electron transfer. The fact appears to be very significant that vanadium changes the activities of NADPH and NADH diaphorases, G6PDH, and Na^+,K^+-ATPase at the same level in the kidney, that is, distal tubules.

The cardiovascular effects following chronic exposure to vanadium in rats and rabbits were similar to those observed following experimental chronic exposure to cadmium chloride; for instance, cadmium reduced or increased both CI and CO in rabbits and rats, respectively (Boscolo and Carmignani, 1986; Carmignani and Boscolo, 1984).

One may suggest that chronic exposure to vanadium leads to the development of arterial hypertension by prevalently increasing arterial resistance. The effects of vanadium on cardiovascular homeostasis appear to be complex and include specific actions on central neurogenic pathways, central and peripheral catecholaminergic mechanisms, autacoidal systems (kallikrein–kinin, renin–angiotensin–aldosterone, and enkephalin systems), and effectors (vessels and heart). Only in part do the levels and times of exposure and the species represent a discriminant factor for the activation (qualitative or quantitative) of these mechanisms by vanadium.

5. EPICRITIC CONCLUSIONS

Since the role of vanadium as essential element in humans is not yet proved, its possible role in the control of cardiovascular function remains hypothetical (French and Jones, 1993). However, the great deal of literature data (Jandhyala and Hom, 1983; Nechay et al., 1985; Nielsen, 1995) and our experimental results (see also Table 6) make plausible the hypothesis that vanadium (mostly as vanadate and/or vanadyl) plays a role in influencing cardiac function and vascular tone.

Our studies show that the key mechanisms of cardiovascular regulation that could be affected by vanadium (following chronic exposure) are the following:

- Release of peripheral catecholamines independently on sympathetic nerve activity.
- Baroreflex and vagal nerve activities.
- Central sympathetic tone.
- Reactivity of the cardiac and vascular $\beta_{1,2}$- and α_2-adrenoreceptors.
- Peripheral renin–angiotensin–aldosterone, kallikrein–kinin, and enkephalin systems.
- Nitric oxide.

Although vanadium was found to be present almost completely in tetravalent form in tissues of rats exposed to 40 ppm of vanadate [V(V)], one cannot exclude that vanadium toxicity to the cardiovascular system depends also on its pentavalent oxidation state. In this regard, experiments on mouse fibroblasts showed that the persistence of vanadium(V) in the cell, as the consequence of a lack of bioreduction to the less toxic vanadium(IV) by intracellular glutathione, would be responsible for the cytotoxic effects of vanadium. Thus, the mechanisms of toxicity of vanadium on the cardiovascular system may be linked to the observed effects induced by vanadate, which include inhibition of enzymes, alteration of protein phosphorylation, and generation of free

Table 6 Vanadium Content in Animal and Human Cardiac Tissue

Compound	Oxidation State	Species	Exposure Type	Exposure Dose	Exposure Length	Mean V Content in Heart (Wet Weight)	Reference
Dietary V: 2.8 μg/L (water); 78 ng/g (food)	Unknown	Rat	Ingestion (ad libitum)	2.44 μg/kg[a]	21 d 77 d 115 d	29.7 ng/g 8.9 ng/g 6.8 ng/g	Edel et al., 1984
$NaVO_3$	+5	Rat	Per os (ig.a.)	500 ng/kg	1 d	0.1 pg/g	Edel and Sabbioni, 1988
$VOSO_4$	+4	Rat	Per os (ig.a.)	500 ng/kg	1 d	0.1 pg/g	
$NaVO_3$	+5	Rat	Intratracheal	200 ng/kg	3 h 1 d 12 d	0.17 ng/g 0.07 ng/g 0.04 ng/g	
$VOSO_4$	+4	Rat	Intratracheal	200 ng/kg	3 h 1 d	0.13 ng/g 0.04 ng/g	
$NaVO_3$	+5	Rat	Iv.i. (mother)	277 ng/kg	1 d before birth	4.3 pg/g (fetus) 102 pg/g (mother)	Edel and Sabbioni, 1989
$NaVO_3$	+5		Iv.i. (mother)	277 ng/kg	10 d after injection 7 d after lactation	0.69 pg/g (sucking rat) 0.23 pg/g (wealing rat)	
$NaVO_3$	+5	Rat	Iv.i.	28 μg/kg	1 d	8 ng/g	Sabbioni and Marafante, 1978
$VOSO_4$	+4		Iv.i.	28 μg/kg	21 d	4 ng/g	
V^{5+}	+5	Rat	Iv.i.	40 μg/kg	1 d	4.3 ng/g	Sabbioni et al., 1978

Compound	Oxidation state	Species	Route	Dose	Duration	Concentration	Reference
V^{4+}	+4					5.8 ng/g	
V^{3+}	+3					6.7 ng/g	
VS_4^{3-}	+5					8.6 ng/g	
Na_3VO_4	+5	Rat	Drinking water (230 μg/L)	6.9 μg/kg	28 d	0.54 μg/g	
$VOSO_4$	+4		Drinking water (230 μg/L)	6.9 μg/kg	28 d	0.6 μg/g	
$NaVO_3$	+5		Drinking water (150 μg/L)	4.5 μg/kg	28 d		
$NaVO_3$	+5	Rat	Drinking water (100 mg/L)	3 mg/kg[a]	7 m	2.71 μg/g (auricle) 0.55 μg/g (ventricle)	Carmignani et al., 1991, 1992b
					12 m	3.19 μg/g (auricle) 0.66 μg/g (ventricle)	
$NaVO_3$	+5	Rabbit	Drinking water (1 mg/L)	26 μg/kg[b]	1 y	22 μg/kg 20 μg/kg (aorta)	Carmignani et al., 1995, 1996
Dietary V: 2.2 μg/L (water); 85 ng/g (food)	Unknown	Rabbit	Ingestion (ad libitum)	1 μg/kg	3 m	1.2 ng/g	Sabbioni, unpublished observations
Diet	Unknown	Human	Dietary intake	Unknown	—	1.1 ng/g	Byrne and Costa, 1978

[a] Estimated assuming a daily consumption of food and water of 30 g and 35 mL, respectively (Sabbioni et al., 1978).
[b] Estimated assuming a daily consumption of food and water of 200 g and 250 mL, respectively (Carmignani et al., 1995).

211

radicals (Sabbioni et al., 1993). Interestingly, the lack of a direct inhibitory effect of vanadium on the activity of the main antioxidant enzymes suggests that many biological and toxicological effects of vanadium may be mediated more by oxidative reactions of the metal or its complexes with physiologically relevant biomolecules than by a direct modulation of enzymatic activities (Serra et al., 1992). Therefore, the mechanisms of vanadium toxicity to the cardiovascular system may be these:

- Inhibition of enzymes (e.g., Na^+,K^+-ATPase, G6PDH, NADPH diaphorase).
- Activation of enzymes (e.g., MAO, NADH diaphorase).
- Alterations in protein phosphorylation.
- Oxidative reactions.
- Inability of glutathione to reduce vanadium(V).

The results of studies on experimental animals show that "subpharmacological" doses of vanadium (such as that of 1 ppm in drinking water) are very active in altering cardiovascular function. Our preliminary data show that the mean vanadium content in kidney, heart, and aorta of rabbits exposed for 1 year to 1 ppm of vanadium in drinking water were 78 ± 3, 22 ± 2, and 20 ± 1 $\mu g/g$ wet weight, respectively (means $\pm$ *SEM; n* $= 4$); these values were about five times higher than those of the control animals. These levels of vanadium in tissues of rabbits appear to be similar to those of humans with environmental or occupational exposure to this element. However, further studies seem to be necessary to extrapolate the data obtained in experimental animals to humans. Analogously, other studies will be able to better clarify the mechanisms of cardiovascular toxicity of vanadium in addition to the reported actions of enzymatic inhibition or activation of this element. The cardiovascular effects of the "pharmacological" doses of vanadium have in any case to be taken into account whenever the insulinomimetic properties of vanadium would be proposed as therapy for the diabetic patient (Brichard and Henquin, 1995; Halberstam et al., 1996; Orvig et al., 1995).

Although the cardiovascular toxic effects of vanadium were not demonstrated in humans, studies on the cardiovascular effects of vanadium may be useful for explaining alterations of the mechanisms of cardiovascular regulation involved in essential hypertension. It is likely that, when the ethiopathogenesis of this disease ultimately is better explained, such a nosographic entity will be divided into different subsets. One subset may be the hypertension induced by (chronic) environmental exposure to toxic agents. In this regard, studies performed in the United States and Great Britain demonstrated a positive correlation between blood lead and BP levels in the normal population (see Lener et al., this volume, Ch. 1). Moreover, recent observations seem to demonstrate a positive relationship between lead in bones and BP levels in populations living in the United States (Landrigham et al., personal communication, 1996). However, there are no studies on the effects of multiple environmental exposures of the cardiovascular system to toxic agents because of the difficulties of epidemiological investigations and the cost of analyses of metals and chemical compounds in biological samples.

Despite the inadequacy of epidemiological investigations, there are several studies on cardiovascular effects of metals (e.g., lead, cadmium, mercury, arsenic). In some of these studies, control animals were fed a diet with a content of toxic metals lower than that of the human diet (Perry and Erlanger, 1976; Petering, 1974). These groups of animals showed reduced BP in relation to other groups treated with low doses of heavy metals.

We found specific alterations in the regulatory mechanisms of cardiovascular function in rats and rabbits exposed not only to vanadium but also to cadmium (Carmignani and Boscolo, 1984), lead (Boscolo et al., 1992; Carmignani et al., 1987), mercury (Carmignani et al., 1983a, 1992c), and arsenic (Carmignani et al., 1983b). This would suggest that not only environmental lead but also multiple exposure to toxic metals, including vanadium, may alter BP regulating mechanisms of humans with a genetic predisposition to essential hypertension. In particular, interactions among toxic metals (e.g., lead, cadmium, vanadium) appear to be highly synergic in altering selective neurogenic, autacoidal, receptor, and transductional mechanisms regulating cardiac activity and vascular tone (Carmignani and Boscolo, 1984; Carmignani et al., 1987).

ACKNOWLEDGMENTS

This research was supported by grants of Italian MURST (National Projects "Nuovi Approcci Valutativi in Tossicologia" and "Patologia da Tossici Ambientali") and CNR (Coordinate Project "Bersagli Farmacologici nelle Interazioni tra Cellule e Molecole Endogene") to Marco Carmignani and Anna Rita Volpe (1994–1995), and by the CNR grant no. 93.03339 ("Innovazioni Tecnologiche nello Studio dell'Esposizione a Metalli sul Sistema Cardiovascolare") to Paolo Boscolo.

The expert technical assistance of Giuseppe Ripanti, Institute of Pharmacology, Catholic University School of Medicine, Rome, is gratefully acknowledged.

REFERENCES

Akera, T., Temma, K., and Takeda, K. (1983). Cardiac actions of vanadium. *Fed. Proc.* **42,** 2984–2988.

Antonaccio, M. J. (1990). *Cardiovascular Pharmacology.* Raven Press, New York, 556 pp.

Bello-Reuss, E. N., Grady, T. P., and Mazumdar, D. C. (1979). Serum vanadium levels in chronic renal disease. *Ann. Int. Med.* **91,** 743–749.

Benabe, J. E., Cruz-Soto, M. A., and Martinez-Maldonado, M. (1984). Critical role of extracellular calcium in vanadate-induced renal vasoconstriction. *Am J. Physiol.* **246,** F317–F322.

Bhoola, K. D., Figueroa, C. O., and Worthy, K. (1992). Bioregulations of kinins: Kallikrein, kininogens and kininases. *Pharmacol. Rev.* **44,** 1–81.

Boscolo, P., and Carmignani, M. (1986). Mechanisms of cardiovascular regulation in male rabbits chronically exposed to cadmium. *Br. J. Ind. Med.* **43,** 605–610.

Boscolo, P., Carmignani, M., Carelli, G., Finelli, V. N., and Giuliano, G. (1992). Zinc and copper in tissues of rats with blood hypertension induced by long-term lead exposure. *Toxicol. Lett.* **63,** 135–139.

Boscolo, P., Carmignani, M., Volpe, A. R., Felaco, M., Del Rosso, G., Porcelli, G., and Giuliano, G. (1994). Renal toxicity and arterial hypertension in rats chronically exposed to vanadate. *Occup. Environ. Med.* **51,** 500–503.

Braunwald, E. (1994). Cellular basis of cardiac contraction. In K. J. Isselbacher, E. Braunwald, and G. H. Williams (Eds.), *Harrison's Principles of Internal Medicine,* 13th ed. McGraw-Hill, New York, Vol. 1, pp. 988–998.

Brichard, S. M., and Henquin, J. C. (1995). The role of vanadium in the management of diabetes. *Trends Pharmacol. Sci.* **16,** 265–270.

Byrne, A. R., and Kosta, L. (1978). Vanadium in foods and human body fluids and tissues. *Sci. Total Environ.* **10,** 17–30.

Carmignani, M., and Boscolo, P. (1984). Cardiovascular responsiveness to physiological agonists of male rats made hypertensive by long-term exposure to cadmium. *Sci. Total Environ.* **34,** 19–33.

Carmignani, M., Boscolo, P., Ripanti, G., Porcelli, G., and Volpe, A. R. (1993). Mechanisms of the vanadate-induced arterial hypertension only in part depend on the levels of exposure. In M. Anke, D. Meissner, and C. F. Mills (Eds.), *Trace Elements in Man and Animals.* Verlag Media Touristik, Gersdorf (Germany), pp. 971–975.

Carmignani, M., Boscolo, P., Volpe, A. R., Togna, G., Masciocco, L., and Preziosi P. (1991). Cardiovascular system and kidney as specific targets of chronic exposure to vanadate in the rat: Functional and morphological findings. *Arch. Toxicol.* **14**(Suppl.), 124–127.

Carmignani, M., Finelli, V. N., and Boscolo, P. (1983a). Mechanisms in cardiovascular regulation following chronic exposure of male rats to inorganic mercury. *Toxicol. Appl. Pharmacol.* **69,** 442–450.

Carmignani, M., Boscolo, P., and Iannaccone, A. (1983b). Effects of chronic exposure to arsenate on the cardiovascular function of rats. *Br. J. Ind. Med.* **40,** 280–284.

Carmignani, M., Finelli, V. N., Boscolo, P., and Preziosi, P. (1987). Sex-related interactions of cadmium and lead in changing cardiovascular homeostasis and tissue metal levels of chronically exposed rats. *Arch. Toxicol.* **11**(Suppl.), 216–219.

Carmignani, M., Preziosi, P., Del Carmine, R., Porcelli, G., and Volpe, A. R. (1992a). Kallikrein–kinin, enkephalin, renin aldosterone and catecholamine systems in the vanadate (as vanadyl)-induced arterial hypertension. *Agents Actions* **38**(3), 243–247.

Carmignani, M., Volpe, A. R., Porcelli, G., Boscolo, P., and Preziosi, P. (1992b). Chronic exposure to vanadate as factor of arterial hypertension in the rat: Toxicodynamic mechanisms. *Arch. Toxicol.* **15,** 117–120.

Carmignani, M., Boscolo, P., Artese, L., Del Rosso, G., Porcelli, G., Felaco, M., Volpe, A. R., and Giuliano, G. (1992c). Renal mechanisms in the cardiovascular effects of chronic exposure to inorganic mercury in rats. *Br. J. Ind. Med.* **49,** 226–232.

Carmignani, M., Sabbioni, E., Boscolo, P., Pietra, E., and Ripanti, G. (1989). Cardiovascular effects of long-term exposure to vanadate in rats. In M. Anke and W. Baumann (Eds.), *Sixth Trace Element Symposium 1989 (Mo, V).* University of Jena Press, Jena, pp. 106–113.

Carmignani, M., and Volpe, A. R. (1996). Nitric oxide and delta-opioid receptors in the tissue interactions among enkephalins, angiotensins and kinins. *Immunopharmacology* **32,** 172–175.

Carmignani, M., Volpe, A. R., Boscolo, P., and Giuliano, G. (1995). Vanadate and cardiovascular system. *G. Ital. Med. Lav.* **17,** 148–162.

Carmignani, M., Volpe, A. R., Masci, O., Boscolo, P., Di Giacomo, F., Grilli, A., Del Rosso, G., and Felaco, M. (1996). Vanadate as factor of cardiovascular regulation by interactions with the catecholamine and nitric oxide systems. *Biol. Trace Elem. Res.,* **51,** 1–12.

Churchill, P. C. (1985). Second messanger in renin secretion. *Am. J. Physiol.* **249**, F175–F180.

Churchill, P. C., Rossi, N. F., Churchill, M. C., and Ellis, V. R. (1990). Vanadate-induced inhibition of renin secretion is unrelated to inhibition of Na,K-ATPase activity. *Life Sci.* **44**, 1953–1959.

Cornelis, R., Versieck, J., Mees, L., Hoste, J., and Barbier, F. (1981). The ultratrace element vanadium in human serum. *Biol. Trace Elem. Res.* **3**, 257–263.

Cruz-Soto, M. A., and Martinez-Maldonado, M. (1984). Modification of the renal response to ouabain by vanadate in vivo. *Clin. Res.* **32**, 532A.

Dafnis, E., Spohn, M., Lonis, B., Kurtzman, N. A., and Sabatini, S. (1992). Vanadate causes hypokalemic distal renal tubular acidosis. *Am. J. Physiol.* **31**, F449–F453.

Day, H., Middendorf, D., Lubert, B., Heinz, A., and Grantham, J. (1980). The renal response to intravenous vanadate in rats. *J. Lab. Clin Med.* **96**, 392.

Domingo, J. L., Gomez, M., Liobet, J. M., Corbella, J., and Keen, C. L. (1991). Improvement of glucose homeostasis by oral vanadyl or vanadate treatment in diabetic rats is accompanied by negative side effects. *Pharmacol. Toxicol.* **66**, 279–287.

Edel, J., Pietra, R., Sabbioni, E., Marafante, E., Springer, A., and Ubertalli, L. (1984). Disposition of vanadium in rat tissue at different age. *Chemosphere* **13**, 87–93.

Edel, J., and Sabbioni, E. (1988). Retention of intratracheally instilled and ingested tetravalent and pentavalent vanadium in the rat. *J. Trace Elem. Electrolytes Health Dis.* **2**, 23–30.

Edel, J., and Sabbioni, E. (1989). Vanadium transport across placenta and milk of rats to the fetus and newborn. *Biol. Trace Elem. Res.* **22**, 265–275.

Frank, A., Madej, A., Galgan, V., and Peterson, R. L. (1996). Vanadium poisoning of cattle with basic slag. Concentrations in tissues from poisoned animals and from a reference, slaughterhouse material. *Sci. Total Environ.* **181**, 73–92.

French, R. J., and Jones, P. J. H. (1993). Role of vanadium in nutrition: Metabolism, essentiality and dietary considerations. *Life Sci.* **52**, 339–346.

Gomez, M., Domingo, J. L., Liobet, J. M., and Corbella, J. (1991). Effectiveness of some chelating agents on distribution and excretion of vanadium in rats after prolonged oral administration. *J. Appl. Toxicol.* **11**, 195–198.

Halberstam, M., Cohen, N., Shlimovich, P., Rossetti, L., and Shamoon, H. (1996). Oral vanadyl sulphate improves insulin sensitivity in NIDDM but not in obese nondiabetic subjects. *Diabetes* **45**, 659–666.

Hansen, O. (1979). Facilitation of ouabain binding to (Na,K)-ATPase by vanadate at in vivo concentrations. *Biochim. Biophys. Acta* **568**, 265–269.

Hardman, J. G., Limbird, L. E., Molinoff, P. B., Ruddon, R. W., and Goodman Gilman, A. (Eds.), (1996). *Goodman Gilman's Pharmacological Basis of Therapeutics.* 9th ed. McGraw-Hill, New York, 1905 pp.

Iannaccone, A., Boscolo, P., and Cavallotti, C. (1978). Glucose-6-phosphate dehydrogenase in kidneys of lead poisoned rats and adrenalectomized rats. *Nephron* **20**, 220–224.

Inciarte, D. J., Steffen, R. P., Dobbins, D. E., Swindall, B. T., Johnston, J., and Haddy, F. J. (1980). Cardiovascular effect of vanadate in the dog. *Am. J. Physiol.* **239**, H47–H56.

Jandhyala, B. S., and Hom, G. J. (1983). Physiological and pharmacological properties of vanadium. *Life Sci.* **33**, 1325–1340.

Kalyani, P., Vijaya, S., and Ramasarma, T. (1992). Characterization of oxigen free radicals generated during vanadate-stimulated NADH oxidation. *Mol. Cell. Biochem.* **111**, 33–40.

Kretzschmar, M., and Braunlich, H. (1990). Role of GSH in vanadate reduction in young and mature rats: Evidence for direct participation of glutathione in vanadate inactivation. *J. Appl. Toxicol.* **10**, 295–300.

Kurtz, T. W., and Spence, M. A. (1993). Genetics of essential hypertension. *Am. J. Med.* **94**, 77–88.

Laniyonu, A., Saifeddine, M., Ahmad, S., and Hollenberg, M. D. (1994). Regulation of vascular and gastric smooth muscle contractility by pervanadate. *Br. J. Pharmacol.* **113**, 403–410.

Larsen, J. A., and Thomsen, O. O. (1980). Vanadate induced oliguria and vasoconstriction in the cat. *Acta Physiol. Scand.* **110**, 367.

Lefkowitz, R. J., Hoffman, B. B., and Taylor, P. (1996). Neurotransmission. In J. G. Hardman, L. E. Limbird, P. B. Molinoff, R. W. Ruddon, and A. Goodman Gilman (Eds.), *Goodman Gilman's Pharmacological Basis of Therapeutics.* 9th ed. McGraw-Hill, New York, pp. 105–139.

Levinsky, N. G. (1994). Sodium and water. In K. J. Isselbacher, E. Braunwald, and G. H. Williams (Eds.), *Harrison's Principles of Internal Medicine.* 13th ed. McGraw-Hill, New York, pp. 242–253.

Lopez-Novoa, J. M., and Garrido, M. C. (1986). Hemodynamic effects of vanadate administration in rats with different levels of sodium intake. *Am. J. Med. Sci.* **291**, 152–156.

Marmé, D. (1987). *Calcium and Cell Physiology.* Springer-Verlag, Berlin, 390 pp.

Moncada, S., and Higgs, A. (1993). The L-arginine nitric oxide pathway. *New Engl. J. Med.* **329**, 2002–2010.

Moncada, S., Palmer, R. M. J., and Higgs, E. A. (1991). Nitric oxide: Physiology, pathophysiology and pharmacology. *Pharmacol. Rev.* **43**, 109–142.

Nechay, B. R. (1984). Mechanisms of action of vanadium. *Ann. Rev. Pharmacol. Toxicol.* **24**, 501–524.

Nechay, B. R., Nanninga, L. B., Nechay, P. S. E., Post, R. L., Grantham, J. J., Macara, J. G., Kubena, L. F., Phillips, T. D., and Nielsen, F. H. (1985). Role of vanadium in biology. *Fed. Proc.* **45**, 123–132.

Nielsen, F. H. (1995). Vanadium in mammalian physiology and nutrition. *Met. Ions Biol. Syst.* **31**, 543–573.

Orvig, C., Thomson, K. H., Battell, M., and Mc Neill, J. H. (1995). Vanadium compounds as insulin mimics. *Met. Ions Biol. Syst.,* **31**, 575–594.

Orvini, E., and Lodola, L. (1979). Determination of the chemical forms of dissolved vanadium in freshwater as determined by [48]V radiotracer experiments and neutron activation analysis. *Science Total Environ.* **13**, 195–207.

Pearse, A. G. E. (1972). *Histochemistry, Theoretical and Applied.* Churchill-Livingstone, Edinburgh, 871 pp.

Perry, H. M. Jr., and Erlanger, M. W. (1976). Mechanism of cadmium-induced hypertension. In D. D. Hemphill (Ed.), *Trace Substances in Environmental Health.* University of Missouri, Columbia, Vol. 10, pp. 339–348.

Petering, H. G. (1974). The effects of cadmium and lead on copper and zinc metabolism. In J. W. Horkstra, J. W. Suttie, H. E. Ganther, and W. Mertz (Eds.), *Trace Element Metabolism in Animals.* University Park Press, Baltimore, pp. 311–325.

Pirkle, J. L., Schwartz, J., Landis, J. R., and Harlan, W. R. (1985). The relationship between blood lead levels and blood pressure and its cardiovascular risk implications. *Am. J. Epidemiol.* **121**, 246–258.

Pocock, S. J., Shaper, A. G., Ashby, D., Delves, T., and Whitehead, T. P. (1984). Blood lead concentration, blood pressure and renal function. *Br. Med. J.* **289**, 872–874.

Sabbioni, E., Goetz, L., and Bignoli, G. (1984). Health and environmental implications of trace metals released from coal-fired power plants: An assessment study of the situation in the European Community. *Sci. Total Environ.* **40**, 141–154.

Sabbioni, E., Kuèera, J., Pietra, R., and Vesterberg , O. (1996). A critical review on normal concentrations of vanadium in human blood, serum, and urine. *Science Total Environ.* **188**, 49–58.

Sabbioni, E., and Marafante, E. (1978). Metabolic patterns of vanadium in the rat. *Bioinorg. Chem.* **9**, 389–407.

Sabbioni, E., and Marafante, E. (1981). Relations between iron and vanadium metabolism: In vivo incorporation of vanadium into iron proteins of the rat. *J. Toxicol. Environ. Health* **8**, 419–429.

Sabbioni, E., Marafante, E., Amantini, L., and Ubertalli, L. (1978). Similarity in metabolic patterns of different chemical species of vanadium in the rat. *Bioinorg. Chem.* **8,** 503–515.

Sabbioni, E., and Maroni, M. (1983). *A study on Vanadium in Workers from Oil Fired Power Plants.* Commission of the European Communities, Luxembourg, 36 pp.

Sabbioni E., Minoia, C., Pietra, R., Fortaner, S., Gallorini, M., and Saltelli, A. (1992). Trace element reference values in tissues from inhabitants of the European Community. II. Examples of strategy adopted and trace element analysis of blood, limph nodes and cerebrospinal fluid of Italian subjects. *Sci. Total Environ.* **120,** 39–50.

Sabbioni, E., Pozzi, G., Devos, S., Pintar, A., Casella, L., and Fischbach, M. (1993). The intensity of vanadium (V)-induced cytotoxicity and morphological transformation in BALB/3T3 cells is dependent on glutathione-mediated bioreduction to vanadium (IV). *Carcinogenesis* **14,** 2565–2568.

Sabbioni, E., Pozzi, G., Pintar, A., Casella, L., and Garattini, S. (1991). Cellular retention, cytotoxicity and morphological transformation by vanadium (IV) in BALB/3T3 cell lines. Carcinogenesis **12,** 47–52.

Sandirasegarane, L., and Gopalakrishnan, V. (1995). Vanadate increases cytosolic free calcium in rat aortic smooth muscle cells. *Life Sci.* **56**(7), PL169–174.

Sauvadet, A., Pecker, F., and Pavoine, C. (1995). Inhibition of the sarcolemmal Ca^{++} pump in embryonic chick heart cells by mini-glucagon. *Cell-Calcium* **18,** 76–85.

Schroeder, H. A., Balassa, J. J., and Tipton, I. H. (1963). Abnormal trace metals in man— vanadium. *J. Chronic Dis.* **16,** 1047–1071.

Scicli, A. G., and Carrettero, A. O. (1986). Renal kallikrein-kinin system. *Kidney Int.* **29,** 120–130.

Serra, M. A., Pintar, A., Casella, L., and Sabbioni, E. (1992). Vanadium effect on the activity of horseradish peroxidase, catalase, glutathione peroxidase, and superoxide dismutase in vitro. *J. Inorg. Biochem.* **46,** 161–174.

Sharma, R. P., Coulumbe, R. A. Jr., and Srisuchart, B. (1985). Effects of dietary vanadium exposure on levels of regional brain neurotransmitters and their metabolites. *Biochem. Pharmacol.* **35,** 451–465.

Simonoff, M., Liabador, Y., Peers, A. M., and Simonoff, G. N. (1984). Vanadium in human serum, as determined by neutron activation analysis. *Clin. Chem.* **30,** 1700–1703.

Steffen, R. P., Pamnani, M. B., Clough, D. L., Huot, S. J., Muldoon, S. M., and Haddy, F. J. (1981). Effects of prolonged dietary administration of vanadate on blood pressure in the rat. *Hypertension* **3,** 1173–1178.

Strauss, J. D., Zeuger, C., Van Eyk, J. E., Bletz, C., Troschka, M., and Ruegg, J. C. (1992). Troponin replacement in permeamilized cardiac muscle. Reversible extraction of troponin I by incubation with vanadate. *FEBS Lett.* **310,** 229–234.

Sundet, W. D., Wang, B. C., Hakumärri, M. O. K., and Goetz, K. L. (1984). Cardiovascular and renin responses to vanadate in the conscious dog: Attenuation after calcium channel blockade (41786). *Proc. Soc. Exp. Biol. Med.* **175,** 185–190.

Todaro, A., Bronzato, R., Buratti, M., and Colombi, A. (1991). Esposizione acuta a polveri contenenti vanadio: Effetti sulla salute e monitoraggio biologico in un gruppo di lavoratori addetti alla manutenzione delle caldaie. *Med. Lav.* **82,** 142–147.

Vesterberg, O., Alessio, L., Brune, D., Gerhardsson, L., Herber, R., Kazantsis, G., Nordberg, G. F., and Sabbioni, E. (1993). International project for producing reference values of concentrations of trace elements in human blood and urine—*TRACY. Scand. J. Work. Environ. Health* **19** (Suppl. 1), 19–26.

Vio, C. P., and Figueroa, D. (1985). Subcellular localization of renal kallikrein by ultrastructural immunocytochemistry. *Kidney Int.* **28,** 36–42.

Vollkel, N. F., and Czartolomna, J. (1991). Vanadate potentiates hypoxic pulmonary vasoconstriction. *J. Pharmacol. Exp. Ther.* **259,** 666–672.

Volpe , A. R., Giardina, B., Preziosi, P., and Carmignani, M. (1996). Biosynthesis of endothelium-derived nitric oxide by bradykinin as endogenous precursor. *Immunopharmacology* **33,** 287–290.

Willebucher, R. F., Xie, Y. N., Eysselein, V. E., and Snape, W. J. Jr. (1992). Mechanisms of cAMP-mediated relaxation of distal circular muscle in rabbit colon. *Am. J. Physiol.* **262,** G154–164.

Williams, G. H. (1994). Hypertensive vascular disease. In K. J. Isselbacher, E. Braunwald, and G. H. Williams (Eds.), *Harrison's Principles of Internal Medicine.* 13th ed. McGraw-Hill, New York, pp. 1116–1131.

11

EFFECTS OF VANADATE IN ADRENAL GLAND OF MAMMALS

Maria da Graça Fauth

Instituto de Biociências, Pontifícia Universidade Católica do Rio Grande do Sul, Porto Alegre, Brasil

Kátia Padilha Barreto, Marcelo de Lacerda Grillo, and Guillermo Federico Wassermann

Instituto de Biociências, Universidade Federal do Rio Grande do Sul, Porto Alegre, Brasil

Vanadium in the Environment. Part 2: Health Effects, Edited by Jerome O. Nriagu.
ISBN 0-471-17776-8. © 1998 John Wiley & Sons, Inc.

1. INTRODUCTION

Vanadium is chemically classified as a transition element, located in the VB group and in the fourth period of the periodic table, with oxidation numbers that vary from $+2$ to $+5$ (Chapter 4, Vol. 30). Vanadium is found in water, rocks, and soils in low concentration and in relatively high concentration in coal and oil deposits. It is the 21st most abundant element in the earth's crust in a concentration of 135 ppm and the 34th in the oceans in a rate of 2 ppb.

Owing to its great industrial importance (Chapter 1, Vol. 30) and to the fact that it is an essential element to some life forms, vanadium's biological actions have gained a great interest in the scientific investigations of the last decades. The diverse biological action of vanadium results from its capacity to function as an oxyanion, oxycation, or prooxidant. Therefore, in corporal fluids, in a pH between 4.0 and 8.0, vanadium's predominant form is the vanadate or metavanadate VO_3^-, with an oxidation number of $+5$. The VO_3^- ion, through an anion transportation system, will be able to get into certain cells and be reduced therein,—not through the actions of enzymes, by glutatione, NADH, or ascorbate—to VO^{2+}, vanadyl ion with an oxidation number of $+4$ (Nechay, 1984).

Various forms of vanadate that differ in their composition are already known. The ones whose biological action is presently known are VO_4^{3-}, ortho-vanadate ion; $V_2O_7^{4-}$, divanadate or pyrovanadate; and VO_3^-, metavanadate. When in solution, the conversion of the ortho, pyro, and meta forms is practically instantaneous, as is the formation of oligomers $(VO_3)_n^{-n}$. In millimolar concentrations and neutral pH, the vanadium solutions contain monomers, dimers, tetramers, and pentamers. Vanadate is electronically as well as structurally similar to phosphate (Cantley et al., 1977). Both present a tetrahedral structure when in solution, the V–O bond being 1.66 Å long, while the P–O bond is 1.55 Å long. Because of this similarity, the VO_3^- ion can replace phosphate in many bond sites (Nechay, 1984).

The vanadyl ion, VO^{2+}, is similar to the Mg^{2+} ion; they present atomic radii of 0.60 and 0.65 Å respectively. In a relation like that of vanadate vis-à-vis phosphate, the vanadyl ion will also be able to replace the Mg^{2+} ion at many cellular and biochemical action sites (Nechay, 1984).

Vanadium distribution over the human body varies according to the tissue and the organ that is being studied. The highest concentrations were found in teeth, bones, hair, and nails, it being accepted that, in the last two cases, the concentrations were functions of environmental exposure. In daily diet, a person consumes from 10 to 60 μg of vanadium. The excretion through urine is approximately 10 μg per day (Nechay, 1984). The estimated concentration of vanadium in mammal tissues is 17 μg/L or 0.34 μM in blood, 26 μg/L or 0.52 μM in erythrocytes, and from 20 to 30 μg/kg of dry weight in adult human liver, spleen, pancreas, and prostate gland (Cantley et al., 1977). The concentrations of vanadium that were determined in different tissues exhibit great dispersion, probably owing to the techniques that were used. Simonoff et al. (1984) determined the concentration of vanadium in human serum by neutron activation analysis, with postirradiation chemical separation. The obtained values were around 670 ng/L. The concentration in plasma, determined by various researchers, vary from nanograms per liter when neutron activation analysis is used, to micrograms per liter when atomic absorption is used. Nechay et al. (1986) refer to vanadium concentrations in liver of around 5 ng/g, and concentrations in lungs and hair of 12–140 ng/g.

2.   INTERACTION WITH ATPases

Vanadium seems to be present in cellular fluids in the oxidation state +5, in the VO_3^- ion form, and in the intracellular medium in the predominant VO^{2+} vanadyl ion form. Both forms interfere in numerous enzymes related to phosphate, in concentrations that vary from nanomolar to millimolar (Nechay, 1984).

The main system of active transport in many animal cells is the pump that removes sodium from, and simultaneously deposits potassium into, the interior of the cells. This process requires ATP and is specifically inhibited by cardiac glycosides like ouabain. The protein that makes up this cellular structure is called Na^+,K^+-ATPase (Fig. 1). The energy necessary to the construction of the sodium and potassium gradient comes from the hydrolysis of the ATP to ADP and Pi. The ATPase activity, associated with the Na^+ and K^+ pump, is the result of the sodium-dependent phosphorylation of the aspartate residue

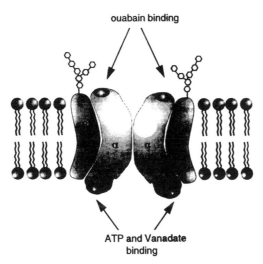

Figure 1. Schematic diagram of the Na^+,K^+-ATPase insert in the lipidic bilayer, showing the binding sites of ouabain on the extracellular side, and the binding sites of vanadate and ATP on the cytoplasmatic side, both in α subunit.

of the pump protein and the subsequent potassium-dependent hydrolysis of the enzyme acyl phosphate group. The initial phosphorylation of the enzyme molecule occurs only after three sodium ions are bonded to the sites on the cytoplasmatic side of the membrane. Thereby the E_1-P configuration of the enzyme can be obtained. The phosphorylation begins a fast transition of the pump to the E_2-P state, whence sodium is liberated in the extracellular medium. Then potassium is bonded to the E_2-P structure, starting the acyl phosphate hydrolysis. This process makes E_2 unstable, which reverts spontaneously to E_1, bringing potassium to the interior of the cell (Siegel, 1989) (Fig. 2).

Vanadium interacts with transport ATPases in many cellular types. In view of its physicochemical properties, its nutritional requirement, distribution in tissues, and interactions with Na^+,K^+-ATPases, Cantley et al., in 1978, and Cantley and Aisen, in 1979, stated that vanadium is a potent physiological regulator of the sodium and potassium pump. Through a similar mechanism vanadium also interferes with Ca^{2+}-ATPases. It is bonded to the Na^+,K^+-ATPase in two sites, one of them having high affinity, in micromolar, concentrations and the other one having low affinity, in millimolar concentrations (Nechay, 1984).

A large body of evidence suggests that the toxicity of vanadium is strictly dependent on its oxidation state. In in vitro studies using BALB/3T3 cells, it was found that pentavalent vanadium is considerably more cytotoxic that the

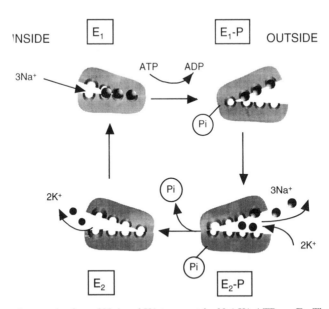

Figure 2. Postulate mechanism of Na^+ and K^+ transport by Na^+,K^+-ATPase. E_1: Three Na^+ ions inside the cell bind to the Na^+ sites of the ATPase. E_1-P: Phosphorylation of an aspartate residue of the ATPase by ATP. E_2-P: The phosphorylation induces a conformational change that results in the release of Na^+ ions into the extracellular medium. Two K^+ ions outside the cell bind to the K^+ sites on the ATPase. E_2: After hydrolysis of the phosphoryl group bound to the ATPase, K^+ ions enter the cell.

tetravalent form (Sabbioni et al., 1991). In contrast with pentavalent vanadium, tetravalent vanadium did not exhibit transforming effects in these cells. It has been proposed that vanadium toxicity is related to the saturation of the biochemical pathways, that reduce the vanadium(V) form to the tetravalent species in the cell. The excess of vanadate that is not reduced could be responsible for the toxic effects (Sabbioni et al., 1991).

The membrane ATPase inhibition by the vanadate (VO^{3-}_4) ion occurs on the cytoplasmatic side, beginning in nanomolar concentrations. Vanadate is an analog of the transition state in phosphoryl transfer reactions because it readily forms a pentacovalent bipyramidal structure like that of phosphate ester undergoing hydrolysis.

The ATPases are also inhibited by ouabain through an inhibition mechanism that is different from the one of the vanadate ion, as ATPases have specific receptors for the bond of this steroid. Ouabain simultaneously blocks ATP hydrolysis and ion translocation. Inhibition of the Na^+,K^+-ATPase by ouabain disrupts the stable osmotic pressure maintained by the pump. This regulation is a physiological mechanism, since the presence and secretion of ouabain in rat and bovine adrenals has been described in recent studies (Hamilton et al., 1994).

Cardiotonic steroids inhibit the dephosphorylation reaction of the Na^+,K^+-ATPase, but only if the cardiotonic steroid is located on the extracellular face of the membrane. Thus, inhibition of dephosphorylation by this steroid has the same sidedness as the activation of dephosphorylation by K^+. Cardiotonic steroid stabilizes the E_2 form of the pump.

3.   INSULIN-LIKE EFFECTS

The vanadate ion presents insulin-like effects in many cellular events. The chemistry of vanadate is complex when in solution and many of its insulin-like effects can be related to the valence state. While in some biological phenomena vanadate shows an action that is similar to that of insulin, in other ones it is different.

Vanadium's insulin-like properties in vitro have been described in hepatocytes (Tamura et al., 1984), adipocytes (Kadota et al., 1987), and erythrocytes (Vareka and Carafoli, 1982).

Vanadium in vivo has been shown to exert insulin-like effects on transport, glucose oxidation, and potassium uptake, as well as on the activity of glycogen synthase in isolated rat adipocytes and skeletal muscle (Roden et al., 1993). Other vanadium effects on the carbohydrate metabolism can be related to the reduction of the plasmatic glucose concentration. Vanadium compounds at much higher concentrations than are typically ingested are being considered in the treatment of diabetes mellitus. The hormonal regulation of glycolysis and of the gluconeogenesis in liver is mediated by fructose 2,6-bisphosphate. The vanadate ion stimulates the 2,6-bisphosphate fructose formation. Consequently, an increase in hepatic glycolysis and an inhibition of hepatic gluconeo-genesis will take place (Pilkis and El-Maghrabi, 1988).

Vanadate is shown by certain researchers to be an agent that interferes in the autophosphorylation of the insulin receptor. The insulin molecule is bonded to the α subunit, stimulating the phosphorylation of the tyrosine residue of the receptor β subunit, which activates the transphosphorylation of exogenous substrates and intracellular proteins (Duckworth et al., 1988).

The vanadate effects were also investigated in the cells of the pancreatic islets, where the ion slightly increased insulin release (Zhang et al., 1991) and inhibited the ATPases of the plasmatic membrane.

These actions of vanadate are probably due to increased dependent inhibi-tion of phosphatases or to increased binding of insulin to its target cells, possibly due to enhanced tyrosine phosphorylation of the insulin receptor by vanadate. The actions of insulin and vanadium on the insulin receptor are similar, but the mechanisms are not identical.

Owing to the similarities of action between vanadate and insulin, this ion has also been used as a tool to gain insight into the mechanism of insulin action (Duckworth et al., 1988).

The major insulin effect in the target tissues is to stimulate the transport of sugars, amino acids, fatty acids, ions, and nucleic acid precursors on the cellular surface (Goldfine, 1981). Insulin, when interacting with its receptor, has a cellular action that deeply affects plasmatic membrane physiology. The hormone activates enzymes, such as Na^+,K^+-ATPase or phosphodiesterase. In many cellular types, it modulates the number of receptors in the cell surface to insulin-like growth factor II or transferrin, also increasing the transport of various metabolites such as glucose and amino acids.

4. AMINO ACID TRANSPORT

The cellular content is separated from the extracellular medium by the plasma-tic membrane, formed by proteins and a lipid bilayer and therefore working as a barrier to most polar molecules. Because of this fact, the cells have special ways to transfer molecules through the membrane. As happens in artificial lipid membranes, cellular membranes allow water and other small polar mole-cules, as well as hydrophobic molecules, to pass through the membrane by simple physical diffusion.

Cellular membranes have specific proteins that are responsible for the transfer of a solute or a group of polar solutes, for example, ions, amino acids, sugars, nucleotides, and many cellular metabolites. These carrier proteins can be found in many forms and in all kinds of biological membranes. Some of them function as pumps that actively control the movement of specific solutes against their electrochemical gradient through the so-called active system, which is firmly bound to a metabolic source of energy. This transport often involves ATP hydrolysis or Na^+ and H^+ countertransport, decreasing their electrochemical gradients.

The transport proteins specifically bind and transfer the soluble molecule through the lipid bilayer. This process is similar to the enzyme–substrate reaction. The carrier molecule has a specific binding site for the solute. When the carrier is saturated and all the sites are occupied, the transport is at maximum rate of reaction. This rate is the V_{max}, a specific characteristic of each carrier molecule, as well as the Michaelis–Menten term, K_m. The solute binding can be blocked by competitive inhibitors, which compete for the binding site, or by noncompetitive inhibitors, which bind to another site, different from the substrate site, and modify the structure of the carrier molecule.

Amino acid transport is an essential phenomenon of cell metabolism. Through amino acid transport, the cell is provided with the fundamental elements that form cell proteins: enzymes, structural and secretion proteins, and others. According to the nutritional condition of the cell, amino acids can also constitute energy sources to ensure the perfect operation of its basic activities.

Ionic involvement in amino acid transport can occur through various mechanisms. For example, the ions can participate as cosubstrates and stimulate the uptake, turn carriers unstable, inhibit transport, change membrane permeability, or may simply not interfere with transport.

Sodium-dependent transport is characteristic of a great number of tissues, as well as cellular organelles like mitochondria and nucleus.

Most amino acids are actively transported to the cell interior, being able to reach concentrations 2–20 times higher than in the extracellular medium. Amino acid transport is related to the sodium electrochemical gradient, determined by the active transport of the ion to the exterior of the cells that provides energy to the active transport of the organic solute. So the amino acid carrier molecule has a binding site for sodium, too. The ion interaction with the carrier molecule increases the site affinity with the amino acid. Since the intracellular sodium concentration is low, the possibility of its interaction with the carrier molecule in the interior of the cell is small. This way, the sodium gradient through the membrane conducts an asymmetry in the affinities of the amino acid carrier, resulting in the active transport of the amino acid to the interior of the cell. Various amino acid transport systems have been already characterized in many cell types, to date. However, the most common and best known is the A system.

The A system of neutral amino acid transport mediates zwitterion amino acid transport. It accepts linear aliphatic amino acids, having small side chains, such as glycine, alanine, threonine, asparagine, glutamine, serine, and methionine and imine acids, such as proline. It accepts the alkylation of the α-amino group and mediates the transport of nonmetabolizable analogs such as the α-aminoisobutyric acid (α-AIB) and its N-methylated derivative, α-methylaminoisobutyric (α-MeAIB) (Gazzola et al., 1980). It is sodium-dependent and its activity is modified by extracellular pH. Transport through this system is subject to transinhibition by intracellular substrates.

The A system regulation is complex; its activity can be modulated by hormones, growth factors, nutritional conditions, and sodium electrochemical gradient. It is the main neutral amino acid transport system present in the various cell types of mammals and birds.

The mediated transport systems are subject to hormonal regulation. Many hormones act on the mediated transport of specific solutes. In some cases, the hormone effect consists in modifying carrier molecule affinity with the transported solute. In other situations, the hormone stimulates the synthesis of new carrier proteins, increasing the maximum transport capacity. The hormonal action on amino acid transport has been extensively studied in the last decades, as amino acid uptake is of essential importance to cellular metabolism.

Insulin may be the most studied hormone involved in phenomenon of amino acid transport. The stimulatory effect of insulin on the A transport system, in rat liver, has been shown to be dependent on protein synthesis and on microtubule function, and to be characterized by an increase in V_{max} (Fehlmann et al., 1979; Prentki et al., 1981).

Amino acid transport regulation by insulin in human fibroblast culture was studied by Longo et al. in 1985. Insulin increases transport activity through an increase in the maximum rate of reaction (V_{max}) without a significant increase of K_m. This fact suggests that the hormone increases the disposability of active carriers in the cell membrane. Insulin can promote the synthesis of new carrier proteins or can convert inactive carrier proteins into active forms or even translocate carrier proteins from other cellular compartments to the membrane (Lienhard, 1983). Long- and short-term hormonal actions on amino acid transport allow the postulation of a direct effect on the synthesis of de novo carrier molecules.

Tovar et al. (1991) investigated the amino acid uptake in soleus muscle of rats. Their studies have shown that MeAIB transport is clearly accumulated in soleus muscle; it is subject to competition by high concentrations of individual small neutral amino acids or by amino acid mixtures simulating plasma amino acid profiles, but not by most of the large neutral amino acids. It can be stimulated by insulin, shows adaptive regulation upon amino acid deprivation and is required for sodium energy source.

5.   EFFECTS OF VANADATE ON AMINO ACID TRANSPORT IN ADRENAL GLANDS

The adrenal glands of mammals consist of two functionally distinct endocrine glands involved in the same capsule. They are richly vascularized: The blood flow, by unit of weight, is one of the highest of the mammal organism (Spector, 1956). They have two different kinds of tissue, forming the medulla and the cortex. The medulla is composed of chromaffinic tissue. It produces adrenaline and noradrenaline. In mammals, the medulla is surrounded by the cortex (Orth et al., 1992).

The cortex is essential for life, being involved in many physiological processes of the cell economy. It produces more than 50 different steroids, being a source of glucocorticoids, mineralocorticoids, and also androgens and estrogens. The steroid hormones are bonded to intracellular receptors and the resulting complex is bonded to the nuclear DNA, regulating the transcription of specific genes whose production mediates the effects of the hormones. Structurally, the cortex is divided in three zones: glomerulosa, fasciculata, and reticularis.

The zona glomerulosa is more external, adjacent to the gland capsule, and its cells are disposed in irregular masses. It is responsible for the synthesis of most mineralocorticoids. Aldosterone, the main mineralocorticoid and main sodium retainer steroid hormone in most mammals, is produced exclusively in the glomerulosa zone. Its production is regulated by the renin–angiotensin system, ACTH, and potassium.

The middle zone is called the zona fasciculata. It is the widest adrenal zone, and its cells present a great number of lipid droplets and an abundant agranular endoplasmic reticulum. The zona fasciculata is the main source of glucocorticoids, cortisol in most mammals, and corticosterone in many rodents; and in mature bovines it produces cortisol as well as corticosterone. The glucocorticoids have numerous and complex actions in practically all cells of the organism, regulating the metabolism of proteins, nucleic acids, fats, and carbohydrates. ACTH stimulates the secretion and release of cortisol as well as other steroids and small amounts of intermediaries of cortisol biosynthesis.

The inner cortical zone is called the zona reticularis, which produces sexual steroids, androgens and estrogens, as well as their precursors.

The adrenal gland of rat must be considered a target organ for insulin (Machado et al., 1982). Insulin binding to rat adrenal was studied in vivo by iv injection of ^{125}I-insulin either alone or together with an excess of unlabeled hormones (insulin, glucagon, prolactin, or growth hormone). In addition, isolated glands from normal or streptozotocin diabetic rats were incubated in vitro with ^{125}I-insulin and varying concentrations of unlabeled insulin. Both experiments showed specific binding sites in the adrenal glands. The insulin stimulation of deoxyglucose uptake was examined in isolated glands from normal and streptozotocin diabetic rats. Insulin induced a stimulation of deoxyglucose transport on both groups. These results indicated that insulin has an important metabolic influence on the adrenal gland (Silva et al., 1984).

In 1992, Ito et al., showed that both the bovine adrenal cortex and medulla present specific insulin receptors and that they are indistinguishable from those found in other mammal tissues. In this regard, several previous reports have demonstrated the effects of insulin on cell growth and function in adrenocortical tissue. Insulin interferes in adrenal glucogen metabolism in rats (Piras et al., 1973), stimulates cell growth in bovine adrenocortical cell culture in physiological concentrations (Simonian et al., 1982; Simonian and Gill, 1979), stimulates glucose transport and protein synthesis in chromaffinic cells (Delicado and Miras Portugal, 1987), and increases catecholamine content in chromaffinic cell culture (Wilson and Kirshner, 1983).

Insulin stimulation of amino acid transport in the adrenal glands of cows and rats is mediated by the sodium- and energy-dependent system A, potentiated by Na^+-K^+-ATPase. This stimulatory action is related to voltage-dependent Ca^{2+} channels, particularly the T-type channels, since insulin action on amino acid transport in bovine adrenals is inhibited by the action of voltage-dependent calcium channel blockers (Wassermann et al., 1994).

Insulin action on the $[^{14}C]\alpha$-AIB transport in the adrenal glands of bovines and rats was demonstrated by Wassermann et al. (1989) and Fauth et al. (1991). The presence of insulin in the incubation medium produced a significant increment in $[^{14}C]\alpha$-AIB uptake in adrenals incubated in Krebs–Ringer–bicarbonate buffer (KRb). In the sodium-free medium (KRb–choline) the uptake of $[^{14}C]\alpha$-AIB was considerably decreased. Similar results were obtained when the system A was blocked by MeAIB (20 nmol/L); under this condition, amino acid uptake was reduced to nearly half of the control values. Under these experimental conditions insulin was also ineffective (Fig. 3, Fig. 4). The adrenal response to insulin indicates that the gland has receptors that recognize the hormone and that insulin probably regulates important metabolic events.

Verapamil (2.5×10^{-4} M), a specific blocker for voltage-dependent Ca^{2+} channels (Goldfraind et al., 1986), produced a reduction in basal $[^{14}C]AIB$ transport and nullified insulin stimulatory action (Fig. 4). It has been shown that the increase in amino acid uptake produced by isoproterenol in mouse kidney cortex slices is abolished by calcium transport inhibitors (Goldstone et al., 1983). It has also been reported that the drug nullified the action of FSH on amino acid transport in the testes (Wassermann and Loss, 1989).

The specific inhibition of Na^+, K^+-ATPase by ouabain (1.0 mM) in the adrenals of cows produced a strong decrease in basal amino acid transport and total inhibition of the stimulatory action of insulin (Fig. 5).

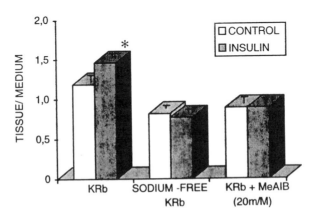

Figure 3. Effect of insulin (3×10^{-6} M) on the uptake of $[^{14}C]AIB$ in rat adrenal glands. Statistical evaluation was carried out using one-way and two-way ANOVA, followed by Duncan's multiple range test. Differences were considered to be significant when $p < 0.05$. *As compared with control, sodium-free KRb and KRb plus MeAIB.

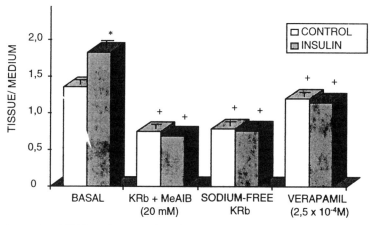

Figure 4. Uptake of [^{14}C]AIB in the bovine adrenal cortex (means ± SEM). Statistical analysis: (a) *Unpaired Student's t-test. Significantly different from the adrenal cortex incubated without insulin ($p < 0.01$). (b) +One-Way ANOVA, Duncan's multiple range test revealed significant differences ($p < 0.05$) between basal vs. MeAIB, KRb-choline, verapamil and insulin vs. MeAIB, KRb-choline, verapamil.

The action of the vanadate ion on the amino acid transport in adrenal glands of bovines and rats was tested at different concentrations of the ion. The chemical compound used in the experiment was sodium orthovanadate (Na_3VO_4). The presence of this salt changes the pH of the solutions as shown in Figure 6. In our experiments, the pH of the incubation media containing the vanadate ion was corrected to 7.4 with carbogen $O_2:CO_2$ (95:5 v/v), in

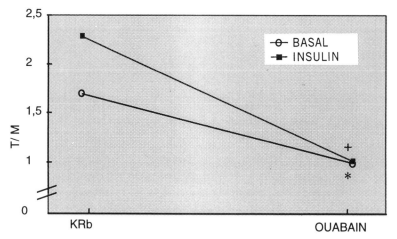

Figure 5. Effect of ouabain (1mM) on basal and insulin-stimulated [^{14}C]MeAIB uptake into the glomerulosa zone of bovine adrenal glands. Statistical evaluation was carried out using one-way and two-way ANOVA, followed by Duncan's multiple range test. +As compared with basal KRb, $p < 0.01$. *As compared with KRb plus insulin, $p < 0.01$.

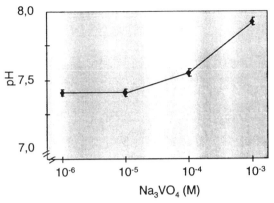

Figure 6. Influence of sodium vanadate concentration on KRb buffer pH (means ± SEM). The addition of vanadate to the KRb buffer increases the pH values. In the experiment with adrenals, the pH of the solutions was maintained by gasification with O_2:CO_2 (95:5 v/v).

order to obtain results within the physiological pH. The basal [^{14}C]MeAIB uptake and insulin-stimulated uptake had their maximum values at pH 7.4. Lower uptake rates were observed at pH 6.0 or 8.0 (Table 1).

Sodium vanadate produced an inhibitory effect on the basal as well as in the insulin-stimulated uptake of [^{14}C]MeAIB transport. At low concentrations of 10^{-6} and 10^{-5} M, insulin action was inhibited in the rat adrenal cortex (Table 2), whereas in cow adrenal cortex, it was reduced only slightly at these concentrations (Table 3). With 10^{-4} and 10^{-3} M vanadate, there was complete inhibition in the case of the cow adrenal. The effect of vanadate is probably due to the binding of the VO_3^- ion to Na^+,K^+-ATPase, slowing the E_2–E_1 conformational change, and so diminishing enzyme activity (Beaugé and Dipolo, 1979).

The perfect operation of ATPases has a great importance in amino acid-mediated transport, since this phenomenon depends on the sodium and electrochemical energy gradient.

Table 1   Influence of pH on [^{14}C]MeAIB Uptake into Slices of Bovine Adrenal Cortex (means ± SEM)[a]

pH	Control	Insulin	Increase, %
6.0	2.05 ± 0.05	2.43 ± 0.03*	18.5
7.4	2.58 ± 0.20	3.44 ± 0.28[+]	34.0
8.0	2.21 ± 0.13	2.94 ± 0.08*	33.0

[a] Statistical evaluation was carried out using one-way and two-way ANOVA, followed by Duncan's multiple range test. Differences were considered to be significant when $p < 0.05$. *As compared with control pH 7.4 ($p < 0.05$). [+]As compared with insulin pH 6 and 8, $p < 0.05$.

Table 2 Vanadate Concentration-Dependent Inhibition of Basal and Insulin-Stimulated [^{14}C]MeAIB Uptake in Rat Adrenal Glands (means ± SEM)[a]

Na$_3$VO$_4$ (M)	Control	Insulin
0	1.07 ± 0.07*	1.30 ± 0.05
10^{-6}	0.86 ± 0.07[+]	0.98 ± 0.07[+]
10^{-5}	0.80 ± 0.04[+]	0.85 ± 0.04[+]
10^{-4}	0.92 ± 0.03[+]	1.04 ± 0.08[+]
10^{-3}	0.84 ± 0.07[+]	0.92 ± 0.09[+]

[a] Statistical evaluation was carried out using one-way and two-way ANOVA, followed by Duncan's multiple range test. Differences were considered to be significant when $p < 0.05$. *As compared with insulin, $p < 0.05$; [+]As compared with control without insulin, $p < 0.05$.

The actions of vanadate and ouabain on the basal and insulin-stimulated amino acid transport showed qualitative and quantitative differences. Ouabain, described recently as a steroid produced by the adrenal gland, presents an accurate regulatory effect on the ATPases and can be considered, like the vanadate ion, a potent physiological regulator or modulator of the sodium–potassium pump.

The vanadate ion and ouabain inhibitory effects on Na$^+$,K$^+$-ATPase are synergic. Ouabain as well as the K$^+$ and Mg^{2+} ions, agents that promote the definition of the E$_2$ conformational structure, stimulate binding of the vanadate ion to the enzyme. In another way, in the absence of ATP, the vanadate ion is able to promote the binding of ouabain to the Na$^+$,K$^+$-ATPase, potentializing the inhibitory effect on enzyme activity. The synergic action of sodium vana-

Table 3 Inhibition of Basal and Insulin-Stimulated [^{14}C]MeAIB Uptake by Different Molar Concentrations of Vanadate in Bovine Adrenal Cortex Slices (means ± SEM)[a]

Na$_3$VO$_4$ (M)	Control	Insulin
0	2.94 ± 0.10*	3.56 ± 0.12
10^{-6}	2.66 ± 0.13*	3.32 ± 0.17
10^{-5}	2.71 ± 0.18*	3.05 ± 0.32*
10^{-4}	2.72 ± 0.08*	2.90 ± 0.22*
10^{-3}	2.02 ± 0.10[+]	2.31 ± 0.11[+]

[a] Statistical evaluation was carried out using one-way and two-way ANOVA, followed by Duncan's multiple range test. Differences were considered to be significant when $p < 0.05$. *As compared with insulin, $p < 0.05$; [+]As compared with control without insulin, $p < 0.05$.

Table 4   Inhibitory Effect of Ouabain, Sodium Vanadate, or Ouabain Plus Sodium Vanadate in [^{14}C]MeAIB Uptake in Rat Adrenal Glands (means ± SEM)[a]

Group	^{14}C(MeAIB) uptake	Inhibition, %
Control	1.10 ± 0.02	
Na$_3$VO$_4$ (10^{-7} M)	0.99 ± 0.05	10
Ouabain (0.1 mM)	0.92 ± 0.02[a]	17
Na$_3$VO$_4$ (10^{-7} M) + Ouabain (0.1 mM)	0.69 ± 0.02[b,c,d]	37

[a] Statistical evaluation was carried out using one-way and two-way ANOVA, followed by Duncan's multiple range test. a, b, c, d $p < 0.05$: (a) as compared with control group, (b) as compared with Na$_3$VO$_4$ (10^{-7} M) group, (c) as compared with ouabain (0.1 mM) group, (d) as compared with control group.

date and ouabain on the inhibition of the transport of methylaminoisobutyric acid through the system A, which is dependent on the Na$^+$,K$^+$-ATPase, has been analyzed in the adrenals of rats (see Table 4). However, the synergic action of these two endogenous substances is not very well known and explored. The important physiological control exerted by them on ATPases has not been fully considered and its significance has not been emphasized.

6.   CONCLUSIONS

According to Cantley et al. (1977) and Nechay (1984), naturally occurring vanadium is among the lowest of trace elements in mammals. It is considered an element essential to the perfect development and function of organisms of various species, including mammals and birds. A diversity of values have been found that constitute physiological concentrations of vanadium. This element interacts with membrane ATPase in concentrations that vary from nanomolar to millimolar, and it inhibits neutral amino acid transport in mammal adrenals in concentrations from micromolar to millimolar. These concentrations can be found in tissues and plasma of animals and humans exposed to contamination sources, which implies that the potential toxic effects of vanadate on these physiological phenomena are a real possibility.

ACKNOWLEDGMENTS

This work was supported by grants from Fundação de Amparo a Pesquisa do Rio Grande do Sul, Pro-Reitoria de Pesquisa UFRGS, Conselho Nacional de Desenvolvimento Científico e

Tecnologico, Financiadora de Estudos e Projetos, Brasil and Capes, Ministério de Educação, Brasil. M. G. F. is indebted to Pontificia Universidade Católica do Rio Grande do Sul, Porto Alegre, Brasil.

REFERENCES

Beaugé, L. A., and Dipolo, R. (1979). Vanadate selectively inhibits the K_0^+-activated Na^+ efflux in squid axon. *Biochim Biophys. Acta.* **551**, 220–223.

Cantley, L. C., Jr., and Aisen, P. (1979). The fate of cytoplasmatic vanadium. Implications on (Na, K)-ATPase inhibition. *J. Biol. Chem.* **254**, 1781–1784.

Cantley, L. C., Jr., Cantley, L. G., and Josephson, L. (1978). A characterization of vanadate interactions with the (Na, K)-ATPase. Mechanistic and regulatory implications. *J. Biol. Chem.* **253**, 7361–7368.

Cantley, L. C. Jr., Josephson, L., Warner, R., Yanagisawa, M., Lechene, C., and Guidotti, G. (1977). Vanadate is a potent (Na, K)-ATPase inhibitor found in ATP derived from muscle. *J. Biol. Chem.* **252**, 7421–7423.

Delicado, E. G., and Miras Portugal, M. T. (1987). Glucose transporters in isolated chromaffin cells. *Biochem. J.* **243**, 541–547.

Duckworth, W. C., Solomon, S. S., Liepnieks, J., Hamel, F. G., Hand, S., and Peavy, D. E. (1988). Insulin-like effects of vanadate in isolated rat adipocytes. *Endocrinology* **122**, 2285–2289.

Fauth, M. G., Barreto, K. P., and Wassermann, G. F. (1991). Insulin stimulatory action on amino acid uptake in bovine adrenal cortex or glomerulosa zone. *Comp. Biochem. Physiol.* **98A**, 513–515.

Fehlmann, M., Le Cam, A., and Freychet, P. (1979). Insulin and glucagon stimulation of amino acid transport in isolated rat hepatocytes. *J. Biol. Chem.* **254**, 10431–10437.

Gazzola, G. C., Dall'Asta, V., and Guidotti, G. G. (1980). The transport of neutral amino acid in cultured human fibroblasts. *J. Biol. Chem.* **255**, 929–936.

Goldfine, I. D. (1981). Interaction of insulin, polypeptide hormones and factors with intracellular membranes. *Biochim. Biophys. Acta,* **650**, 53–67.

Goldfraind, T., Miller, R., and Wibo, M. (1986). Calcium antagonism and calcium entry blockade. *Pharmacol. Rev.* **38**, 321–416.

Goldstone, A. D., Koenig, H., Lu, C. Y., and Trout, J. J. (1983). β-Adrenergic stimulation evokes a rapid, Ca^{2+}-dependent stimulation of endocytosis, hexose and amino acid transport associated with increased Ca^{2+} fluxes in mouse kidney cortex. *Biochem. Biophys. Res. Commun.* **114**, 913–921.

Hamilton, B. P., Manunta, P., Lardo, J., Hamilton, J. H., and Hamlyn, J. M. (1994). The new adrenal steroid hormone ouabain. *Curr. Opinion Endocrinol. Diabetes,* **1**, 123–131.

Ito, Y., Yasuda, K., and Takeda, N. (1992). Characterization of insulin receptors in bovine adrenal cortex and medulla. *Endocrinol. Jpn.* **39**, 217–222.

Kadota, S., Fantus, G., and Deragnon, G. (1987). Peroxide(s) of vanadium: A movel and potent insulin-mimetic agent which activates the insulin receptor kinase. *Biochem. Biophys. Res. Commun.* **147**, 259.

Lienhard, G. E. (1983). Regulation of cellular membrane transport by the exocytotic ansertion and endocytic retrieval of transporters. *Trends Biochem. Sci.* **8**, 125–127.

Longo, N., Franchi-Gazzola, R., Bussolati, O., Dall'Asta, V., Foá, P. P., Guidotti, G. G., and Gazzola, G. C. (1985). Effect of insulin on the activity of amino acid transport systems in cultured human fibroblasts. *Biochim. Biophys. Acta* **844**, 216–223.

Machado, V. L. A., Silva, R. S. M., Marques, M., and Wassermann, G. F. (1982). Insulin specific uptake and action on deoxiglucose transport in rat adrenal glands. *Med. Sci. Res.* **10**, 459–460.

Nechay, B. R. (1984). Mechanisms of action of vanadium. *Ann. Rev. Pharmacol. Toxicol.* **24**, 501–524.

Nechay, B. R., Nanninga, L. B., Nechay, P. S. E., Post, R. L., Grantham, J. J., Macara, I. G., Kubena, L. F., Phillips, T. D., and Nielsen, F. H. (1986). Role of vanadium in biology. *Fed. Proc.* **45**, 123–132.

Orth, N. D., Kovacs, J. W., and Debold, C. R. (1992). The adrenal cortex. In *Williams Textbook of Endocrinology.* Philadelphia: Saunders, pp. 489–619.

Pilkis, S. J., and El-Maghrabi, M. R. (1988). Hormonal regulation of hepatic gluconeogenesis and glycolysis. *Ann. Rev. Biochem.* **57**, 755–783.

Piras, M. M., Bindstein, E., and Piras, R. (1973). Regulation of glycogen metabolism in the adrenal gland. IV. The effect of insulin on glycogen synthetase, phosphorylase and related metabolites. *Arch. Biochem. Biophys.* **154**, 263–269.

Prentki, M., Crettaz, M., and Jeanrenaud, B. (1981). Role of microtubules in insulin and glucagon stimulation of amino acid transport in isolated rat hepatocytes. *J. Biol. Chem.* **256**, 4336–4340.

Roden, M., Liener, K., Fürnsinn, C., Prskavec, M., Nowotny, P., Steffan, I., Vierhapper, H., and Waldhäusl, W. (1993). Non-insulin-like action of sodium orthovanadate in the isolated perfused liver of fed, non-diabetic rats. *Diabetologia,* **36**, 602–607.

Sabbioni, E., Pozzi, G., Pintar, A., Casella, L., and Garattini, S. (1991). Cellular retention, cytotoxicity and morphological transformation by vanadium(IV) and vanadium(V) in BALB/3T3 cells lines. *Carcinogenesis* **12**, 47–52.

Siegel, G. (1989). *Basic Neurochemistry.* Raven Press, New York, pp. 64–67.

Silva, R. S. M., Machado, V. L. A., Marques, M., and Wassermann, G. F. (1984). Insulin binding sites and action in the adrenal glands from normal and streptozotocin diabetic rats. *Horm. Metabol. Res.* **16**, 77–81.

Simonian, M. H., and Gill, G. N. (1979). Regulation of deoxyribonucleic acid synthesis in bovine adrenocortical cells in culture. *Endocrinology* **104**, 588–595.

Simonian, M. H., White, M. L., and Gill, G. N. (1982). Growth and function of cultured bovine adrenocortical cells in a serum-free defined medium. *Endocrinology* **111**, 919–927.

Simonoff, M., Llabador, Y., Peers, A. M., and Simonoff, G. N. (1984). Vanadium in human serum, as determined by neutron activation analysis. *Clin. Chem.* **30**, 1700–1703.

Spector, W. S. (1956). *Handbook of Biological Data.* Saunders, Philadelphia, Table 273.

Tamura, S., Brown, T. A., Whipple, J. H., Fujita-Yamaguchi, Y., Bubler, R. E., Cheng, K., and Larner, J. (1984). A novel mechanism for the insulin-like effect of vanadate on glycogen synthase in rat adipocytes. *J. Biol. Chem.* **259**, 6650–6658.

Tovar, R. A., Tews, J. K., Torres, N., and Harper, A. E. (1991). Neutral amino acid transport into rat skeletal muscle: Competition, adaptive regulation, and effects of insulin. *Metabolism* **40**, 410–419.

Varecka, L., and Carafoli, E. (1982). Vanadate-induced movements of Ca^{2+} and K^+ in human red blood cells. *J. Biol. Chem.* **257**, 7414–7421.

Wassermann, G. F., Barreto, K. P., and Fauth, M. G. (1994). Ca^{2+} channels and insulin action on amino acid uptake by bovine adrenal cortex. *Med. Sci. Res.* **22**, 601–602.

Wassermann, G. F., Fauth, M. G., and Machado, V. L. A. (1989). Insulin action on amino acid uptake by rat adrenal glands in vitro. *Med. Sci. Res.* **17**, 675–676.

Wassermann, G. F., and Loss, E. S. (1989). Effect of calcium channel blocker, verapamil, on amino acid uptake stimulated by FSH in rat testes. *Med. Sci. Res.* **17**, 779–780.

Wilson, S. P., and Kirshner, N. (1983). Effects of ascorbic acid, dexamethasone and insulin on the catecholamine and opioid peptide stores of cultured adrenal medullary chromaffin cells. *J. Neurosci.* **3**, 1971–1978.

Zhang, A., Gao, Z. Y., Gilon, P., Nenquin, M., Drews, G., and Henquin, J. C. (1991). Vanadate stimulation of insulin release in normal mouse islets. *J. Biol. Chem.* **266**, 21649–2165.

12

OXIDATIVE STRESS AND PRO-OXIDANT BIOLOGICAL EFFECTS OF VANADIUM

Janusz Z. Byczkowski

ManTech Environmental Technology, Inc., Dayton, Ohio 45437-0009

Arun P. Kulkarni

Toxicology Program, College of Public Health MDC-56, University of South Florida, Tampa, FL 33612

Vanadium in the Environment. Part 2: Health Effects, Edited by Jerome O. Nriagu.
ISBN 0-471-17776-8. © 1998 John Wiley & Sons, Inc.

1. INTRODUCTION

1.1. Vanadium as a Main Inorganic Pollutant of Petroleum

Fossil fuels, especially petroleum (crude mineral oil), contain significant amounts of vanadium compounds (Eckardt, 1971; Sokolov, 1986). Vanadium and other trace metals, unlike organic pollutants, are not biodegradable in the environment. Therefore, inorganic vanadium compounds redistributed by human activity tend to build up in the ecosystem to levels that may be toxic to living organisms. It was estimated that as much as 66,000 tons of vanadium are released and redistributed into the atmosphere each year (Nriagu and Pacyna, 1988). Devastating effects on the environment may result from massive incidental and/or intentional burning of vanadium-containing crude oil and its spilling into the sea (Sadiq and Zaidi, 1984; Kalogeropoulos et al., 1989; Vasquez et al., 1991; Madany and Raveendran, 1992; Moeller et al., 1994). Crude oil from certain locations may be especially rich in vanadium. For example, fly ash resulting from combustion of Venezuelan oil may contain up to 80% of vanadium compounds (Hudson, 1964).

In addition to fossil fuels, some ores may contain significant amounts of this metal, and thus, occupational exposure to vanadium is quite common in modern petrochemical, mining, and steel industries (Goldsmith et al., 1976; Lees, 1980; Fisher et al., 1983; Schiff and Graham, 1984; White et al., 1987; Karimov et al., 1988, 1989, 1991; Sarsebekov et al., 1994). Vanadium is often associated with uranium ore and may contribute to the increased occupational risk to those employed in uranium mines (Paschoa et al., 1987). Huge amounts of vanadium are usually deposited in the smokestacks and exhaust systems of engines, boilers, and generators powered by heavy fuel oils (e.g., mazut). Vanadium may also be found in diesel fuel exhaust particles (Kleinman et al., 1977; Levy et al., 1984; Rossi et al., 1986; Pisteli et al., 1991; Todaro et al., 1991; Hauser et al., 1995). Vanadium pentoxide (V_2O_5) is the most ubiquitous vanadium compound (Troppens, 1969; Nechay, 1984), although the natural oil-fired fly ash may also contain as many as a dozen other vanadium compounds (Bowden et al., 1953).

Vanadium toxicity is a true concern for industrial workers and military personnel exposed to its compounds on land and sea. In addition to vanadium exposure at the work place (Zychlinski, 1980), the general population is also increasingly exposed to this metal (Flyger et al., 1976), mostly as a result of increased utilization of vanadium-containing natural oil (Schiff and Graham, 1984). In humans, vanadium-bearing particles may persist in the lungs for many years (Paschoa et al., 1987). Thus, among the inhabitants of U.S. cities, vanadium deposits in lungs are markedly increasing with age as a result of its accumulation (Tipton and Shaffer, 1964).

Acute inhalation of dust containing high concentrations of vanadium can cause harmful health effects in humans, mostly in the respiratory tract, including lung irritation, coughing, wheezing, chest pain, runny nose, and sore throat. No comprehensive human studies are available on the health effects of chronic exposure or the carcinogenicity of vanadium; thus, the U.S. Department of Health and Human Services and the U.S. Environmental Protection Agency have not classified vanadium as to its human carcinogenicity (Agency for Toxic Substances and Disease Registry, 1992). A focused issue, containing invited reviews and original research papers on mechanisms and biochemical effects of vanadium, edited by Srivastava and Chiasson (1995), was published in *Molecular and Cellular Biochemistry.*

1.2. Oxidative Stress

Animal studies have shown that vanadium compounds induce oxidative stress and lipid peroxidation in vivo (Stohs and Bagchi, 1995). An exposure of murine C3H/10T1/2 cells in culture to as low as 5–20 μM vanadium(V) was found to induce up to 100-fold expression of the proliferin gene family, which is indicative of a cellular state of oxidative stress (Parfett and Pilon, 1995). Even concentrations 10 times lower (0.5–2 μM vanadium(V)) stimulated growth of MCF-7 cells in vitro. The proliferative effect of vanadium(V) on these cells reached a plateau at 1 μM, declined at 3 μM, and disappeared at 5 μM (Auricchio et al., 1996).

The significance and biological implications of chemically induced oxidative stress have been reviewed extensively by Byczkowski and Channel (1996). In summary, oxidative stress is a pathophysiological process in which intracellular balance between endogenous as well as exogenous pro-oxidants and antioxidants is shifted towards pro-oxidants, leaving cells unprotected from free-radical attack. Ultimate cellular pro-oxidants are mostly free radicals (Fig. 1), defined as molecules or groups of atoms that have one or more unpaired electrons and are capable of independent existence. These free radicals derive

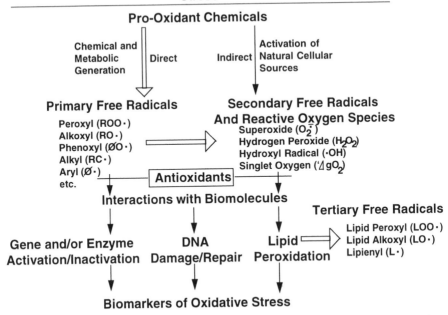

Figure 1. Generation of primary, secondary, and tertiary free radicals and reactive oxygen species involved in the oxidative stress induced by pro-oxidant chemicals (modified from Byczkowski and Channel, 1996).

directly from pro-oxidant chemicals (Kehrer, 1993) or may arise indirectly from stimulated natural functions in aerobic cells (Byczkowski and Gessner, 1988); for instance, free radicals may be produced by stimulated lipid peroxidation (Gower, 1988; Finley and Otterburn, 1993; Haegele et al., 1994). Thus, from the pro-oxidant xenobiotics, further chemical and/or metabolic reactions may generate primary free radicals (Fig. 1). Then, in an avalanche-type process, secondary free radicals and reactive oxygen species may be released (for review see Roberfroid and Calderon, 1994). Both primary and secondary free radicals may initiate lipid peroxidation, leading to the generation of tertiary free radicals (Fig. 1). Additional factors, such as aging (Stadtman et al., 1993), dietary deficiencies, or the presence of transition metals (Kulkarni and Byczkowski, 1994a,b), may augment the oxidative stress. A depletion of cellular antioxidants such as vitamin E, C or reduced glutathione (GSH), and a defective enzymatic scavenging by catalase, selenium-containing GSH peroxidase, and superoxide dismutase, further increase the level of free radicals and may enhance damage to cellular components. Oxidative stress can be reversed by natural and synthetic antioxidants (Williams, 1993; Papas, 1993; Pratt, 1993). Oxidative stress triggers biological reactions that counteract cellular damage. It may induce enzymes with free-radical scavenging and repair activities and/or activate an oxidative-stress-responsive nuclear transcription factor κB (NFκB; Meyer et al., 1993). The importance of oxidative stress and the published literature detailing methodology employed for its measurement have been reviewed in three recent publications by Pryor (1993), Sies (1994), and Jaeschke (1995), and in a multiple-author book (Spatz and Bloom, 1992).

Chemically induced oxidative stress causes derangement of antioxidant mechanisms in tissues (Videla et al., 1990), may lead to lipid peroxidation (Comporti, 1985), and may cause stimulation of cellular proliferation and/or apoptosis (Corcoran et al., 1994) that may finally result in cell injury (de Groot and Littauer, 1989). A study by Biasi et al. (1995) demonstrated that lipid peroxidation is a cause rather than an effect of necrotic tissue damage. Figure 2 summarizes dose dependency observed in the oxidative stress induced by pro-oxidant chemicals.

Chen and Chan (1993), using 3T3-L1 cells cultured in a serum-free medium, demonstrated that vanadium(V) increases [^{3}H]thymidine incorporation into DNA and enhances expression of the *c-fos* gene in a manner analogous to a redox cycling naphthoquinone (DMNQ). The authors suggested that both pro-oxidant compounds, orthovanadate and DMNQ, increased tyrosine protein phosphorylation early in the signal transduction cascade of growth factor receptors, leading to augmentation of cell proliferation. Apparently the common factor in the mode of action of these two completely dissimilar chemicals was an oxidative stress caused as a result of intracellular redox cycling (Chen and Chan, 1993).

2.   PRO-OXIDANT PROPERTIES OF VANADIUM

A pro-oxidant effect of vanadium in tissue preparations was first described by Bernheim and Bernheim (1938). Since then, the biologically relevant redox properties of vanadium have stirred both extensive research activity and controversy. Especially vanadium-stimulated NAD(P)H oxidation, reported for the first time by Byczkowski and Zychlinski (1978) in mitochondrial preparations, attracted much attention and became a subject of passionate discussions (Erdmann et al., 1981; Liochev and Fridovich, 1990, 1996). Because the progress in this field has been rapid, it was not possible, in the format of only one chapter, to review all of the work and cite each relevant publication. Instead, only selected experimental and review papers that we thought were key to understanding and substantiating the phenomenon of oxidative stress caused by vanadium are discussed. We apologize to those whose work is not cited here.

2.1.   Compartmentalization of Vanadium(V) and Vanadium(IV) in the Organism

In oxygenated blood, the absorbed vanadium circulates as polyvanadate (vanadium(V), isopolyanions containing vanadium in the +5 oxidation state). Dif-

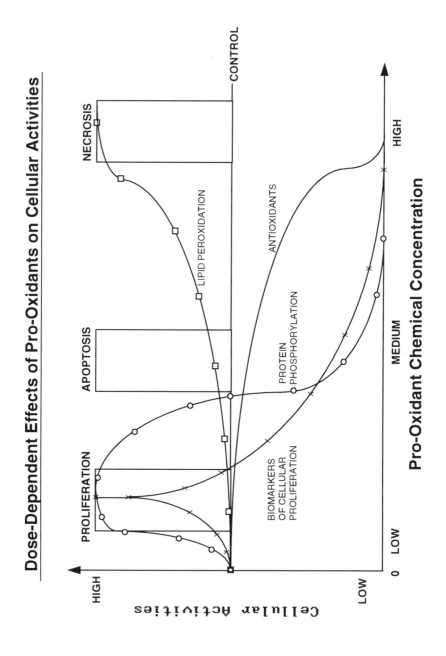

Dose-Dependent Effects of Pro-Oxidants on Cellular Activities

Cellular Activities

HIGH

LOW

PROLIFERATION

APOPTOSIS

NECROSIS

CONTROL

LIPID PEROXIDATION

ANTIOXIDANTS

PROTEIN PHOSPHORYLATION

BIOMARKERS OF CELLULAR PROLIFERATION

0 LOW   MEDIUM   HIGH

Pro-Oxidant Chemical Concentration

ferent tissues retain vanadium mainly as vanadyl (vanadium(IV), cationic form of vanadium in the +4 oxidation state) (Erdmann et al., 1984) in the presence of endogenous reducing compounds (such as glutathione-SH; Bruech et al., 1984). The highest concentration of vanadium in rats fed either vanadium(V) or vanadium(IV) was found in the kidneys, while the liver and spleen contained about three times the lower concentrations. The lowest accumulations of vanadium were found in the lung, blood plasma, and blood cells. The half-life for vanadium elimination from the body was about 12 days in rats (Ramanadham et al., 1991). To date, apparently, there is no systematic pharmacokinetic study conducted on vanadium absorption, distribution, metabolism, and disposition, and no pharmacokinetic model is available describing comparative kinetics and toxicity of vanadium administered by different routes.

A successful attempt to isolate an inhibitor of (Na^+,K^+)-ATPase, inherent in the Sigma Grade commercial ATP preparations isolated from muscle, led Cantley et al. (1977) to the discovery that sodium orthovanadate (Na_3VO_4, a salt containing vanadium(V)) inhibits this activity at a very low concentration. In rat cardiac subcellular preparations, vanadium(V) was an even more potent inhibitor of the (Na^+,K^+)-ATPase than cardiac glycoside (ouabain), which is used as a "specific inhibitor" of the sodium pump (Erdmann et al., 1984). However, a high degree of inhibition of (Na^+,K^+)-ATPase activity was observed only in the purified enzyme systems or when damaged subcellular preparations were employed. Since no evidence of inhibition was noted in intact cells or in vivo, it has been proposed that some kind of compartmentalization and/or "detoxification" of vanadium exists in the undamaged tissue. It was suggested that the inhibitory vanadium(V) might be intracellularly reduced to vanadyl (vanadium(IV)), which was believed to be noninhibitory (Erdmann et al., 1984). Inside the cell, vanadium(IV) may further bind to several endogenous ligands, and be protected against reoxidation (Nechay et al., 1986). Although, as shown in the hog kidney (North and Post, 1984) and rat brain (Svoboda et al., 1984) microsomal preparations, vanadyl may also act as an inhibitor of (Na^+,K^+)-ATPase.

Using the inhibition of substrate oxidation in liver mitochondria (Byczkowski et al., 1979) as an indicator of vanadium(V) accumulation, it was possible

---

Figure 2. Effects of different concentrations of pro-oxidant chemicals on cellular function and activities. □—□, lipid peroxidation; ———, vitamin E-type antioxidant level; ○—○, protein tyrosine phosphorylation; ×—×, ornithine decarboxylase and S-adenosylmethionine decarboxylase activities. The curves are computer-generated results of simulation of lipid peroxidation and antioxidant depletion (BBPD computer program developed by Byczkowski et al., 1996, based on data by Tappel et al., 1989); protein tyrosine phosphorylation (BBPD computer program developed by Byczkowski and Flemming, 1996, based on data by Vroegop et al., 1995, and experimental results of Heffetz et al., 1990); and ornithine and S-adenosylmethinine decarboxylation (BBPD computer program developed by Byczkowski and Flemming, 1996, based on Corcoran et al., 1994).

Oxidation of Vanadyl Depending on O₂ Availability

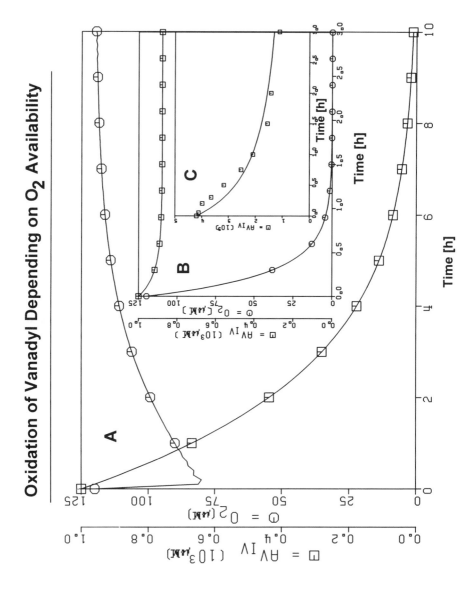

to demonstrate a significant amount of vanadate accumulation in the liver during experimental administration of V_2O_5 to rats (Zychlinski and Byczkowski, 1990). Although under physiological conditions, the partial oxygen pressure in liver is low (pO_2 from about 30 mm Hg in the perivenous zone to about 56 mm Hg in the periportal zone), the actual O_2 concentration in different compartments across the hepatic acinus may vary more than twofold (Kulkarni and Byczkowski, 1994b). Owing to preferential partitioning, it may be even three times higher in a lipid phase of biomembranes than in the water phase (Antunes et al., 1994). This suggests that the process of "detoxification" by reduction of vanadium(V) to vanadium(IV) in the rapidly perfused liver tissue may be insufficient to prevent its deleterious effects on biomembranes and bioenergetic functions, especially under conditions of chronic exposure to V_2O_5.

These apparent discrepancies may be explained by means of computer simulations of vanadyl oxidation with a biologically based pharmacokinetic (BBPK) model, considering the oxygen availability (Fig. 3). The model was calibrated with the experimental data from North and Post (1984), as shown in Figure 3C. Under conditions of chronic oxidative stress tissue reserves of GSH may be depleted. Under these circumstances, for example, 1 mM vanadyl may be completely reoxidized to vanadium(V) by the available oxygen within 10 h (Fig. 3A), provided that the tissue, stripped of antioxidants, is receiving a constant supply of oxygen (assumed to be initially no more than 124 μM around arterial vessels in a rapidly perfused tissue). However, the same initial concentration of oxygen may be completely depleted within 1 h in the presence of the same 1 mM vanadyl when the diffusion of fresh oxygen is restricted (Fig. 3B). This process may be drastically accelerated by the presence of traces of catalytically active transition metals, for example, copper (Byczkowski et al., 1988). Moreover, in the perfused rat liver, vanadium(V) caused significant vasoconstriction, which resulted in a progressive decrease in perfusion rate (Younes and Strubelt, 1991) that was irreversible by antioxidants (Younes et al., 1991). Thus, under the conditions of reduced rate of perfusion or hypoxia, at best up to 12% of vanadyl may be oxidized, assuming that vanadium(IV) oxidation is the only process using oxygen. In reality, the tissues become anaerobic much faster (for instance, an average estimated O_2 concentration in the liver was found to be only about 35 μM; Antunes et al., 1994), which

---

Figure 3. Oxidation of dissolved vanadyl, at nearly neutral pH, depending on oxygen availability. The curves are computer-generated results of simulation with a biologically based pharmacokinetic model (BBPK computer program based on North and Post, 1984, and Byczkowski et al., 1988), assuming replenishment of oxygen by diffusion from the air (C) or from perfusing blood (A), or with oxygen supply turned off (B). The curves marked with squares show vanadyl concentration (AV_{IV}, μM), and curves with circles show oxygen concentration (O_2, μM). The squares in Figure 3 C depict experimental data used for model calibration, from North and Post (1984).

may inhibit further reoxidation of intracellular vanadyl in poorly perfused or hypoxic tissues, even without the contribution of GSH and other cellular reductants.

2.2. Generation of Free Radicals

Autooxidation of vanadium(IV) with O_2 from the air is a potentially deleterious but rather a slow process, producing superoxide anion radical ($O_2^{\cdot-}$) and vanadium(V):

$$V_{(IV)} + O_2 \leftrightarrows V_{(V)} + O_2^{\cdot-} \qquad (a)$$

This reaction may either be drastically accelerated by the presence of catalytic amounts of transition metal cations (e.g., Cu^{2+}) or slowed down by chelators (e.g., EDTA; Byczkowski et al., 1988). Superoxide anion radicals may dismutate spontaneously or enzymatically (by superoxide dismutase, SOD), producing hydrogen peroxide (H_2O_2). It may, in turn, react with remaining vanadium(IV), producing hydroxyl free radical ($HO^{\cdot}$) in a Fenton-type reaction:

$$V_{(IV)} + H_2O_2 \rightarrow V_{(V)} + HO^{\cdot}$$

Theoretically, $HO^{\cdot}$ may react with H_2O_2 to generate $O_2^{\cdot-}$ in a Haber–Weiss reaction. While the Fenton-type oxidation of vanadium(IV) can easily be demonstrated under anaerobic conditions in vitro (Stankiewicz et al., 1991), its biological significance seems to be rather low (Byczkowski and Kulkarni, 1992b). On the other hand, in a chelating environment (Crans et al., 1989), superoxide may remain complexed within the vanadium moiety, forming a peroxy-vanadyl-type intermediate:

$$V_{(V)} + O_2^{\cdot-} \leftrightarrows [V_{(IV)} - OO^{\cdot}] \qquad (b)$$

It seems that amino acids, peptides, and other natural ligands (analogous to Tris buffer) may stabilize the product of one-electron reduction of vanadium(IV) (Byczkowski and Kulkarni, 1992b). In contrast to superoxide, this hypothetical peroxy-vanadyl intermediate probably disappears very quickly under the biologically relevant conditions (Liochev and Fridovich, 1990) via hydrogen abstraction from biomolecules:

$$[V_{(IV)} - OO^{\cdot}] + H^{\cdot} \rightarrow [V_{(IV)} - OOH] \leftrightarrows V_{(V)} + H_2O_2 \quad (c)$$

This reaction, through the pervanadate (vanadyl hydroperoxide) intermediate, may regenerate vanadium(IV), which is necessary for continuous operation of the redox cycle. The vanadium redox cycling, initiated by one-electron oxidation of vanadyl or one-electron reduction of vanadate with superoxide anion radical, may operate as a chain reaction providing primary, secondary, and tertiary free radicals and reactive oxygen species as long as $O_2^{\cdot-}$ and/or biomolecules capable of reducing vanadium(V) back to vanadium(IV) are available. A summary of the reactions involved in the vanadium redox cycling is shown schematically in Figure 4.

2.3. NAD(P)H Oxidation

The stimulatory effect of vanadate on oxidation of NADH was originally described by Byczkowski and Zychlinski (1978) in sonicated rat liver mitochondrial preparations (Zychlinski and Byczkowski, 1978; Byczkowski et al., 1979). Subsequently, a similar vanadium(V)-stimulated NADH oxidation was reported in cardiac cell membranes (Erdmann et al., 1979), followed by many claims for NADH and NADPH oxidation in several other subcellular preparations (Ramasarma et al., 1981) and enzymatic systems (Darr and Fridovich, 1984).

Vanadate can be directly reduced by NAD(P)H a biologically relevant pH, but this reaction is very slow (Byczkowski et al., 1979; Vyskocil et al., 1980):

$$V_{(V)} + NAD(P)H \rightarrow V_{(IV)} + NAD(P) \qquad (d)$$

In the presence of oxygen and an appropriate biological preparation, this reaction can be increased by a factor of several hundred. This reaction exhibits

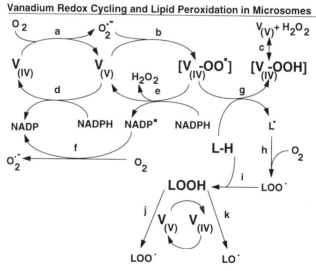

Figure 4. Summary scheme of the postulated reactions participating in vanadium redox cycling, NAD(P)H oxidation, and lipid peroxidation (according to Byczkowski et al., 1988; Byczkowski and Kulkarni 1992b).

specific pH optima (Vyskocil et al., 1980) and is sensitive to specific inhibitors and boiling (Ramasarma et al., 1981). No wonder that many researchers thought the reaction (d) to be catalyzed by either "NAD(P)H : vanadate reductase" or "vanadium-dependent NAD(P)H oxidase." In fact, this reaction, which is insensitive to inhibition by SOD or catalase, is a "general property of endomembranes" (Ramasarma et al., 1981) containing flavin coenzymes capable of accepting electrons from NAD(P)H. Vanadate can accept electrons in a way similar to ferricyanide (Ramasarma et al., 1981). The oxidation–reduction potential for one-electron reduction of ferricyanide under standard conditions (E'_0) is $+0.26$ V, whereas the E'_0 for one-electron reduction of vanadium(V) is even more positive (about $+1.0$ V). The biological importance of reaction (d) lies in its ability to provide reduced vanadium, vanadium(IV).

Under aerobic conditions, autooxidation of vanadium(IV) may generate $O_2^{\cdot-}$ (reaction a), which, reacting with vanadium(V), may turn on a fast and destructive one-electron oxidation of NAD(P)H:

$$[V_{(IV)} - OO^{\cdot}] + NAD(P)H \rightarrow V_{(V)} + NAD(P)^{\cdot} + H_2O_2 \quad (e)$$

As demonstrated by Liochev and Fridovich (1988), $O_2^{\cdot-}$ generated in the system containing vanadium(V) and NADH initiates a free-radical reaction and causes a rapid oxidation of NADH with the estimated chain length of 15 NADH oxidized per $O_2^{\cdot-}$. As expected, the overall process was inhibited by SOD (because of reaction b), but not by catalase. Under aerobic conditions, this process may be self-sustaining, as the next step can regenerate $O_2^{\cdot-}$. Thus, in the presence of oxygen, the free-radical intermediate NAD(P) may pass its unpaired electron on dioxygen, generating more $O_2^{\cdot-}$.

$$NAD(P)^{\cdot} + O_2 \rightarrow NAD(P) + O_2^{\cdot-} \quad (f)$$

Obviously, in a well-perfused tissue any endogenous compound (e.g., GSH) capable of reducing the initial amount of vanadate (generated endogenously or introduced) to vanadium(IV) may also initiate the vanadium redox cycling and depletion of NAD(P)H. This deleterious redox cycling may be broken by anoxia or by chain-breaking antioxidants. A summary of the main reactions involved in the vanadium-mediated NAD(P)H oxidation is shown in Figure 4.

2.4. Depletion of Cellular Antioxidants

Both water- and lipid-soluble antioxidants, along with enzymatic scavengers, represent the cellular defense mechanisms that allow, for instance, the liver tissue to detoxify as much as $60-72$ μmol $O_2^{\cdot-}$ per gram liver without significant injury (Jaeschke, 1995). However, significant differences exist in the tissue levels of antioxidant protection and the ratio of water-soluble versus lipid-soluble antioxidants in different species. For instance, human liver contains only 1–2 μmol GSH per gram as compared with 7–8 μmol GSH per gram in rat liver (Purucker and Wernze, 1990). On the other hand, physiological levels of lipid-soluble antioxidants and natural free-radical scavengers in a human liver are within an order of magnitude higher than in a mouse liver (Cutler, 1991; Sohal, 1993).

It was shown by Bruech et al. (1984) that exposure to vanadium(V) results in the depletion of cellular thiol pool. GSH is the main intracellular thiol compound that also plays an important role as a main water-soluble cellular antioxidant. GSH is involved in the homeostasis of the intracellular redox state and is coupled to the oxidation state of cysteine residues in proteins (Ziegler, 1985). Tissue GSH was shown to participate directly in vanadate inactivation (Kretzschmar and Braunlich, 1990). As was observed in vitro, vanadium(V) caused the oxidation of thiols, including GSH and cysteine, and the formation of thiyl radicals in this reaction was suggested (Shi et al., 1990). Depletion of GSH not only decreases the antioxidant defense in the cytosol, but also prevents regeneration of a vital lipid-soluble antioxidant, α-tocopherol (vitamin E), increasing the vulnerability of phospholipid-rich biomembranes to oxidative stress and lipid peroxidation.

Experimental treatment of rats with vanadium(V) in drinking water (0.15 mg of ammonium metavanadate/1 mL) for 14 days resulted in a decrease in the activity of two essential "antioxidant enzymes," catalase and glutathione peroxidase, in the liver and kidneys. This decrease was linked with the increased spontaneous lipid peroxidation measured in liver and kidney homogenates. However, there was no change in the activity of cytosolic and mitochondrial superoxide dismutase (Russanov et al., 1994).

3. EFFECTS ON MITOCHONDRIA

Experiments performed on rat liver and wheat seedling mitochondria showed that, depending on the localization of vanadate, in the intermembrane space or at the inner side of the inner mitochondrial membrane, either the inhibition of the respiration with NAD-linked substrates and succinate (but not ascorbate) or short-circuiting of the respiratory chain can be observed (Fig. 5; Byczkowski et al., 1979).

3.1. Inhibition of the Respiratory Chain by Vanadium(V)

In intact mitochondria vanadate accumulates in the intermembrane space and blocks electron transfer through the respiratory chain between cytochrome c_1 and cytochrome c, which causes inhibition of NADH-linked substrates and succinate (Byczkowski and Zychlinski, 1978; Zychlinski and Byczkowski, 1978; Byczkowski et al., 1979). In the isolated undamaged rat liver mitochondria, the respiration with glutamate and succinate at state 3 was more sensitive to inhibition by vanadium(V) than at state 4, whereas the efficiency of oxidative

Mode of Action of Vanadium in Mitochondria

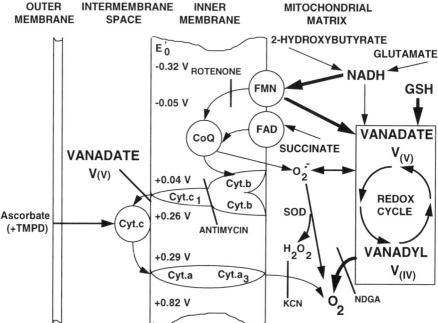

Figure 5. Summary scheme of the mode of action of vanadium in mitochondria (according to Byczkowski and Sorenson, 1984).

phosphorylation (measured as ratio ADP to O) was not affected significantly up to the 0.1 mM concentration of vanadium(V). Similarly, mitochondrial respiration with ascorbate (+N,N,N′,N′-tetramethyl-p-phenylenediamine, TMPD), which supplies electrons to cytochrome c, was not inhibited (Zychlinski and Byczkowski, 1990). The oxygen uptake by mitochondria isolated from the livers of rats treated intratracheally with a massive single dose of 5 mg V_2O_5/kg was significantly inhibited for up to 48 h with glutamate and succinate, but not with ascorbate. Even more pronounced inhibition was observed after chronic treatment of rats with 0.56 mg V_2O_5/kg monthly for 12 months (Zychlinski and Byczkowski, 1990). Other significant changes in the rats from the chronic treatment group included decreased blood glucose and cholesterol concentrations, and increased (almost twofold) average hydroxyproline content in the lungs (Zychlinski et al., 1991).

3.2. Short-Circuiting of the Respiratory Chain

In rat liver mitochondria damaged by sonication, vanadium(V) caused a dramatic stimulation of oxygen uptake with NADH-dependent substrates (Byczkowski and Zychlinski, 1978; Zychlinski and Byczkowski, 1978; Byczkowski et al., 1979). It seems that when the inner mitochondrial membrane permeability barrier to vanadate polyanions is broken, vanadium(V) undergoes a redox cycling. Most likely, vanadium(V) at the inner side of the inner mitochondrial membrane is reduced by NADH at the level of the flavoprotein center of the dehydrogenase (reaction d). Although reduction of vanadium(V) by NADH may proceed nonenzymatically, it is much slower than the reduction with dihydroflavin in flavoprotein containing flavin mononucleotide (FMN) (Byczkowski et al., 1979). The oxidation–reduction potential of flavoprotein under standard conditions (E'_0) is less negative than that of NADH (-0.05 V vs. -0.32 V, respectively; Lehninger, 1965); by comparison, the E'_0 for one-electron reduction of vanadium(V) is about $+1.0$ V, but may be further affected by chelation with amino acids, Tris, and other buffers.

It should be emphasized that any reaction capable of reducing vanadium(V) to vanadium(IV) (with NADH, GSH, sugars, etc.) under aerobic conditions would initiate vanadium redox cycling and eventually lead to the destructive reaction (5) and $O_2^{\cdot -}$ generation (reaction 6) (Fig. 4). Similarly, even trace amounts of $O_2^{\cdot -}$ may initiate vanadium redox cycling and produce even more $O_2^{\cdot -}$ in a chain reaction (Liochev and Fridovich, 1988). Accordingly, it was proposed that NAD(P)H oxidation in the presence of vanadium(V) amplifies the initial generation of $O_2^{\cdot -}$ and thus may be used as a sensitive assay method for vanadium(V) (Liochev and Fridovich, 1990). While the idea of $O_2^{\cdot -}$ "amplification" appears attractive, its application to the results observed with mitochondrial preparations is very difficult. First of all, mitochondria contain a highly active SOD, and even in isolated submitochondrial sonic particles, washed and recentrifuged, a residual SOD activity enriched the reaction mixture with O_2 (from the dismutated $O_2^{\cdot -}$). The additional amount of O_2 initially caused a substantial delay in the recorded oxygen uptake compared with the NADH oxidation measured spectrophotometrically (Byczkowski et al., 1979). Eventually the traces of SOD became overwhelmed and, finally, destroyed by free radicals, and only then did the oxygen uptake achieve its fastest rate. Moreover, when NADH oxidation was initiated by high concentrations of vanadium(IV), the reaction was supported by boiled mitochondria, purified cytochrome c oxidase, or even inorganic Cu^{2+} (Crane, 1975; Byczkowski and Zychlinski, 1978; Zychlinski and Byczkowski, 1978; E'_0 potential for cytochrome a is $+0.29$ V, and for Cu^{2+} is $+0.15$ V).

The proposed mode of action of vanadium on mitochondria is schematically shown in Figure 5. This scheme is based on the results of our experiments and the evidence presented by other investigators (Byczkowski and Sorenson, 1984). The overall process may be triggered either with vanadium(V) by one-electron reduction or with vanadium(IV) by one-electron oxidation. However, Liochev and Fridovich (1996) have recently proposed that the role of a biological membrane in this process is to produce $O_2^{\cdot -}$, implying that the vanadium redox cycling may be initiated only by $O_2^{\cdot -}$. Their arguement does not explain why, in intact plant mitochondria and sonicated rat liver mitochondria, the vanadium(V)-stimulated oxygen uptake is insensitive to rotenone with malate and/or glutamate (+NAD). The insensitivity of this process to rotenone (which in mitochondria blocks the respiratory chain at the site before ubiquinone, and keeps flavin-dependent dehydrogenases in the reduced state but prevents the leakage of $O_2^{\cdot-}$) rules out the involvement of $O_2^{\cdot-}$ in the vanadium(V)-stimulated mitochondrial oxygen uptake, at least in liver and plant mitochondria (Byczkowski et al., 1979).

4.   EFFECTS ON MICROSOMES

In a process analogous to that described above for mitochondria, the addition of vanadium(V) to the isolated rat hepatic or human placental microsomes, in the presence of NADPH, also evoked a significant increase in the oxygen uptake and NADPH oxidation rate (Byczkowski at al., 1988; Zychlinski et al., 1991). The experiments performed with microsomes isolated from the human placenta showed that vanadium(V) and vanadium(IV) trigger lipid peroxidation due to redox cycling, increase oxygen uptake, and cause the depletion of NADPH (Byczkowski et al., 1988). Although the mechanism proposed for this phenomenon was criticized by Liochev and Fridovich (1990), to date it still remains the best possible explanation for the results obtained (Byczkowski and Kulkarni, 1992b).

4.1.   Microsomal Lipid Peroxidation

In microsomes from human tissue, vanadyl and vanadate triggered lipid peroxidation due to redox cycling and formation of the reactive peroxy–vanadyl complex with superoxide (Byczkowski et al., 1988). On the basis of our experiments and the evidence presented by other workers, the following mechanism was postulated (Byczkowski et al., 1988): (1) Vanadyl(IV) oxidizes non-enzymatically, generating vanadium(V) and superoxide ($O_2^{\cdot-}$) (reaction a). (2) vanadium(V) with superoxide instantaneously forms a peroxy-vanadyl complex (vanadium(V)–OO$^{\cdot}$) (reaction b). (3) The peroxy-vanadyl complex attacks microsomal polyunsaturated lipid (L-H), abstracting a hydrogen atom and initiating lipid peroxidation (reaction g). The overall process (Fig. 4) leads to vanadium redox cycling, consumption of O_2, depletion of NAD(P)H, and destruction of microsomal lipids, and represents a source of reactive oxygen species.

5.   CARCINOGEN CO-OXYGENATION

In several target tissues, vanadium is very likely to undergo one-electron redox cycling and initiate lipid peroxidation (Byczkowski et al., 1988). Under these pro-oxidant conditions, aromatic hydrocarbon copollutants such as benzo(a)-

pyrene (B(a)P), may be subjected to co-oxygenation and activation to their reactive intermediates, which bind to the vital macromolecules (Dix and Marnett, 1983). This process of co-oxygenation was observed earlier by Byczkowski and Gessner (1987, a,b,c,d) during the interaction of B(a)P with asbestos and/or catalytically reactive iron in mouse liver microsomes. The results of the experiments with isolated lipoxygenase supplemented with linoleic acid suggested an involvement of lipid peroxyl and other free-radical products of lipid peroxidation in activation of B(a)P-7,8-dihydrodiol to the ultimate mutagenic and carcinogenic epoxide (Byczkowski and Kulkarni, 1989, 1990a, 1992a). Therefore, co-occurrence of vanadium and benzo(a)pyrene as environmental pollutants raises a major concern for the possibility of synergistic interaction between them (Byczkowski and Kulkarni, 1990b).

5.1. Polyunsaturated Fatty Acid Peroxidation

In an air-saturated incubation medium containing polyunsaturated fatty acid, the autooxidation of vanadium(IV) initiates lipid peroxidation, in a process analogous to that described above for microsomes (Zychlinski et al., 1991). The reaction can be partially inhibited by chain-breaking antioxidant (nordihydroguaiaretic acid, NDGA) and accelerated by preformed hydroperoxides (Byczkowski and Kulkarni, 1990b). Under identical conditions, vanadium(V) had no effect. Apparently, the reactive intermediate responsible for initiation of lipid peroxidation was not a pentavalent vanadium itself, but rather its complex with $O_2^{\cdot-}$ (peroxy-vanadyl):

$$[V_{(IV)} - OO^{\cdot}] + L - H \rightarrow [V_{(IV)} - OOH] + L^{\cdot} \qquad (g)$$

The generation of lipid-derived free radicals was sufficient to propagate lipid peroxidation in the linoleic acid system in vitro:

$$L^{\cdot} + O_2 \rightarrow LOO^{\cdot} \qquad (h)$$

The next step in the process is expected to yield lipid hydroperoxide:

$$LOO^{\cdot} + LH \rightarrow LOOH + L^{\cdot} \qquad (i)$$

However, the end products of vanadium(IV)-initiated peroxidation of linoleic acid did not contain linoleate hydroperoxide (Byczkowski and Kulkarni, 1992b). Therefore, it was postulated that vanadium depletes lipid hydroperoxides as shown in the following reactions:

$$LOOH + V_{(V)} \rightarrow LOO^{\cdot} + V_{(IV)} \qquad (j)$$

$$LOOH + V_{(IV)} \rightarrow LO^{\cdot} + V_{(V)} \qquad (k)$$

Both reactions (j) and (k) are not dependent on superoxide ion radicals; they generate reactive lipid alkoxyl and peroxyl radicals and may further propagate lipid peroxidation (Fig. 6). It is believed that lipid-derived epoxy-peroxyl radicals serve as the ultimate oxidants in the co-oxygenation of B(a)P-7,8-dihydrodiol (Hughes et al., 1989). Understanding of these reactions is crucial for explanation of the limited sensitivity of the vanadium-initiated B(a)P and its 7,8-dihydrodiol co-oxygenation to inhibition by SOD (Byczkowski and Kulkarni, 1992b).

5.2 Benzo(a)pyrene Co-oxygenation

Several transition metal cations affect oxygen activation and influence B(a)P metabolism in mammalian tissues (Byczkowski and Gessner, 1987a,b,c,d). For instance, there is a growing body of evidence that iron-initiated lipid peroxidation may enhance the formation of the ultimate carcinogenic B(a)P metabolite (Byczkowski and Kulkarni, 1994). Byczkowski and Gessner (1987a,c) have shown that reactive oxygen species involved in the peroxidation

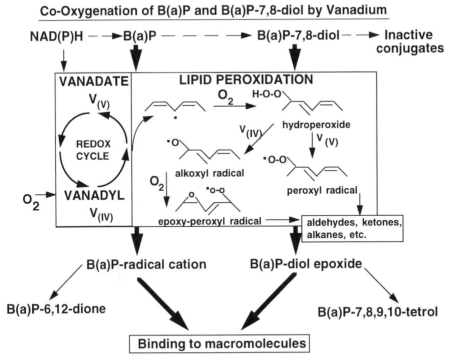

Figure 6. Summary scheme of the coupling between vanadium-initiated lipid peroxidation and benzo(a)pyrene (B(a)P) metabolism. Broken lines depict normal microsomal B(a)P metabolism, which may be inhibited by lipid peroxidation and vanadium redox cycling. Thick arrows depict activation of B(a)P and its 7,8-dihydrodiol epoxide by co-oxygenation (according to Byczkowski and Kulkarni, 1992b; 1994).

of hepatic microsomal lipids (reviewed by Byczkowski and Gessner, 1988) change the balance between bioactivation and conjugation of B(a)P metabolites, causing an accumulation of the B(a)P-7,8-diol epoxide and other intermediates capable of binding to macromolecules (Fig. 6). Further studies conducted on mouse liver microsomes (Byczkowski and Gessner, 1987c,d) suggested that even the trace amount of iron present in asbestos fibers can activate B(a)P under conditions of oxidative stress. Experimental data also showed that vanadium redox cycling effected a dose-dependent co-oxygenation of B(a)P and its 7,8-dihydrodiol, thereby increasing the generation of metabolites capable of interaction with macromolecules (Byczkowski and Kulkarni, 1992b).

6. EFFECTS ON SIGNAL TRANSDUCTION

At the tissue level the extracellular signals conveyed by hormones, growth factors, cytokines, and so forth, are transmitted to the cellular nucleus by a network of protein intermediates that are sequentially phosphorylated in response to the extracellular signaling molecule. The cellular system involved in signal transduction consists of transmembrane receptors, protein kinases, phosphoprotein phosphatases, transcription factors, and the regulatory or promoter regions of genes. Several oxidants can cause deregulation of signal transduction and simultaneous activation of protein kinase C and some transcription factors (e.g., activation protein-1 or AP-1, and NFκB; Klaassen, 1996). In this regard, it is interesting that the oxidative stress induced by redox cycling of naphthoquinones or orthovanadate was found to result in the activation of a membrane-associated phosphatidylinositol kinase through tyrosine-protein phosphorylation (Chen et al., 1990).

The regulation of protein phosphorylation represents a balance between protein kinase and protein phosphatase activities (Fischer et al., 1991). The SH2 domain (src homology 2 domain) binds to phosphotyrosine residues. The SH2 domain, containing phosphotyrosyl protein phosphatases (PTPases), causes down-regulation of growth factor receptor-initiated tyrosine phosphorylation and down-stream signal transduction pathways. The PTPases have dual specificity, hydrolyzing not only phosphotyrosyl esters, but also phosphoseryl and phosphothreonyl esters. These PTPases are involved in the regulation of cell proliferation and modulation of cell cycle (Byczkowski and Channel, 1996). Inhibition of receptor dephosphorylation and stimulation of MAP kinase activity by pro-oxidant chemicals may lead to activation of p21^{ras} and the transmission of the signal to the nucleus. This mechanism was enhanced significantly by the depletion of intracellular GSH (Lander et al., 1995).

6.1 Inhibition of Protein Tyrosine Phosphatase

Vanadium has been shown to inhibit, in vitro, several enzyme systems: ATPases, adenylate kinase, phosphofructokinase, glucose-6-phosphate dehydrogenase, ribonuclease, squalene synthetase, and PTPases (see review by Nechay, 1984, and references therein). Among these, the inhibition of (Na^+,K^+)-ATPase by vanadate (Rifkin, 1965) has drawn the most attention. Vanadate can replace phosphate as a substrate for glyceraldehyde 3-phosphate dehydrogenase, which leads to the formation of the unstable analog of 1,3-diphosphoglycerate (Simons, 1979). As a phosphate analog, vanadate at (sub)-micromolar concentrations inhibits several other phosphatases and ATPases, such as (Ca^{2+},Mg^{2+})-ATPase, (H^+,K^+)-ATPase, dynein ATPase, and so forth. (Nechay, 1984; Jandhyala and Hom, 1983; Shimizu, 1995), and thus interferes with the phosphate transfer or release reactions. On the other hand, vanadyl vanadium$_{(IV)}$ seems not to affect the (Na^+,K^+)-ATPase, but it inhibits alkaline phosphatase activity (Simons, 1979).

Vanadate and pervanadate (vanadyl hydroperoxide) are potent protein tyrosine phosphatase inhibitors (Fig. 7). Pervanadate may be produced, in vitro, in the reaction of orthovanadate with hydrogen peroxide (reaction c) (Ianzu et al., 1990). However, in the intracellular compartments with reductive

Dose-Dependent Effects of Pervanadate on PTPase Activity

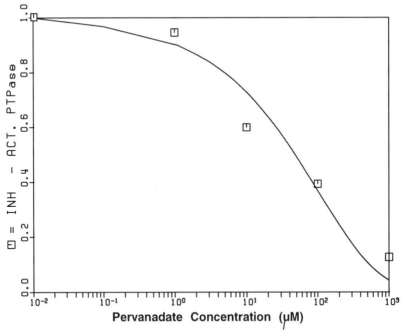

Figure 7. Dose-dependent effects of vanadate in the presence of excess H_2O_2 (pervanadate) on PTPase activity in intact rat hepatoma (Fao) cells. The curve is a computer-generated result of simulation with a computer program (BBPD computer program developed by Byczkowski and Flemming, 1996, based on Heffetz et al. 1990). The squares depict experimental data from Heffetz et al. (1990).

environments, it is very likely to undergo the redox cycling, yielding a family of reactive and free-radical compounds (Fig. 4). It was demonstrated that an alteration of the intracellular redox balance by decreasing the GSH levels selectively increases protein tyrosine phosphorylation, while it does not alter the serine/threonine phosphorylation (Staal et al., 1994). This effect was accounted for by selective effects on redox-sensitive protein tyrosine phosphatase. Typically, PTPases contain nucleophilic cysteinyl residue in their catalytic center (Stone and Dixon, 1994). Their enzymatic activities are rapidly inhibited by small disulfides (Ziegler, 1985). It seems that the cysteinyl residue must be kept in the reduced -SH form; therefore, thiol-directed reagents that oxidize it cause inhibition of PTPases (Fischer et al., 1991). It is therefore possible that, at millimolar concentrations, vanadium can affect intracellular redox potential and can lead to the inhibition of PTPase activity by causing oxidation of essential cysteinyl thiols. This pro-oxidant action of vanadium(V) is independent of its interfernce with phospate groups; thus, it seems that at least two independent sites of inhibition should exist within the protein tyrosine phosphatase for vanadium(V) and other pro-oxidants (Hecht and Zick, 1992). Accordingly, it was suggested that vanadium(V) induced the expression of two categories of genes in mouse C127 cells by two separate mechanisms (Yin et al., 1992). The stimulation of the synthesis of mRNA of *c-jun* was dramatically enhanced by either NADH or H_2O_2 and partly inhibited by catalase, suggesting the involvement of vanadium redox cycling, while the vanadate-stimulated synthesis of mRNA of actin and *c-Ha-ras* was unaffected by oxidants, reductants, or antioxidants.

6.2. Inhibition of Receptor and Phosphoprotein Dephosphorylation

Increased phosphorylation of several soluble proteins and membrane receptors by vanadate and pervanadate was reported in diverse biological systems (Heffetz et al., 1990; Purushotham et al., 1995; Rokhlin and Cohen, 1995; Haque et al., 1995; Lin and Grinnell, 1995; Yamaguchi et al., 1995). Similar stimulatory effects on receptor phosphorylation status and enzymatic protein phosphorylation also were reported for other pro-oxidants (Tan et al., 1995). On the other hand, at high concentrations free radicals generated by the pro-oxidants caused nonspecific cross-linking, polymerization, and/or fragmentation of proteins, changes in membrane fluidity, and a decrease in the number of binding sites and/or binding affinity for specific ligands. In macrophages, vanadium(V) reduced their ability to interact with interferon γ as well as other cytokines, while it increased their basal spontaneous release of $O_2^{\cdot-}$ and H_2O_2 (Cohen et al., 1996).

6.3. Nonspecific Interactions

Paradoxically, under certain conditions vanadate was shown to stimulate Ca^{2+}-ATPase from pig heart sarcoplasmic reticulum (Erdmann et al., 1984). Reportedly, toxic doses of vanadium stimulated activity of monoamine oxidase (see review by Jandhyala and Hom, 1983, and references therein) and also the secretion of pancreatic enzymes (Proffitt and Case, 1984a,b).

Vanadate, at reasonably low concentrations, enhanced adenyl cyclase activity in several cell membrane preparations, thereby increasing the levels of cyclic AMP (Erdmann et al., 1984). This mechanism seems responsible for the secretion of Cl^-, Na^+, and water in rat jejunum (Hajjar et al., 1986). However, in some isolated multicellular model systems like toad skin or bladder, vanadate inhibited the effects of cyclic AMP on osmotic water flow (Nechay, 1984). Also, both vanadate and vanadyl, in vitro, decreased cyclic AMP production in the luteinizing hormone-treated rat corpora lutea, and involvement of vanadium-stimulated tyrosine phosphorylation in this process was proposed (Lahav et al., 1986). Involvement of tyrosine kinase stimulation was also suggested in pervanadate stimulation of respiratory burst in neutrophils (Yamaguchi et al., 1995). Therefore, it seems that stimulation of tyrosine phosporylation of phospholipase C, activation of phospholipase C, increased production of diacyl-glycerols (DAG), and possibly protein kinase C and/or tyrosine kinase activation (Yamaguchi et al., 1995) may be all involved in nonspecific effects of high concentrations of vanadium$_{(V)}$ on several cellular systems that are analogous to the more specific action exerted, for instance, by a classical tumor promoter, phorbol acetate (Byczkowski and Channel, 1996). On the other hand, drastic oxidation of purified protein kinase C damaged both regulatory and catalytic domains (Ramasarma, 1990), which may cause inactivation of protein kinase C-dependent signal transduction in vivo.

7. CONCLUSIONS

Although in a trace amount vanadium may be essential to the normal metabolic functions of the organism, excessive exposure to this element is potentially deleterious, and may even be toxic. Owing its complex chemistry and ability to undergo a free-radical redox cycling, vanadium may adversely affect several biochemical functions and is capable of causing oxidative stress, co-oxygenating carcinogens, and interfering with signal transduction. The results reported in the literature are sometimes based on a treatment with a single dose or on in vitro experiments in the presence of just one concentration of vanadium. A very few systematic dose-response studies have been conducted with vanadium to date. Clearly, more experimental data are necessary to assess fully its toxicological potential and the possibility of health hazard from exposure to vanadium compounds.

7.1. Future Research Needs

A systematic pharmacokinetic study should be conducted on vanadium absorption, distribution, metabolism, and disposition. The development of a pharmacokinetic/dynamic model should be very helpful in understanding the kinetics and toxicity of vanadium administered by different routes (analogous to the pharmacokinetic model described for chromium by O'Flaherty, 1993, 1996). Currently, there is no information available about the microcompartmentalization and the redox state of vanadium in different subcellular fractions of the target tissues. More experimental data are necessary to understand the mode of action of vanadium on the cellular signal transduction pathway and DNA transcription, as well as carcinogenicity, if any.

ACKNOWLEDGMENTS

This study was supported in part by Department of the Air Force contract no. F41624-96-C-9010.

ABBREVIATIONS USED:
AP-1: activator protein-1;

B(a)P: benzo(a)pyrene;

BBPD: biologically based pharmacodynamics;

BBPK: biologically based pharmacokinetics;

DMNQ: 2,3-dimethoxy-1,4-naphthoquinone;

FMN: flavin mononucleotide;

GSH: reduced glutathione;

NAD(P)H: reduced nicotinamide adenine dinucleotide (and its phosphate);

NFκB: nuclear factor κB;

NDGA: nor-dihydroguaiaretic acid;

$O_2^{\cdot-}$: superoxide anion radical;

PBPK: physiologically based pharmacokinetics;

PTPase: phosphotyrosyl protein phosphatase;

SH2: src homology 2 domain, a domain binding to phosphotyrosine-containing proteins;

SOD: superoxide dismutase;

TMPD: N,N,N',N'-tetramethyl-p-phenylenediamine;

$V_{(V)}$: pentavalent vanadium;

$V_{(IV)}$: tetravalent vanadium;

REFERENCES

Agency for Toxic Substances and Disease Registry (ATSDR). (1992). *Toxicological Profile for Vanadium.* U.S. Department of Health and Human Services, Public Health Service, Atlanta, GA.

Antunes, F., Salvador, A., Marinho H. S., and Pinto, R. E. (1994). A mathematical model for lipid peroxidation in inner mitochondrial membranes. *Trav. Lab.* **33**(Suppl. T-1), 1–52.

Auricchio, F., Migliaccio, A., Castoria, G., Di Domenico, M., Bilancio, A., and Rotondi, A. (1996). Protein tyrosine phosphorylation and estradiol action. *Ann. NY Acad. Sci.* **784,** 149–170.

Bernheim, F., and Bernheim, M. L. C. (1938). Action of vanadium on tissue oxidations. *Science* **88,** 481–482.

Biasi, F., Bosco, M., Lafranco, G., and Poli, G. (1995). Cytolysis does not per se induce lipid peroxidation: Evidence in man. *Free Radical Biol. Med.* **18,** 909–912.

Bowden, A. T., Draper, P., and Rawling, H. (1953). The problem of fuel oil deposition in open-cycle gas turbines. *Proc. Inst. Mech. Eng.* **167,** 291–313.

Bruech, M., Quintanilla, M. E., Legrum, W. Koch, J., Netter, K. J., and Fuhrmann, G. F. (1984). Effects of vanadate on intracellular reduction equivalents in mouse liver and the fate of vanadium in plasma, erythrocytes and liver. *Toxicology* **31,** 283–295.

Byczkowski, J. Z., and Channel, S. R. (1996). Chemically induced oxidative stress and tumorigenesis: Effects on signal transduction and cell proliferation. *Toxic Substance Mechan.* **15,** 101–128.

Byczkowski, J. Z., Channel, S. R., Pravecek, T. L., and Miller, C. R. (1996). Mathematical model for chemically induced lipid peroxidation in precision-cut liver slices: Computer simulation and experimental calibration. *Computer Methods Prog. Biomed.* **50,** 73–84.

Byczkowski, J. Z., and Flemming, C. D. (1996). Computer-aided dose-response characteristics of chemically initiated oxidative stress in vitro. *Toxicologist* **30,** 240(1227).

Byczkowski, J. Z., and Gessner, T. (1987a). Effects of superoxide generated in vitro on glucuronidation of benzo(a)pyrene metabolites by mouse liver microsomes. *Int. J. Biochem.* **19,** 531–537.

Byczkowski, J. Z., and Gessner, T. (1987b). Interaction between vitamin K3 and benzo(a)pyrene metabolism in uninduced microsomes. *Int. J. Biochem.* **19,** 1173–1179.

Byczkowski, J. Z., and Gessner, T. (1987c). Action of xanthine–xanthine oxidase system on microsomal benzo(a)pyrene metabolism in vitro. *Gen. Pharmacol.* **18,** 385–395.

Byczkowski, J. Z., and Gessner, T. (1987d). Asbestos-catalyzed oxidation of benzo(a)pyrene by superoxide-peroxidized microsomes. *Bull. Environ. Contam. Toxicol.* **39,** 312–317.

Byczkowski, J. Z., and Gessner, T. (1988). Biological role of superoxide ion-radical. *Int. J. Biochem.* **20,** 569–580.

Byczkowski, J. Z., and Kulkarni, A. P. (1989). Lipoxygenase-catalyzed epoxidation of benzo(a)pyrene-7,8-dihydrodiol. *Biochem. Biophys. Res. Commun.* **159,** 1190–1205.

Byczkowski, J. Z., and Kulkarni A. P. (1990a). Activation of benzo(a)pyrene-7,8-dihydrodiol in rat uterus: An in vitro study. *J. Biochem. Toxicol.* **5,** 139–145.

Byczkowski, J. Z., and Kulkarni, A. P. (1990b). Lipid peroxidation and benzo(a)pyrene derivative co-oxygenation by environmental pollutants. *Bull. Environ. Contam. Toxicol.* **45,** 633–640.

Byczkowski, J. Z., and Kulkarni, A. P. (1992a). Linoleate-dependent co-oxygenation of benzo(a)-pyrene and benzo(a)pyrene-7,8-dihydrodiol by rat cytosolic lipoxygenase. *Xenobiotica* **22,** 609–618.

Byczkowski, J. Z., and Kulkarni, A. P. (1992b). Vanadium redox cycling, lipid peroxidation and co-oxygenation of benzo(a)pyrene-7,8-dihydrodiol. *Biochim. Biophys. Acta* **1125,** 134–141.

Byczkowski, J. Z., and Kulkarni, A. P. (1994). Oxidative stress and asbestos. In J. O. Nriagu and M. S. Simmons (Eds.), *Environmental Oxidants.* John Wiley and Sons, New York, pp. 459–474.

Byczkowski, J. Z., and Sorenson, J. R. J. (1984). Effects of metal compounds on mitochondrial function: A review. *Sci. Total. Environ.* **37,** 133–162.

Byczkowski, J. Z., Wan, B., and Kulkarni, A. P. (1988). Vanadium-mediated lipid peroxidation in microsomes from human term placenta. *Bull. Environ. Contam. Toxicol.* **41,** 696–703.

Byczkowski, J. Z., and Zychlinski, L. (1978). Interaction of vanadium with electron-transport chain in rat liver mitochondria in vitro. In *16th Meeting of the Polish Biochemical Society.* Abstracts, p. 78 (SH-4). Lodz, Poland.

Byczkowski, J. Z., Zychlinski, L., and Tluczkiewicz, J. (1979). Interaction of vanadate with respiratory chain of rat liver and wheat seedling mitochondria. *Int. J. Biochem.* **10,** 1007–1011.

Cantley, L. C. Jr., Josephson, L., Warner, R., Yangisawa, M., Lechene, C., and Guidotti, G. (1977). Vanadate is a potent (Na,K)-ATPase inhibitor found in ATP derived from muscle. *J. Biol. Chem.* **252,** 7421–7423.

Chen, Y., and Chan, T. M. (1993). Orthovanadate and 2,3-dimethoxy-1,4-naphthoquinone augment growth factor-induced cell proliferation and c-fos gene expression in 3T3-L1 cells. *Arch. Biochem. Biophys.* **305,** 9–16.

Chen, Y. X., Yang, D. C., Brown, A. B., Jeng, Y., Tatoyan, A., and Chan, T. M. (1990). Activation of a membrane-associated phosphatidylinositol kinase through tyrosine-protein phosphorylation by naphthoquinones and orthovanadate. *Arch. Biochem. Biophys.* **238,** 184–192.

Cohen, M. D., McManus, T. P., Yang, Z., Qu, Q., Schlesinger, R. B., and Zelikoff, J. T. (1996). Vanadium affects macrophage interferon-γ-binding and -inducible responses. *Toxicol. Appl. Pharmacol.* **138,** 110–120.

Comporti, M. (1985). Biology of disease, lipid peroxidation and cellular damage in toxic liver injury. *Lab. Invest.* **53,** 599–623.

Corcoran, G. B., Fix, L., Jones, D. P., Moslen, M. T., Nicotera, P., Oberhaimmer, F. Z., and Buttyan, R. (1994). Apoptosis: Molecular control point in toxicity. *Toxicol. Appl. Pharmacol.* **128,** 169–181.

Crane, F. L. (1975). Oxidation of vanadium IV by cytochrome c oxidase: Evidence for a terminal copper pathway. *Biochem. Biophys. Res. Commun.* **63,** 355–361.

Crans, D. C., Bunch, R. L., and Theisen, L. A. (1989). Interaction of trace levels of vanadium(IV) and vanadium(V) in biological systems. *J. Am. Chem. Soc.* **111,** 7597–7607.

Cutler, R. G. (1991). Antioxidants and aging. *Am. J. Clin. Nutr.* **53,** 373S–379S.

Darr, D., and Fridovich, I. (1984). Vanadate and molybdate stimulate the oxidation of NADH by superoxide radical. *Arch. Biochem. Biophys.* **232,** 562–565.

de Groot, H., and Littauer, A. (1989). Hypoxia, reactive oxygen, and cell injury. *Free Radical Biol. Med.* **6,** 541–551.

Dix, T. A., and Marnett, L. J. (1983). Metabolism of polycyclic aromatic hydrocarbon derivatives to ultimate carcinogens during lipid peroxidation. *Science* **221,** 77–79.

Eckardt, R. E. (1971). Petroleum fuel and airborne metals. *Arch. Environ. Health.* **23,** 166–170.

Erdmann, E., Krawietz, W., Philipp, G., Hackbarth, I., Schmitz, W., and Scholz, H. (1979). Purified cardiac cell membranes with high (Na$^+$+K$^+$)ATPase activity contain significant NADH-vanadate reductase activity. *Nature* **282,** 335–3336.

Erdmann, E., Krawietz, W., Vyskocil, F., Dlouha, H., and Teisinger, J. (1981). Importance of cardiac cell membranes in vanadate-induced NADH oxidation. *Nature* **294,** 288.

Erdmann, E., Werdan, K., Krawietz, W., Schmitz, W., and Scholtz, H. (1984). Vanadate and its significance in biochemistry and pharmacology. *Biochem. Pharmacol.* **33,** 945–950.

Finley, J. W., and Otterburn, M. S. (1993). The consequences of free radicals in foods. *Toxicol. Ind. Health* **9,** 77–91.

Fischer, E. H., Charbonneau, H., and Tonks, N. K. (1991) Protein tyrosine phosphatases: A diverse family of intracellular and transmembrane enzymes. *Science* **253,** 401–406.

Fisher, G. L., McNeill, K. L., Prentice, B. A., and McFarland, A. R. (1983). Physical and biological studies of coal and oil fly ash. *Environ. Health Perspect.* **51,** 181–186.

Flyger, H., Jensen, F. P., and Kemp, K. (1976). Air pollution in Copenhagen. Part I. Element analysis and size distribution of aerosols. *RISO Report No. 338,* pp. 1–70.

Goldsmith, A. H., Vorpahl, K. W., French, K. A., Jordan, P. T., and Jurinski, N. B. (1976). Health hazards from oil, soot and metals at a hot forging operation. *Am. Ind. Hyg. Assoc. J.* **37,** 217–226.

Gower, J. D. (1988). A role for dietary lipids and antioxidants in the activation of carcinogens. *Free Radical Biol. Med.* **5**, 95–111.

Haegele, A. D., Briggs, S. P., and Thompson, H. J. (1994). Antioxidant status and dietary lipid unsaturation modulate oxidative DNA damage. *Free Radical Biol. Med.* **16**, 111–115.

Hajjar, J. J., Zakko, S., and Tomicic, T. K. (1986). Effect of vanadate on water and electrolyte transport in rat jejunum. *Biochim. Biophys. Acta* **863**, 325–330.

Haque, S. J., Flati, V., Deb, A., and Williams, B. R. G. (1995). Roles of protein-tyrosine phosphatases in stat1α-mediated cell signaling. *J. Biol. Chem.* **270**, 25709–25714.

Hauser, R., Elreedy, S., Hoppin, J. A., and Christiani, D. C. (1995). Upper airway response in workers exposed to fuel oil ash: Nasal lavage analysis. *Occup. Environ. Med.* **52**, 353–358.

Hecht, D., and Zick, Y. (1992). Selective inhibition of protein tyrosine phosphatase activities by H_2O_2 and vanadate in vitro. *Biochem. Biophys. Res. Commun.* **188**, 773–779.

Heffetz, D., Bushkin, I., Dror, R., and Zick, Y. (1990). The insulinomimetic agents H_2O_2 and vanadate stimulate protein tyrosine phosphorylation in intact cells. *J. Biol. Chem.* **265**, 2896–2902.

Hudson, T. G. F. (1964). *Vanadium Toxicology and Biological Significance.* Elsevier Publishing Co., Amsterdam.

Hughes, M. F., Chamulitrat, W., Manson, R. P., and Eling, T. E. (1989). Epoxidation of 7,8-dihydroxydroxy-7,8-dihydrobenzo(a)pyrene via a hydroperoxide-dependent mechanism catalyzed by lipoxygenases. *Carcinogenesis* **10**, 2075–2080.

Ianzu, T., Taniguchi, T., Yanagi, S., and Yamamura, H. (1990). Protein-tyrosine phosphorylation and aggregation of intact human platelets by vanadate with H_2O_2. *Biochem. Biophys. Res. Commun.* **170**, 259–263.

Jaeschke, H. (1995). Mechanisms of oxidant stress-induced acute tissue injury. *Proc. Soc. Exp. Biol. Med.* **209**, 104–111.

Jandhyala, B. S., and Hom, G. J. (1983). Physiological and pharmacological properties of vanadium. *Life Sci.* **33**, 1325–1340.

Kalogeropoulos, N., Scoullos, M., Vassilaki-Grimani, M., and Grimanis, A. P. (1989). Vanadium in particles and sediments of the northern Saronikos Gulf, Greece. *Sci. Total Environ.* **79**, 241–252.

Karimov, M. A., Doskeeva, R. A., and Sarsebekov, E. K. (1991). Blastomogenic action of heavy vanadium-rich petroleum and its processing products [in Russian]. *Gig. Tr. Prof. Zabol.*, No. 1, pp. 18–20.

Karimov, M. A., Doskeeva, R. A., Sarsebekov, E. K., Egorova, Z. D., and Orlova, G. V. (1989). Vanadium level in the urine of workers on a shift-work schedule during drilling of oil wells for high-vanadium petroleum [in Russian]. *Gig. Tr. Prof. Zabol.*, No. 3, pp. 9–12.

Karimov, M. A., Doskeeva, R. A., Sarsebekov, E. K., Rakhimov, K. D., and Bel'khodzhaeva, A. A. (1988). Blastomogenic and nephrotoxic effects of a petroleum product obtained from high-vanadium oil at a temperature of 300–350 degrees C [in Russian]. *Gig. Tr. Prof. Zabol.*, No. 11, pp. 53–55.

Kehrer, J. P. (1993). Free radicals as mediators of tissue injury and disease. *Crit. Rev. Toxicol.* **23**, 21–48.

Klaassen, C. D. (Ed). (1996). *Casarett and Doull's Toxicology: The Basic Science of Poisons.* McGraw-Hill, New York.

Kleinman, M. T., Bernstein, D. M., and Kneip, T. J. (1977). An apparent effect of the oil embargo on total suspended particulate matter and vanadium in New York City air. *J. Air Pollut. Control. Assoc.* **27**, 65–67.

Kretzschmar, M., and Braunlich, H. (1990). Role of glutathione in vanadate reduction in young and mature rats: Evidence for direct participation of glutathione in vanadate inactivation. *J. Appl. Toxicol.* **10**, 295–300.

Kulkarni, A. P., and Byczkowski, J. Z. (1994a). Effects of transition metals on biological oxidations. In J. O. Nriagu and M. S. Simmons (Eds.), *Environmental Oxidants*. John Wiley and Sons, New York, pp. 475–496.

Kulkarni, A. P., and Byczkowski, J. Z. (1994b). Hepatotoxicity. In E. Hodgson and P. E. Levi (Eds.), *Introduction to Biochemical Toxicology*. Appleton & Lange, Norwalk, CT, pp. 459–490.

Lahav, M., Rennert, H., and Barzilai, D. (1986). Inhibition by vanadate of cyclic AMP production in rat corpora lutea incubated in vitro. *Life Sci.* **39**, 2557–2560.

Lander, H. M., Ogiste, J. S., Teng, K. K., and Novogrodsky, A. (1995). p21^{ras} as a common signaling target of reactive free radicals and cellular redox stress. *J. Biol. Chem.* **270**, 21195–21198.

Lees, R. E. (1980). Changes in lung function after exposure to vanadium compounds in fuel oil ash. *Br. J. Ind. Med.* **37**, 253–256.

Lehninger, A. L. (1965). *The Mitochondrion: Molecular Basis of Structure and Function.* W. A. Benjamin, New York, pp. 109–114.

Levy, B. S., Hoffman, L., and Gottsegen, S. (1984). Boilermakers' bronchitis. Respiratory tract irritation associated with vanadium pentoxide exposure during oil-to-coal conversion of a power plant. *J. Occup. Med.* **26**, 567–570.

Lin, Y-C., and Grinnell, F. (1995). Treatment of human fibroblasts with vanadate and platelet-derived growth factor in the presence of serum inhibits collagen matrix contraction. *Exp. Cell Res.* **221**, 73–82.

Liochev, S., and Fridovich, I. (1988). Superoxide is responsible for the vanadate stimulation of NAD(P)H oxidation by biological membranes. *Arch. Biochem. Biophys.* **263**, 299–304.

Liochev, S. I., and Fridovich, I. (1990). Vanadate-stimulated oxidation of NAD(P)H in the presence of biological membranes and other sources of O$^-_2$. *Arch Biochem. Biophys.* **279**, 1–7.

Liochev, S. I., and Fridovich, I. (1996). Comments on vanadate chemistry. *Free Radical Biol. Med.* **20**, 157.

Madany, I. M., and Raveendran, E. (1992). Polycyclic aromatic hydrocarbons, nickel and vanadium in air particulate matter in Bahrain during the burning of oil fields in Kuwait. *Sci. Total. Environ.* **116**, 281–289.

Meyer, M., Schreck, R., and Baeuerle, P. A. (1993). H$_2$O$_2$ and antioxidants have opposite effects on activation of NF-κB and AP-1 in intact cells: AP-1 as secondary antioxidant-responsive factor. *EMBO J.* **12**, 2005–2015.

Moeller, R. B. Jr., Kalasinsky, V. F., Razzaque, M., Centeno, J. A., Dick, E. J., Abdal, M., Petrov, I. I., DeWitt, T. W., al Attar, M., Pletcher, J. M., and others. (1994). Assessment of the histopathological lesions and chemical analysis of feral cats to the smoke from the Kuwait oil fires. *J. Environ. Pathol. Toxicol. Oncol.* **13**, 137–149.

Nechay, B. R. (1984). Mechanisms of action of vanadium. *Annu. Rev. Pharmacol. Toxicol.* **24**, 501–524.

Nechay, B. R., Nanninga, L. B., and Nechay, P. S. E. (1986). Vanadyl (IV) and vanadate (V) binding to selected endogenous phosphate, carboxyl, and amino ligands; calculations of cellular vanadium species distribution. *Arch. Biochem. Biophys.* **251**, 128–138.

North, P., and Post, R. L. (1984). Inhibition of (Na,K)-ATPase by tetravalent vanadium. *J. Biol. Chem.* **259**, 4971–4978.

Nriagu, J. O., and Pacyna, J. M. (1988). Quantitative assessment of world-wide contamination of air, water and soil by trace metals. *Nature* (*Lond.*) **333**, 134–139.

O'Flaherty, E. J. (1993). A pharmacokinetic model for chromium. *Toxicol. Lett.* **68**, 145–158.

O'Flaherty, E. J. (1996). A physiologically based model of chromium kinetics in the rat. *Toxicol. Appl. Pharmacol.* **138**, 54–64.

Papas, A. M. (1993). Oil-soluble antioxidants in foods. *Toxicol. Ind. Health* **9**, 123–149.

Parfett, C. L., and Pilon, R. (1995). Oxidative stress-regulated gene expression and promotion of morphological transformation induced in C3H/10T1/2 cells by ammonium metavanadate. *Food Chem. Toxicol.* **33**, 301–308.

Paschoa, A. S., Warenn, M. E., Singh, N. P., Bruenger, F. W., Miller, S. C., Cholewa, M., and Jones, K. W. (1987). Localization of vanadium-containing particles in the lungs of uranium/vanadium miners. *Biol. Trace Element Res.* **13,** 275–282.

Pistelli, R., Pupp, N., Forastiere, F., Agabiti, N., Corbo, G. M., Tidei, F., and Perucci, C. A. (1991). Increase of nonspecific bronchial reactivity after occupational exposure to vanadium [in Italian]. *Med. Lav.,* **82,** 270–275.

Pratt, D. E. (1993). Antioxidants indigenous to foods. *Toxicol. Ind. Health* **9,** 63–75.

Proffitt, R., and Case, R. M. (1984a). The effects of vanadate on ^{45}Ca exchange and enzyme secretion in the rat exocrine pancreas. *Cell Calcium* **5,** 321–334.

Proffitt, R., and Case, R. M. (1984b). Vanadate stimulates rat pancreatic enzyme secretion through the release of calcium from an intracellular store. *Cell Calcium* **5,** 335–340.

Pryor, W. A. (1993). Oxidative stress status measurement in humans and their use in clinical trials. In K. Yagi (Ed.), *Active Oxygen, Lipid Peroxides, and Antioxidants.* CRC Press, Boca Raton, FL, pp. 117–126.

Purucker, E., and Wernze, H. (1990). Hepatic efflux and renal extraction of plasma glutathione: Marked differences between healthy subjects and the rat. *Klin. Wochenschr.* **68,** 1008–1012.

Purushotham, K. R., Wang, P., and Humphreys-Beher M. G. (1995). Effect of vanadate on amylase secretion and protein tyrosine phosphatase activity in the rat parotid gland. *Mol. Cell. Biochem.* **152,** 87–94.

Ramanadham, S., Heyliger, C., Gresser, M. J., Tracey, A. S., and McNeil, J. H. (1991). The distribution and half-life for retention of vanadium in the organs of normal and diabetic rats orally fed vanadium(IV) and vanadium(V). *Biol. Trace Elem. Res.* **30,** 119–124.

Ramasarma, T. (1990). H_2O_2 has a role in cellular regulation. *Ind. J. Biochem. Biophys.* **27,** 269–274.

Ramasarma, T. MacKellar, W. C., and Crane, F. L. (1981). Vanadate-stimulated NADH oxidation in plasma membrane. *Biochim. Biophys. Acta* **646,** 88–98.

Rifkin, R. J. (1965). In vitro inhibition of Na$^+$-K$^+$ and Mg2$^+$ ATPases by mono, di, and trivalent cations. *Proc. Soc. Exp. Biol. Med.* **120,** 802–803.

Roberfroid, M., and Calderon, P. B. (1994). *Free Radicals and Oxidation Phenomena in Biological Systems.* Marcel Dekker, New York.

Rokhlin, O., and Cohen, M. B. (1995). Differential sensitivity of human prostatic cancer cell lines to the effects of protein kinase and phosphate inhibitors. *Cancer Lett.* **98,** 103–110.

Rossi, B., Siciliano, G., Giraldi, C., Angelini, C., Marchetti, A., and Paggiaro, P. L. (1986). Toxic myopathy induced by industrial mineral oils: Clinical and histopathological features. *Ital. J. Neurol. Sci.* **7,** 599–604.

Russanov, E., Zaporowska, H., Ivancheva, E., Kirkova, M., and Konstantinova, S. (1994). Lipid peroxidation and antioxidant enzymes in vanadate-treated rats. *Comp. Biochem. Physiol. Pharmacol. Toxicol. Endocrinol.* **107,** 415–421.

Sadiq, M., and Zaidi, T. H. (1984). Vanadium and nickel content of Nowruz spill tar flakes on the Saudi Arabian coastline and their probable environmental impact. *Bull. Environ. Contam. Toxicol.* **32,** 635–639.

Sarsebekov, E. K., Dzharbusynov, B. U., and Doskeeva, R. A. (1994). The nephrotoxic action of heavy crude with a high vanadium content and of its refinery products [in Russian]. *Urol. Nefrol. Mosk.,* No. 3, pp. 35–36.

Schiff, L. J., and Graham, J. A. (1984). Cytotoxic effect of vanadium and oil-fired fly ash on hamster tracheal epithelium. *Environ. Res.* **34,** 390–402.

Shi, X., Sun, X., and Dalal, N. S. (1990). Reaction of vanadium(V) with thiols generates vanadium(IV) and thiyl radicals. *FEBS Lett.* **271,** 185–188.

Shimizu. T. (1995). Inhibitors of the dynein ATPase and ciliary or flagellar motility. *Methods Cell Biol.* **47,** 497-501.

Sies, H. (1994). Oxidative stress: From basic research to clinical medicine. In A. E. Favier, J. Neve, and P. Faure (Eds.), *Trace Elements and Free Radicals in Oxidative Diseases.* AOCS Press, Champaign, IL, pp. 1–7.

Simons, T. J. B. (1979). Vanadate—A new tool for biologists. *Nature* (Lond.) **281,** 337–338.

Sohal, R. S. (1993). The free radical hypothesis of aging: An appraisal of the current status. *Aging Clin. Exp. Res.* **5,** 3–17.

Sokolov, S. M. (1986). Methodological aspects of assessing atmospheric contamination with metal aerosols in the vicinity of thermal power complexes. *J. Hyg. Epidemiol. Microbiol. Immunol.* **30,** 249–254.

Spatz, L., and Bloom, A. D. (1992). *Biological Consequences of Oxidative Stress: Implications for Cardiovascular Disease and Carcinogenesis.* Oxford University Press, New York.

Srivastava, A. K., and Chiasson, J.-L. (1995). Vanadium compounds: Biochemical and therapeutic applications. *Mol. Cel. Biochem.* **153,** 1–240.

Staal, F. J. T., Anderson, M. T., Staal, G. E. J., Herzenberg, L. A., Gitler, C., and Herzenberg, L. A. (1994). Redox regulation of signal transduction: Tyrosine phosphorylation and calcium influx. *Proc. Natl. Acad. Sci. USA* **91,** 3619–3622.

Stadtman, E. R., Oliver, C. N., Starke-Reed, P. E., and Rhee, S. G. (1993). Age-related oxidation reaction in proteins. *Toxicol. Ind. Health* **9,** 187–196.

Stankiewicz, P. J., Stern, A., and Davison, A. J. (1991). Oxidation of NADH by vanadium: Kinetics, effects of ligands and role of H_2O_2 or O_2. *Arch. Biochem. Biophys.* **287,** 8–17.

Stohs, S. J., and Bagchi, D. (1995). Oxidative mechanisms in the toxicity of metal ions. *Free Radical Biol. Med.* **18,** 321–336.

Stone, R. L., and Dixon, J. E. (1994). Protein-tyrosine phosphatases. *J. Biol. Chem.* **269,** 31323–31326.

Svoboda, P., Teisinger, J. Pilar, J., and Vyskocil, F. (1984). Vanadyl (VO^{2+}) and vanadate (VO_3^-) ions inhibit the brain microsomal Na,K-ATPase with similar affinities; protection by transferrin and noradrenaline. *Biochem. Pharmacol.* **33,** 2485–2497.

Tan, C. M., Xenoyannis, S., and Feldman, R. D. (1995). Oxidant stress enhances adenyl cyclase activation. *Circ. Res.* **77,** 710–717.

Tappel, A. L., Tappel, A. A., and Fraga, C. G. (1989). Application of simulation modeling to lipid peroxidation process. *Free Radical Biol. Med.* **7,** 361–368.

Tipton, I. H., and Shaffer, J. J. (1964). Statistical analysis of lung trace element levels. *Arch. Environ. Health* **8,** 56–67.

Todaro, A., Bronzato, R., Buratti, M., and Colombi, A. (1991). Acute exposure to vanadium-containing dusts: The health effects and biological monitoring in a group of workers employed in boiler maintenance [in Italian]. *Med. Lav.* **82,** 142–147.

Troppens, H. (1969). Vanadium pentoxide poisoning. [in German]. *Dtsch Gesundheitsw.* **24,** 1089–1092.

Vazquez, F., Sanchez, M., Alexander, H., and Delgado, D. (1991). Distribution of Ni, V, and petroleum hydrocarbons in recent sediments from the Veracruz coast, Mexico. *Bull. Environ. Contam. Toxicol.* **46,** 774–781.

Videla, L. A., Barros, S. B. M., and Junquiera, V. B. C. (1990). Lindane-induced liver oxidative stress. *Free Radical Biol. Med.* **9,** 169–179.

Vroegop, S. M., Decker, D. E., and Buxser, S. E. (1995). Localization of damage induced by reactive oxygen species in cultured cells. *Free Radical Biol. Med.* **18,** 141–151.

Vyskocil, F., Teisinger, J., and Dlouha, H. (1980). A specific enzyme is not necessary for vanadate-induced oxidation of NADH. *Nature* **286,** 516–517.

White, M. A., Reeves, G. D., Moore, S. Chandler, H. A., and Holden, H. J. (1987). Sensitive determination of urinary vanadium as a measure of occupational exposure during cleaning of oil fired boilers. *Ann. Occup. Hyg.* **31,** 339–343.

Williams, G. M. (1993). Inhibition of chemical-induced experimental cancer by synthetic phenolic antioxidants. *Toxicol. Ind. Health* **9,** 303–308.

Yamaguchi, M., Oishi, H., Araki, S., Saeki, S., Yamane, H., Okamura, N., and Ishibashi, S. (1995). Respiratory burst and tyrosine phosphorylation by vanadate. *Arch. Biochem. Biophys.* **323,** 382–386.

Yin, X., Davison, A. J., and Tsang, S. S. (1992). Vanadate-induced gene expression in mouse C127 cells: Roles of oxygen derived active species. *Mol. Cell. Biochem.* **115,** 85–96.

Younes, M., Kayser, E., and Strubelt, O. (1991). Effect of antioxidants on vanadate-induced toxicity towards isolated perfused rat livers. *Toxicology* **70,** 141–149.

Younes, M., and Stubelt, O. (1991). Vanadate-induced toxicity towards isolated perfused rat livers: The role of lipid peroxidation. *Toxicology* **66,** 63–74.

Ziegler, D. M. (1985). Role of reversible oxidation–reduction of enzyme thiols-disulfides in metabolic regulation. *Annu. Rev. Biochem.* **54,** 305–329.

Zychlinski, L. (1980). Toxicological appraisal of work-places exposed to the dust containing vanadium pentoxide. [in Polish]. *Bromat. Chem. Toksykol.* **8,** 195–199.

Zychlinski, L., and Byczkowski, J. Z. (1978). Oxido-reduction properties of vanadium at rat liver mitochondria. *Twelfth Federation of European Biochemical Societies Meeting.* Abstracts, p. 1720. Dresden, Germany.

Zychlinski, L., and Byczkowski, J. Z. (1990). Inhibitory effects of vanadium pentoxide on respiration of rat liver mitochondria. *Arch. Environ. Contam. Toxicol.* **19,** 138–142.

Zychlinski, L., Byczkowski, J. Z., and Kulkarni, A. P. (1991). Toxic effects of long-term intratracheal administration of vanadium pentoxide in rats. *Arch. Environ. Contam. Toxicol.* **20,** 295–298.

13

ENDOCRINE CONTROL OF VANADIUM ACCUMULATION

Frederick G. Hamel

Research Service, Veterans Affairs Medical Center, and Departments of Internal Medicine and Pharmacology, University of Nebraska Medical Center, Omaha, NE

Vanadium in the Environment. Part 2: Health Effects, Edited by Jerome O. Nriagu.
ISBN 0-471-17776-8. © 1998 John Wiley & Sons, Inc.

1. INTRODUCTION

The idea that trace element metabolism may be under endocrine control is not novel (Neve, 1992, Allain and Leblondel, 1992). Chromium, copper, fluorine, iodine, manganese, selenium, zinc, and vanadium have various roles in endocrine functions involving the adrenals, gonads, thyroid, bone, and glucose homeostasis (Neve, 1992), so it is not surprising that they should be under hormonal control. Hypophysectomy, thyroparathyroidectomy, and adrenalectomy result in changes in tissue levels of copper, iron, rubidium, strontium, and zinc (Allain and Leblondel, 1992). In this chapter we will examine some of the data concerning endocrine control of vanadium metabolism. We will review some of the literature regarding the essentiality of vanadium in mammals as it potentially relates to endocrine function, and examine possible roles of vanadium in regulating signal transduction mechanisms. Finally we will look at specific endocrine systems that are involved.

2. ESSENTIALITY

While a definitive deficiency syndrome has not been identified, the current consensus is that vanadium is very probably an essential element (Nielsen and Uthus, 1990; Nielsen, 1990; French and Jones, 1993). As early as 1949 it was suggested that vanadium might be essential because it stimulated the mineralization of bone and teeth in rodents (Rygh, 1949). During the 1970s, a number of studies suggested that decreased dietary vanadium decreased growth, altered proper development, decreased perinatal survival, and altered lipid metabolism in rat and chickens (Hopkins and Mohr, 1971; Hopkins and Mohr, 1974; Myron et al., 1975; Nielsen and Ollerich, 1973; Nielsen and Myron, 1976; Nielsen and Uthus, 1977; Nielsen et al., 1978; Schwarz and Milne 1971; Strasia, 1971; Uthus et al., 1978; Nielsen, 1980). These studies, however, failed to produce a clear requirement for vanadium, since they were complicated by the difficulty in excluding the ubiquitous metal from the diets used (Nielsen and Uthus, 1990). Furthermore, variability in the diets and the interaction of vanadium with other dietary components such as chlorine (Hill, 1985), protein (Hill, 1979b), iron (Nielsen, 1985), riboflavin (Hill, 1988), and ascorbic acid (Hill, 1979a) makes the interpretation of the results specious.

More recent studies indicate that vanadium deprivation in goats causes decreased reproduction, growth, and survival, as well as causing bone deformation (Anke et al., 1986). Uthus and Nielsen have done a series of studies (Uthus and Nielsen, 1988; Uthus and Nielsen, 1989; Uthus and Nielsen, 1990) that suggest that vanadium is required for growth and may be related to thyroid function and iodine metabolism. These results are made more interesting by the discovery that vanadium is required for the activity of some haloperxidases (Krenn et al., 1989; van-Schijndlel et al., 1993; Weaver and Krenn, 1990) and nitrogen fixation (Eady, 1990) in lower species. While this does not mean that vanadium is required for similar enzymes in mammals, it does indicate that vanadium can have a vital physiological function, something that had been missing in previous studies of vanadium's essentiality.

The relationship of vanadium to growth and development and possible connection with growth factors is strengthened by the fact that vanadium has been shown to have growth factor-like effects. McKeehan et al. (1977) first showed that vanadium was required for optimal growth of fibroblasts in culture. Vanadium has been shown to have mitogenic responses in human fibroblasts, and in 3T3 and 3T6 cell lines (Carpenter, 1981; Mountjoy and Flier, 1990; Smith, 1983), including the insulin-like regulation of GLUT-1 glucose transporters (Mountjoy and Flier, 1990). Vanadium can have effects similar to fibroblast growth factor, epidermal growth factor, and insulin (Canalis, 1985; Kato et al., 1987; Lau et al., 1988; Nechay et al., 1989). In antigen-transformed 3T3 T cells, both insulin and vanadate can induce mitogenesis, but the specific genes involved differ between the two agents (Wang et al., 1991; Wang and Scott, 1992). While insulin and vanadate are both protein kinase C-independent, insulin is G protein- and polyamine-dependent, whereas vanadium but not insulin was inhibited by genistein (a tyrosine kinase inhibitor). All of these peptides act at least in part by altering the phosphorylation state of proteins within the cell.

Thus it appears likely that vanadium is required for an as yet unknown function dealing with proper growth and development in mammals. Insulin, growth hormone, insulin-like growth factors I and II, and thyroxin are all required for proper growth development. While it would be presumptive to suggest that any of these hormones' effects on growth require or can be mediated by a vanadium-dependent enzyme or metabolic step, the complex interaction of cellular events that make up signal transduction and the promotion of growth allows for a possible interaction of any of these agents with vanadium. Indeed, the discovery that vanadium is required for proper growth and development might well have been the first clue that vanadium is related to insulin and related growth factor activities.

3. PHYSIOLOGICAL FUNCTION

This discovery has led to two proposals for the role of endogenous vanadium in modulating protein phosphorylation. The first, suggested by Nielsen (1991), is that vanadate within the cell reacts with hydrogen peroxide to produce one or more peroxides of vanadium. Peroxides of vanadium, or pervanadates, have been shown to be potent inhibitors of tyrosine phosphatases, and can mimic many of the actions of insulin in vitro and in vivo. These effects are probably mediated by inhibition of dephosphorylation of the insulin receptor, thereby keeping its intrinsic tyrosine kinase in the active state. Furthermore, insulin and other growth factors have been shown to induce hydrogen peroxide production in cells. These factor would work together to prolong the action of insulin, and allow control of insulin's effects. Other factors, such as the oxidation state of the cell, could control the amount and duration of hydrogen peroxide production and therefore the generation of pervanadate. Thus other metabolic events could alter insulin's signal transduction mechanisms and ultimately its action.

The second, closely related theory, proposed by Schechter's group (Elberg et al., 1984), is based on the differing effects of vanadium's two predominate intracellular oxidation states, vanadyl (4+) and vanadate (5+) and is illustrated in Figure 1. They have shown that vanadate stimulates a cytosolic protein tyrosine kinase (cytPTK) by increasing its phosphorylation (the active state). This kinase has a mass of 53,000 Da, is staurosporine-inhibitable, and is distinct from the insulin receptor tyrosine kinase. Vanadate stimulation occurs by inhibiting the protein tyrosine phosphatase that removes the phosphate from cytPTK. Oxidation of vanadyl to vanadate stimulates this pathway.

Shechter's group has further shown that vanadyl can have similar effects working through a membrane-associated protein tyrosine phosphatase (Li et

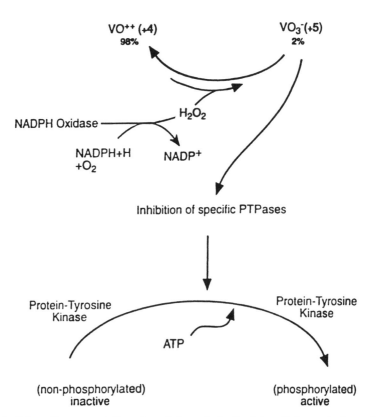

Figure 1. Schematic representation of a putative role for the intracellular pool of vanadium in modulating the activity of nonreceptor protein tyrosine kinases. Reproduced from Elberg et al. (1994), with permission.

al., 1996). Thus there are two insulin receptor kinase-independent systems, one that is vanadate-sensitive and vanadyl-insensitive, and one that is vanadyl-sensitive and vanadate-insensitive. Thus, depending on the oxidation state of the cell and the predominate vanadium species present, one or the other of these systems will be relatively stimulated. While the details of the two systems have not yet been worked out, one can speculate that subsequent steps in the phosphorylation cascade will be somewhat different, depending on the activating vanadium species. Taken together with the previous proposal that pervanadate also endogenously stimulates protein tyrosine phosphorylation, but through the insulin receptor, vanadium can play a central role in modulating the phosphorylation state of various proteins in the cell, especially ones related to the signal transduction mechanisms of insulin and growth factors. This could explain why vanadium is required for proper growth and development of mammalian organisms, but that a deficiency syndrome is difficult to identify. No single specific enzyme or cellular activity is affected, but rather the coordination of various hormonal signals with the metabolic (or oxidative) state of the cell is disrupted when vanadium is absent. If vanadium does play such a central role, then it stands to reason that its metabolism and cellular level would be regulated. There is evidence that this is the case, and that a number of hormones are involved.

4. ENDOCRINE CONTROL

4.1. Organ Half-lives

If vanadium tissue concentrations are hormonally controlled, then uptake and/ or release would have to regulated, and this would vary from organ to organ depending on the hormone involved and the tissue's sensitivity to that hormone. The tissue half-life is a measure of the metabolic turnover of the element, and the shorter the half-life is in an organ, the greater the control possible. We determined the half-lives of vanadium in a number of organs in rats (Hamel and Duckworth, 1995). Rats were fed a liquid diet supplemented with either sodium orthovanadate or vanadyl sulfate for 1 week, and then the supplement was withdrawn. Organ vanadium levels were determined at selected days thereafter, and the loss from the organs was curve-fit to an exponential decay to determine the half-lives of the various organs (shown in Table 1). Shorter half-lives are found in organs that tend to be generally more insulin-sensitive, and organs that retain vanadium tend to be less insulin-sensitive. Thus liver, fat, and muscle, three tissues in the body that are generally considered to be the most insulin-responsive, have relatively short half-lives, whereas spleen, brain, and testes have long half-lives. This indicates that vanadium levels are normally regulated in the body and suggests that insulin, or a related growth factor, may be involved.

Table 1 Vanadium Half-Lives[a]

| Supplement[b] | "Insulin-Sensitive" <------- | | | | | | ------> "Insulin-Insensitive" | | |
	Liver	Kidney	Fat	Muscle	Lung	Heart	Spleen	Brain	Testes
SOV	3.57	3.93	4.06	6.11	5.52	7.03	9.13	11.17	15.95
VS	3.18	3.27	5.04	4.49	4.45	5.05	5.15	9.17	13.50

[a] All values are given in days. Numbers determined by curve fitting to a first-order exponential decay equation. Correlation coefficient for all values ≥0.983.
[b] SOV, sodium orthovanadate; VS, vanadyl sulfate.

Source: Hamel and Duckworth (1995).

270

4.2. Growth Hormone and Thyoxin

The first studies of endocrine control of tissue vanadium levels used hypophysectomized rats to examine the effects of hormone deficiency (Peabody et al., 1976). The Peabody group looked at vanadium metabolism by injecting radioactive [48]V and measuring the isotope in various organs 7 and 14 days later. Hypophysectomized animals showed increases in [48]V content (on a per gram tissue basis) in serum, liver, pancreas, testes, lung, heart, and bone compared with control animals. Since hypophysectomy will affect a wide range of endocrine organ functions, thyroidectomized-parathyroidectomized animals were also examined. These animals showed an increase in vanadium content in pancreas and kidney, suggesting that vanadium content in these organs is, at least in part, controlled by thyroxin or parathyroid hormone.

In an effort to determine which hormone(s) were controlling vanadium, this same group looked at hormone replacement treatment in hypophysectomized rats (Peabody et al., 1977). Bovine growth hormone, thyroxin, and the combination of the two were given daily starting 6 days before assaying vanadium levels by the same radioactive element technique. As previously seen, serum, liver, pancreas, testes, kidney, lung, heart, and bone all showed elevated levels of vanadium in hypophysectomized untreated animals. Bovine growth hormone reduced vanadium levels in all these organs, and pancreas, kidney, lung, heart, and bone levels were statistically significantly different from levels of organs from untreated hypophysectomized rats. However, growth hormone alone did not return any organs to normal vanadium content. Thyroxin treatment returned vanadium levels to normal in serum and pancreas, and significantly reduced the hypophysectomy-induced elevated levels in kidney, lung, and bone. When the two hormones were given concomitantly, all organs were at or below normal levels except lung and testes, and these were down from the levels of untreated hypophysectomized animals. Thus it appears that both growth hormone and thyroxin can affect vanadium metabolism, but the extent of the effect and the organs involved are different for these two hormones. It should also be noted that the effects of growth hormone may be direct on be mediated through the insulin-like growth factors I and II (somatomedins).

4.3. Luteinizing Hormone and Follicle-Stimulating Hormone

The idea that different organs may have their vanadium levels controlled by different hormones is supported by another study by this group (Peabody et al., 1980). Luteinizing hormone (LH), follicle-stimulating hormone (FSH), and the two in combination, were administered subcutaneously to hypophysectomized rats, and vanadium metabolism was assessed again by radioactive [48]V injection. The elevated levels of vanadium in the tissues of hypophysectomized rats were unaffected by LH, FSH, or LH + FSH except in testes. Both LH and FSH alone caused a decrease in vanadium content per gram of tissue.

The combination of LH and FSH was more effective than either alone. Thus vanadium levels in testes, a known specific target tissue of the gonadotropins, are under the control of LH and FSH, although the effect of other factors cannot be excluded.

These studies provide ample evidence that vanadium metabolism is under endocrine control. Thyroxin, luteinizing hormone, and follicle-stimulating hormone appear to have some specific target organs where they regulate vanadium levels. Growth hormone has more general effects, and these studies do not define whether it is a direct effect of growth hormone or the result of some secondary action, such as the release of insulin-like growth factors (IGF) I and II, that mediates much of growth hormone's effects. All three of these hormones/growth factors control a variety of metabolic functions similar to insulin's actions. This suggests that insulin too may exert some effect on vanadium's metabolism. We have evidence that supports that concept.

4.4. Insulin

To determine if insulin is involved, we have examined the levels of vanadium in normal, streptozotocin, and insulin-treated streptozotocin diabetic rats to determine if the tissue levels are altered (Hamel, 1996). We also fed the animals supplemental vanadium to see if the accumulation of vanadium was affected in diabetic animals. Table 2 shows some of the results. Muscle and liver had higher levels of vanadium in diabetic animals than in controls, and this effect was reversed by insulin treatment. This is consistent with insulin stimulation of the loss of vanadium from a specific organ. Kidney, the primary route of excretion, had significantly less vanadium in the diabetic rats, possibly reflecting the retention of the element in the other tissues of the body. Note that the spleen also had elevated vanadium levels, but the insulin treatment did not reduce the vanadium content of the organ. We speculate that this is because spleen vanadium content is not acutely regulated by insulin, as are the more sensitive organs muscle and liver.

In diabetic animals fed supplemental vanadium, however, all the organs had less vanadium than the organs of similarly treated control rats. If intestinal absorption is also under control of insulin, and its transport into the blood

Table 2   Vanadium Levels in Control and Diabetic Rats[a]

Organ	Control	Insulin-Treated Diabetic	Diabetic
Muscle	7.0 ± 0.5	6.2 ± 1.5	10.2 ± 0.8*
Liver	25.0 ± 1.0	23.8 ± 1.8	28.4 ± 2.2[†]
Kidney	95.5 ± 4.3	82.3 ± 4.4	72.4 ± 3.0***
Spleen	41.2 ± 6.6	63.0 ± 7.2*	61.2 ± 6.7*

[a] Values are ng V/g dry tissue weight (mean $\pm$ *SEM*). Value different from control: [†] $p < 0.10$; *$p < 0.05$; ***$p < 0.001$.

Source: Hamel (1996).

stream is stimulated by insulin, then the lower levels seen in all the tissues would be the result of lower intake in the diabetic animals. Alternatively, diabetic animals tend to suffer from diarrhea due to increased intestinal transit of their food. It is possible that this increased transit allows less time for absorption of the dietary vanadium.

To determine how vanadium metabolism is altered in diabetic animals we determined the half-lives of vanadium in organs from diabetic and control animals. (unpublished data). The loss of vanadium from liver is shown in Figure 2. The loss from the organ was curve-fit to either a first-order exponential decay or linear regression. The loss from the control animal showed a better fit to the exponential decay, similar to what we had seen previously. However, the diabetic rat liver showed a better fit to the linear regression line, indicating a zero-order rate loss. Similar findings were seen in muscle tissue. These findings indicate that vanadium metabolism is under the influence of insulin and may be mediated by control of the loss of vanadium from tissue.

Our findings indicate that vanadium content in at least some organs is under the control of insulin. The fact that insulin-sensitive tissues have shorter half-lives, and the fact that the kinetics of the loss of vanadium changes in organs from diabetic animals, indicate that insulin may stimulate the loss of vanadium from tissues. Whether this is a direct effect of insulin on a cellular transport or storage system, or reflects a general effect of stimulating cellular metabolism is not clear. Also, as indicated above, various components of the endocrine system may act together to control vanadium levels.

5. CONCLUSION

Vanadium is an ultratrace metal that is very probably an essential element for proper growth and development of mammals. If so, it is required in very

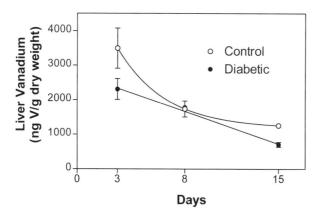

Figure 2. Loss of vanadium from livers of normal and diabetic rats after withdrawal of supplemental dietary vanadyl sulfate. Livers from normal animals show a first-order exponential loss, while livers from diabetic rats are zero-order.

minute quantities, since a deficiency syndrome has not been established. It may function as a cofactor for an as yet unidentified enzyme, but it more likely acts to modulate tyrosine phosphorylation of a variety of cellular proteins. It does this by inhibiting protein tyrosine phosphatases. The various forms of vanadium inhibit different phosphatases with different potencies. In this way the oxidative or metabolic state of the cell, by altering the form of vanadium and therefore which phosphatases are inhibited, can have an influence on the activity of a variety of enzymes that are regulated by tyrosine phosphorylation.

Growth-promoting hormones work in part by altering the phosphorylation state of cellular proteins. Thus it is logical that these hormones may have an effect on the cellular content of vanadium in order to "fine-tune" their actions to the metabolic state of the cell. Thyroxin, growth hormone, luteinizing hormone, follicle-stimulating hormone, and insulin have all been shown to affect vanadium levels in various tissues. In the absence of these hormones, vanadium levels rise in tissues. The increase in vanadium concentration would act to inhibit tyrosine phosphatases, thereby increasing the amount or duration of tyrosine phosphorylation in the cell. This increase would then act to counter the hormone deficiency. Replacement therapy lowers the vanadium concentration back towards normal and reestablishes normal control. The specificity of the hormonal action generally correlates with established target tissues for a given hormone (e.g., LH and FSH work on testes).

The mechanism by which vanadium levels are controlled is still unclear. At least with respect to insulin, the hormone appears to stimulate vanadium's loss from an organ. Whether this action is a specific effect on an ion transporter, or a general effect of stimulating cellular metabolic turnover, remains to be determined. While additional studies are warranted to determine the full extent of hormonal control of trace metals, the interaction between vanadium metabolism and the endocrine system has been established.

REFERENCES

Allain, P., and Leblondel, G. (1992). Endocrine regulation of trace element homeostasis in the rat. *Biol. Trace Elem. Res.* **32,** 187–199.

Anke, M., Groppel, B., Gruhn, K., Kosla, T., and Szilagy, M. (1986). New research on vanadate deficiency in ruminants. In M. Anke, W, Baumann, H. Braunlich, C. Bruckner, and B. Groppel (Eds.), *Spurenelement-Symposium: New Trace Elements.* Jena DDR, Friedrich-Schiller-Universitat, pp. 1266–1275.

Canalis, E. (1985). The effect of sodium vanadate on deoxyribonucleic acid and protein synthesis in cultured rat calvaraie. *Endocrinology,* **116,** 855–862.

Carpenter, G. (1981). Vanadate, EGF, and the stimulation of DNA synthesis. *Biochem. Biophys. Res. Commun.* **102,** 1115–1121.

Eady, R. R. (1990). Vanadium nitrogenases. In M. D. Chasteen (Ed.), *Vanadium in Biological Systems.* Kluwer Academic Publishers, Dordrecht, pp. 99–127.

Elberg, G., Li, J., and Shechter, Y. (1994). Vanadium activates or inhibits receptor and non-receptor protein tryrosin kinasees in cell-free experiments, depending on its oxidation state;

possible role of endogenous vanadium in controlling cellular protein tyrosine kinase activity. *J. Biol. Chem.* **269,** 9521–9527.

French, R. J., and Jones, P. J. H. (1993). Role of vanadium in nutrition: Metabolism, essentiality and dietary considerations. *Life Sci.* **52,** 339–346.

Hamel, F. G. (1996). Alteration of vanadium metabolism in streptozotocin diabetic rats. *Metabolism* (submitted for publication).

Hamel, F. G., and Duckworth, W. C. (1995). The relationship between insulin and vanadium metabolism in insulin target tissues. *Mol. Cell. Biochem.* **153,** 95–102.

Hill, C. H. (1979a). Studies on the ameliorating effect of ascorbic acid on mineral toxicities in the chick. *J. Nutr.* **109**(1), 84–90.

Hill, C. H. (1979b). The effect of dietary protein levels on mineral toxicity in chicks. *J. Nutr.* **109**(3), 501–507.

Hill, C. H. (1985). In M. Abdulla, B. M. Nair, and R. K. Chandra (Eds.), *Nutrient Research,* Suppl 1: *Health Effects and Interactions of Essential and Toxic Elements.* Pergamon Press, New York, pp, 555–559.

Hill, C. H. (1988). In L. S. Hurley, C. L. Keen, B. Lonnerdal, and R. B. Rucker (Eds.), *Trace Elements in Man and Animals 6.* Plenum, New York, pp. 585–587.

Hopkins, L. L., Jr., and Mohr, H. E. (1971). The biological essentiality of vanadium. In W. Mertz, and W. E. Cornatzer (Eds.), *Newer Trace Elements in Nutrition.* Marcel Dekker, New York, pp. 195–213.

Hopkins, L. L., and Mohr, H. E. (1974). Vanadium as an essential element. *Fed. Proc.* **33,** 1773–1775.

Kato, Y., Iwamoto, M., Koike, T., and Suzuki, F. (1987). Effect of vanadate on cartilage-matrix proteoglycan synthesis in rabbit costal chondrocyte cultures. *J. Cell Biol.* **104,** 311–319.

Krenn, B. E., Tromp, M. G. M., and Wever R. (1989). The brown algae *Ascophyllum nodosum* contains two different vanadium bromoperoxidases. *J. Biol. Chem.* **264,** 19207–19212.

Lau, K. H. W., Tanimoto, H., and Baylink, D. J. (1988). Vanadate stimulates bone cell proliferation and bone collagen synthesis in vitro. *Endocrinology* **123,** 2858–2867.

Li, J., Elberg, G., Crans, D. C., and Shechter, Y. (1996). Evidence for the distinct vanadyl (+4)-dependent activating system for manifesting insulin-like effects. *Biochemistry* **35,** 8314–8318.

McKeehan, W. L., McKeehan, K. A., Hammond, S. L., and Ham, R. G. (1977). Improved medium for clonal growth of human diploid fibroblasts at low concentrations of serum protein. *In Vitro* **3,** 399–416.

Mountjoy, K. G., and Flier, J. S. (1990). Vanadate regulates glucose transporter (Glut-1) expression in NIH 3T3 mouse fibroblasts. *Endocrinology* **127,** 2025–2034.

Myron, D. R., Givand, S. H., Hopkins, L. L., and Nielsen, F. H. (1975). Studies on vanadium deficiency in the rat. *Fed. Proc.* **34,** A923.

Nechay, B. R., Norcross-Nechay, K., and Nechay, P. S. E. (1989). *Spurenelement-Symposium: Molybdenum, Vanadium, and other Trace Elements.* Jena, DDR: Friedrich-Schiller-Universitat.

Neve, J. (1992). Clinical implications of trace elements in endocrinology. *Biol. Trace Element Res.* **32,** 173–185.

Nielsen, F. H. (1980). In H. H. Draper (Ed.), *Advances in Nutritional Research,* Vol. 3. Plenum, New York, pp. 157–172.

Nielsen, F. H. (1985). In M. Abdulla, B. M. Nair, and R. K. Chandra (Eds.), *Nutrient Research,* Suppl. 1: *Health Effects and Interactions of Essential and Toxic Elements.* Pergamon Press, New York, pp. 527–530.

Nielsen, F. H. (1990). Nutritional requirements for boron, silicon, vanadium, nickel, and arsenic: Current knowledge and speculation. *Biol. Trace Element Res.* **26**(27), 599–611.

Nielsen, F. H. (1991). Nutritional requirements for boron, silicon, vanadium, nickel, and arsenic: Current knowledge and speculation. *FASEB J.* **5**, 2661–2667.

Nielsen, F. H, and Myron, D. R. (1976). Evidence which indicates a role for vanadium in labile methyl metabolism in chicks. *Fed. Proc.* **35**, A689.

Nielsen, F. H., Myron, D. R., and Uthus, E. O. (1978). In M. Kirchgessner (Ed.), *Trace Element Metabolism in Man and Animals.* Freising-Weihenstephan, BRD, Technical University of Munchen, pp. 244–247.

Nielsen, F. H., and Ollerich, D. A. (1973). Studies on a vanadate deficiency in chicks. *Fed. Proc.* **32**, 929.

Nielsen, F. H., and Uthus, E. O. (1977). The effect of vanadium deficiency on the activity of some enzymes involved with labile methyl and methionine metabolism in the chick. *Fed. Proc.* **36a**, 1123.

Nielsen, F. H., and Uthus, E. O. (1980). The essentiality and metabolism of vanadium. In N. D. Chasteen (Ed.), *Vanadium in Biological Systems.* Kluwer Academic Publishers, Dordrechlt, pp. 51–62.

Peabody, R. A., Wallach, S., Verch, R. L., and Kraszeski, J. (1976). Metabolism of vanadium-48 in normal and endocrine deficient rats. In D. D. Hemphill (Ed.), *Trace Substances in Environmental Health,* Vol. 9. Columbia, University of Missouri Press, pp. 441–450.

Peabody, R. A., Wallach, S., Verch, R. L., and Lifschitz, M. L. (1977), Effect of thyroxin and growth hormone replacement on vanadium metabolism in hypophysectomized rats. In D. D. Hemphill (Ed.), *Trace Substances in Environmental Health,* Vol. 9. Columbia: University of Missouri Press, pp. 297–304.

Peabody, R. A., Wallach, S., Verch, R. L., and Lifschitz, M. L. (1980). Effect of LH and FSH on vanadium distribution in hypophysectomized rats. *Proc. Soc. Exp. Biol. Med.* **165**, 349–353.

Rygh, O. (1949). *Bull. Ste. Chem. Biol.* **31**, 1403–1407.

Schwarz, K., and Milne, D. B. (1971). Growth effects of vanadium in the rat. *Science* **174**, 426–428.

Smith, J. B. (1983). Vandium ions stimulate DNA synthesis in Swiss mouse 3T3 and 3T6 cells. *Proc. Natl. Acad. Sci. USA* **80**, 6162–6166.

Strasia, C. A. (1971). Vanadium: Essentiality and Toxicity in the Laboratory Rat. Ph. D. thesis, Purdue University.

Uthus, E. O., and Nielsen, F. H. (1988). The effect of vanadium, iodine and their interaction on thyroid status indices. *FASEB J.* **2**, A841.

Uthus, E. O., and Nielsen, F. H. (1989). *Spurenelement-Symposium: Molybdenum, Vanadium, and Other Trace Elements.* Friedrich-Schiller-Universität, Jena.

Uthus, E. O., and Nielsen, F. H. (1990). Effect of vanadium, iodine, and their interaction on growth, blood variables, liver trace elements, and thyroid status indices in rats *Magnesium Trace Elem.* **9**(4), 219–226.

Uthus, E. O., Nielsen, F. H., and Myron, D. R. (1978). Studies on the effect of vanadium deficiency on growth and methionine metabolism in rats. *Fed. Proc.* **37**, 893.

van-Schijndel, J. W., Wollenbroek, E. G., and Wever R. (1993). The chloroperoxidase from the fungus *Curvularia inaequalis;* a novel vanadium enzyme. *Biochim Biophys. Acta* **1161**, 249–256.

Wang, H., and Scott, R. E. (1992). Induction of c-jun independent of PKC, pertussis toxin sensitive G protein and polyamines in quiescent SV40-transformed 3T3 T cells. *Exp. Cell Res.* **203**, 47–55.

Wang H. , Wang, J. Y., Johnson, L. R., and Scott, R. E. (1991). Selective induction of c-jun and jun-B but not c-fos or c-myc during mitogenesis in SV40-transformed cells at predifferentiated growth arrest state. *Cell Growth Differ.* **2**, 645–652.

Weaver, R., and Krenn, B. E. (1990). Vanadium haloperoxidases. In N. D. Chasteen (Ed.), *Vanadium in Biological Systems.* Kluwer Academic Publishers, Dordrecht, pp. 81–97.

14

MECHANISMS OF ACTIONS OF VANADIUM IN MEDIATING THE BIOLOGICAL EFFECTS OF INSULIN

Gerard Elberg,* Jinping Li, and Yoram Shechter

Department of Biochemistry, The Weizmann Institute of Science, Rehovot 76100, Israel

Present address: Department of Cell Biology, Baylor College of Medicine, Houston, Texas 77030.

Vanadium in the Environment. Part 2: Health Effects, Edited by Jerome O. Nriagu.
ISBN 0-471-17776-8.   © 1998 John Wiley & Sons, Inc.

---

1. INTRODUCTION

Vanadium is a nutritional element present in mammalian tissues. Low quantities of dietary vanadium may be required for normal metabolism of higher animals. Vanadium exhibits a complex chemistry, fluctuating between several different oxidation forms, depending on the prevailing conditions. Intracellular forms of vanadium fluctuate between the anionic vanadate (VO_3^-) and the cationic vanadyl (VO^{2+}). Although intensive research has been performed on vanadium compounds during the last two decades, its physiological function is still obscure (Simons, 1979; Macara, 1980). A remarkable finding was the induction of normoglycemia in diabetic rodents following oral vanadium therapy (Heyliger et al., 1985; Meyerovitch et al., 1987; Brichard et al., 1988; Venkatesan et al., 1991). In vitro studies revealed that vanadium salts virtually mimic most of the effects of insulin (Tolman et al., 1979; Dubyak and Kleinzeller, 1980; Shechter and Karlish, 1980). A number of reviews have been recently published on the potential use of vanadium as a therapeutic antidiabetic agent (Shechter and Shisheva, 1993; Shechter et al., 1995; Brichard and Henquin, 1995; Orvig et al., 1995). This review focuses on the mechanisms by which vanadate facilitates its insulin-like effects at the molecular level.

2. INSULIN-LIKE EFFECTS OF VANADATE

A major role of insulin in mammals is the regulation of glucose homeostasis by modulating glucose production in the liver and glucose utilization in muscle and fat (Fig. 1). Reduced potency of insulin to arrest hepatic glucose output or to enhance glucose utilization in peripheral tissues leads to insulin resistance. This phenomenon is accompanied with a compensatory oversecretion of insulin by the pancreas (hyperinsulinemia). A failure of the pancreas to secrete insulin in sufficient quantities leads to hyperglycemia (reviewed by Kahn, 1994).

The insulin-like effects of vanadium were originally found in intact cells. Subsequently, vanadium was administrated orally to diabetic animals. In

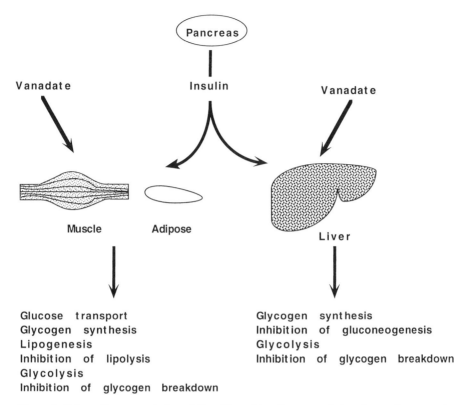

Figure 1. Schematic representation of the effect of insulin on regulating glucose homeostasis. Insulin is secreted by the pancreas and acts on the liver, increasing glucose storage as glycogen and regulating glucose output. Insulin stimulates the entry of glucose into peripheral tissues, adipose, and muscle, acting on glucose metabolism and formation of glycogen and triglycerides for energy storage. Insulinomimetic effects have been observed in animals and intact cells treated with vanadate.

vanadate-treated STZ rats, hyperglycemia was reduced to normal glucose levels within 2–5 days. Sodium metavanadate in drinking water was found to be optimal at a concentration of 0.2 mg/ml, and in some instances normoglycemia persisted after termination of vanadium therapy (Heyliger et al., 1985; Brichard et al., 1988; Meyerovich et al., 1987). The effect of vanadate was also determined in animal models representing insulin-resistance phenotypes. In ob/ob and in db/db mice and in fa/fa rats, administration of vanadate lowered plasma insulin concentrations and led to a better tolerance to glucose absorption (Brichard et al., 1989; 1990, 1992a,b; Ferber et al., 1994). How does vanadate induce these beneficial effects in diabetic rodents? Figure 1 summarizes the putative sites at which vanadate can exert its biological effects in liver, muscle, and fat. However, intact cells (or cell lines) are needed for elucidating the mechanisms involved at the molecular level. Vanadate activates glucose transport in myocytes and adipocytes and glucose metabolism and glycogen synthesis in myocytes, adipocytes, and hepatocytes. Vanadate also inhibits lipolysis, an insulin-like effect that differs mechanistically from the hormonal effect in facilitating glucose metabolism (Tolman et al., 1979; Dubyak and Kleinzeller, 1980; Shechter and Karlish, 1980; Degani et al., 1981; Tamura et al., 1984; Shechter and Ron, 1986; Milrapeix et al., 1991). In diabetic animals, vanadate therapy corrected the expression of genes encoding key enzymes of glucose metabolism such as glucokinase, 6-phosphofructokinase, acetyl coA carboxylase, fatty acid synthetase, phosphoenol carboxykinase, and pyruvate kinase (Bosch et al., 1990; Milrapeix et al., 1991; Saxena et al., 1992; Ferber et al., 1994; Brichard et al., 1994).

3. MOLECULAR BASIS FOR THE INSULINOMIMETIC EFFECTS OF VANADIUM

3.1. General Considerations

The effects of vanadate are likely to be mediated by intracellular events that ultimately lead to evoking the biological responses of insulin. To understand the mechanisms involved, it is necessary to overview the major events known to be involved in insulin's intracellular signaling. For the sake of simplicity, insulin signaling pathways can be divided into three levels (Shechter and Shisheva, 1993; Kahn, 1994): (1) insulin binding to its membranal receptor and tyrosine phosphorylation; (2) cascades involving ser/thr phosphorylation and dephosphorylation; and (3) attenuation of effector molecules such as the glucose transporters, key enzymes of glycogen and lipid metabolism, and transcription factors. A schematic view of insulin transduction pathways indicating the potential sites at which vanadate may exert its insulin-like effects is presented in Figure 2. The specific blockers and the experimental approaches used to determine the site(s) of actions of vanadate are considered in Sections 3.2, 3.3, 3.4, 4 and in Table 1.

3.2. Insulin Receptor and Insulin Receptor Substrate-1 (IRS-1)

The first step in mediating the insulin-dependent cascades consists of insulin binding to its cell surface receptor. The insulin receptor is a transmembrane protein, composed of two α subunits and two β subunits. Binding of insulin to the extracellularly located α subunits activates the intrinsic tyrosine kinase, located intracellularly at the β subunits. This initial event leads to tyrosine phosphorylation of the receptor itself, as well as the phosphorylation of intracellular substrates. The major insulin receptor tyrosine kinase (insRTK) substrate is insulin receptor substrate-1 (IRS-1), a cytosolic docking protein that, following tyrosine phosphorylation at several sites, associates with several proteins. This association may activate, deactivate, or orient these proteins toward their receptive sites. Documented signaling molecules known so far to

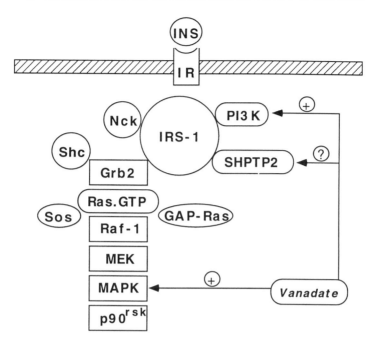

Figure 2. Simplified scheme for signaling pathways of insulin. Insulin (INS) binds extracellulary to the insulin receptor (IR). The active receptor now induces insulin action via activation of intracellular molecules. In the center is IRS-1, a docking protein that binds and activates diverse transduction pathways, Ras-MAP kinase, PI3-kinase, and SHPTP2, a protein phosphotyrosine phosphatase. Putative sites where vanadate may interfere by activation (+) or inhibition (+) are indicated by arrows.

associate with IRS-1 are phosphatidyl inositol-3-kinase, the adaptor proteins Grb2 and Nck, and the PTPase SHPTP2 (reviewed by Czech, 1985; Kahn, 1994; Waters and Pessin, 1996).

Vanadate by itself exerts negligible phosphorylation (and receptor activation) in various cells and tissues (Mooney et al., 1989; Strout et al., 1989; Fantus et al., 1994; D'Onofrio et al., 1994; Wilden and Broadway, 1995). Accordingly, IRS-1 is negligibly phosphorylated as well. Additional evidence that the insRTK is not involved in mediating the rapid metabolic effects of insulin by vanadate have been provided by cell-permeable inhibitors of the insulin receptor. In intact adipocytes, inhibition of the phosphotransferase activity of the insulin receptor blocked the biological effects of insulin on glucose transport and glucose metabolism but did not block the same bioeffects when triggered by vanadate (Shisheva and Shechter, 1992). We conclude that the insulin receptor tyrosine kinase activity and the subsequent phosphorylation of IRS-1 are not required for mediating the rapid effects of insulin by vanadate.

Table 1 Effect of Various Inhibitors on Manifesting the Rapid Biological Effects of Insulin and of Vanadate in Intact Rat Adipocytes

Inhibitor Used	Inhibition Site	Biological Effect in Rat Adipocytes				References
		Hexose Transport and Glucose Metabolism		Inhibition of Isoproterenol-Mediated Lipolysis		
		Triggered by Insulin	Triggered by Vanadate	Triggered by Insulin	Triggered by Vanadate	
Quercetin	insRTK[a]	Inhibited	Not inhibited	Not reversed	Not reversed	Shisheva and Shechter, 1992, 1993a
Wortmannin	PI3-kinase	Inhibited	Inhibited	Reversed	Not reversed	Kahn, 1994; Shisheva and Shechter, 1992; Li et al. 1997
PD98059	MEK	Not inhibited	Not inhibited	Not reversed	Not reversed	Kahn, 1994; D'Onofrio et al., 1994; Lazar et al., 1995
Staurosporine <0.3 μM	Vanadate-activatable cytPTK	Not inhibited	Inhibited[b]	Not reversed	Not reversed	Shisheva and Shechter, 1992, 1993a
Staurosporine >6 μM	Vanadate-activatable memb-PTK	Inhibited	Inhibited	Partially reversed	Reversed	Elberg et al., 1997

[a] Except for autophosporylation.
[b] Glucose metabolism (but not activation of hexose transport) is inhibited.

3.3. PI3-Kinase

Insulin activates PI3-kinase (Folli et al., 1992; Okada et al., 1994). This enzyme phosphorylates the D3 position of the inositol ring phosphatidyl inositol (PI) and generates phospholipid products. PI3-kinase is composed of binding and catalytic subunits of 85 kDa and 110 kDa, respectively (Klippel et al., 1994). Association of PI3-kinase with proteins having two phosphorylated YMXM (or very related) motifs activates PI3-kinase (Piccione et al., 1993; Williams and Shoelson, 1993; Kahn, 1994). In addition to IRS-1, PI3-kinase can associate with several tyrosine-phosphorylated protein tyrosine kinases of receptor and nonreceptor origin such as Src, a cytosolic protein tyrosine kinase (Klippel et al., 1996; Auger et al., 1992; Piccione et al., 1993). Binding of PI3-kinase through the p85-kDa subunit activates the catalytic activity located within the P110 subunit. Activation of PI3-kinase is followed by the activation of several ser/thr protein kinases such as pp70 S6 kinase, Akt, and c-Jun N-terminal kinase (Cross et al., 1995; Klippel et al., 1996). Wortmannin, a potent inhibitor of PI3-kinase (Ui et al., 1995), blocked all the rapid metabolic effects of insulin. These included the activation of glucose uptake and glucose metabolism in rat adipocytes, and the activation of glycogen synthase in skeletal muscle cells. The antilipolytic effect of insulin was also reversed by wortmannin (Okada et al., 1994; Cross et al., 1994). As with insulin, vanadate activates PI3-kinase in rat adipocytes as well. Wortmannin blocks the effects of vanadate in stimulating glucose transport and glucose metabolism. Unlike insulin, however, the antilipolytic effect of vanadate is not quenched by wortmannin (Li et al., 1997).

3.4. Ras-Mitogen-Activated Protein (MAP) Kinase Pathway

Insulin as well as several growth factors activate SOS and Ras by increasing the formation of Ras-GTP, thereby activating cascades of ser/thr protein phosphorylation in the MAP kinase pathway (Fig. 2). MAP kinase phosphorylates and activates several regulatory proteins including the ser/thr kinase p90[rsk] (Cobb et al., 1991; Kozma et al., 1993; Fingar and Birnbaum, 1994). The role of this transduction pathway in insulin signaling has been studied by transfecting cells with dominant negative Ras and by using a specific inhibitor of MEK. The dominant negative Ras largely reduced MAP kinase activation, without affecting activation of glycogen synthase by insulin (Yamamoto-Honda et al., 1995; Gabbay et al., 1996; Dorrestijn et al., 1996). Also, inhibition of the MAP kinase pathway by the MEK inhibitor had no effect on the potency of insulin to activate glucose uptake, lipogenesis, and glycogen synthase in 3T3 L1 adipocytes (Lazar et al., 1995). Vanadate also resembles insulin in activating MAP kinase in rat adipocytes. Interestingly, the effect of vanadate is more intensive and prolonged compared with insulin. The differential pattern of activation of MAP kinase by EGF (rapid activation) and NGF (persistent activation) has been proposed to explain the differential effects of EGF on growth and of NGF on differentiation in PC12, neuronal cells (Peraldi et al., 1993; Nguyen et al., 1993). Further studies revealed that the activation of MAP kinase by insulin and by vanadate occurs through two distinct pathways (manuscript in preparation). As with insulin, activation of MAP kinase is not required for manifesting the metabolic effects of vanadate.

3.5.   SHPTP2: An Insulin-Activated PTPase Species

Following insulin stimulation, SHPTP2, a cytosolic protein phosphotyrosine phosphatase (PTPase), associates with IRS-1 and undergoes activation (Yamauchi et al., 1995). Hausdorf et al. (1995) disrupted this association and demonstrated that SHPTP2 is involved in mitogenicity and in expressing glucose transporters type 1 (GLUT1) stimulated by insulin. Activation of SHPTP2, however, is not involved in enhancing glucose transport via GLUT4 or manifesting any other metabolic effect of insulin. Whether vanadate exerts any effect on this insulin-activated PTPase species remains to be elucidated.

4.   ROLE OF NONRECEPTOR PROTEIN TYROSINE KINASES IN MEDIATING THE INSULIN-LIKE EFFECTS OF VANADATE

The effect of vanadate in modulating tyrosine phosphorylation of intracellular proteins has been studied. Figure 3 shows the tyrosine phosphorylation of intracellular proteins in intact adipocytes treated either with vanadate or with vanadyl. Compared with insulin, phosphrylation of intracellular proteins occurs to a lesser extent. Tyrosine phosphorylation of a 53-kDa protein, however, is considerably more intense in vanadate-treated adipocytes (Mooney et al., 1989). With the notion that endogeneous tyrosine phosphorylation is an early prerequiste event for manifesting the metabolic effects of insulin, we have searched for a protein tyrosine kinase (PTK) that is activated by vanadate. A cytosolic protein tyrosine kinase (cytPTK) with apparent molecular weight of 53 kDa on gel filtration chromatography has been identified. cytPTK activity is elevated 3–5 fold upon treatment of rat adipocytes with vanadate. cytPTK differs from the insRTK in its molecular weight, in preferring CO^{2+} rather than Mn^{2+} as the divalent metal ion cofactor, and in exhibiting different sensitivity to protein tyrosine kinase inhibitors, such as quercetin and staurosporin. Unlike the insRTK, cytPTK is resistant to inactivation by N-ethymaleimide (Shisheva and Shechter, 1991, 1993a; Elberg et al., 1994). The role of cytPTK in mediating the insulin-like effects of vanadate at the intact cell level has been demonstrated by using staurosporine, a potent inhibitor of cytPTK and a weak inhibitor of the insRTK ($K_i \sim 2$ nM and ~ 2 μM, respectively; Shisheva and Shechter, 1993a). Staurosporine can be utilized as a general marker for cytosolic protein tyrosine kinases, inhibiting the latter at nanomolar concentrations, whereas membranal PTKs, including the insRTK, are inhibited at micromolar concentrations of staurosporine (Elberg et al., 1995). While staurosporine was found to inhibit the effect of vanadate on lipogenesis and

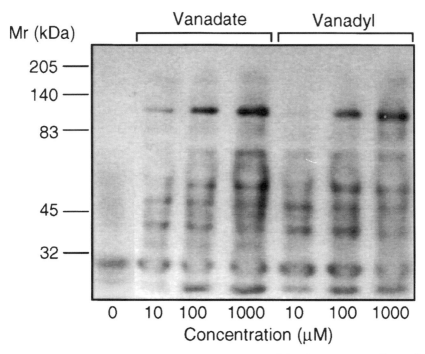

Figure 3. Effect of different concentrations of vanadate or vanadyl on the appearance of intracellular phosphotyrosine-containing proteins. Freshly isolated rat adipocytes were exposed to the indicated concentration of either vanadate or vanadyl. Following cell lysis, phosphotyrosine-containing proteins were analyzed by Western blot.

glucose oxidation at low concentration, the same bioeffects stimulated by insulin were not affected. Also, other insulinomimetic effects of vanadate such as the activation of glucose transport and inhibition of lipolysis were not inhibited by low staurosporine concentrations (Shisheva and Shechter, 1993a). These results suggest that the biological effects stimulated by vanadate, such as lipogenesis and glucose oxidation, are mediated by cytPTK in an insRTK-independent fashion. However, additional mechanisms are lacking to explain vanadium's effects in inhibiting lipolysis and in activating hexose uptake. Recently, we have identified an additional vanadate-activatable nonreceptor PTK, exclusively located at the plasma membrane (Elberg et al., 1997). This membPTK seems to participate in those vanadate effects not mediated by cytPTK and to activate PI3-kinase as well. Insulin itself stimulates the tyrosine phosphorylation of at least two different 60-kDa proteins. One is a direct substrate of the insRTK and consists of a Ras-GAP-associated protein involved in the Ras-MAP kinase pathway, and the other is associated with PI3-kinase (Lavan and Lienhard, 1993; Hosomi et al., 1994; Milarski et al., 1995; Zhang-Sun et al., 1996). It has been recently suggested that two nonreceptor PTKs, such as Fyn and JAK2, may be involved in regulating insulin signaling (Sun et al., 1996; Saad et al., 1996). The relations between the activation of these PTKs and the biological action of insulin or vanadate remain to be determined. In summary, vanadate mediates its insulin-like effects at level 1 of the insulin signaling scheme (Section 3.1), in an insRTK-independent fashion. Two nonreceptor PTKs appear to be involved in manifesting the bioeffects of vanadate.

5.   INHIBITION OF PROTEIN PHOSPHOTYROSINE PHOSPHATASES AND THE INSULIN-LIKE EFFECTS OF VANADATE

5.1.   Mechanism of Inhibition of PTPases

Vanadate is a well-known inhibitor of protein phosphotyrosine phosphatases (PTPases) (Swarup et al., 1982). All members of this PTPase family contain a signature motif in their active site: [I/V] HCXAGXXR[S/T]G (Denu et al., 1996; Tonks and Neel, 1996). The nucleophylic cysteinyl moiety harbored in the active site attacks the tyrosyl phosphate bond to form an enzyme phosphate intermediate during catalysis. Several residues, including the motif's histidyl moiety, create the hydrophobic "pocket" interacting with phosphotyrosine. Because of the resemblance of vanadate to phosphate, it was suggested that a cysteinyl-vanadate intermediate is formed, thereby inhibiting the formation of the enzyme phosphate intermediate. In the presence of nucleophiles, such as hydroxylamine, vanadate fails to inhibit protein phosphotyrosine phosphatases (Shechter et al., 1995; Denu et al., 1996). Thus, from an enzymological standpoint, vanadate appears to inhibit the rate-limiting dephosphorylation step of the phosphoenzyme intermediate (Li et al., 1995a,b).

The general linkage between inhibition of PTPases and activation of nonreceptor PTKs on the one hand, and manifesting insulin effects through receptor-independent pathways on the other hand, was further supported by additional inhibitors of PTPases. Those include molybdate, tungstate, and several vanadium (+4) compounds, which do not necessarily share a common mechanistic pathway in inhibiting this class of enzymes. For example, tungstate and molybdate represent competitive inhibitors of PTPases and do not interfere with the catalytic site of the enzyme (Barford et al., 1994). Vanadyl (+4) differs from vanadate (+5) in its oxidation state and in being cationic and a Mg^{2+} analog (Brichard and Henquin, 1995). The capacity of all those PTPase inhibitors to evoke insulin-like effects in in vitro systems has been systematically analyzed (subsequent paragraphs).

As discussed in the preceding section, we have concluded that vanadate by itself manifests its insulin-like metabolic effects through the activation of nonreceptor PTKs, in an insRTK-independent manner. Our cell-free experiments strongly suggest that at the intact cell level vanadate activates the nonreceptor PTKs (by autophosphorylation on tyrosine moieties) concomitantly with the inhibition of cellular PTPases. The insRTK is only negligibly autophosphorylated and is not activated in vanadate-treated adipocytes (Fantus et al., 1994; Mooney et al., 1989; Strout et al., 1989; Shisheva and Shechter, 1992). Our previous attempts to tackle these experimentally were based on the assumption that the adipose tissue contains several members of PTPases, of which the still obscured insRTK PTPase is insensitive to vanadate inhibition (Shechter et al., 1995). We are currently studying an alternative interpretation, namely, that the insRTK PTPase is sensitive to vanadate inhibition, although at the basal state the active-site receptor's tyrosine moieties are hindered and are therefore insusceptible to undergoing autophosphorylation, even though the specific receptor-PTPase activity is arrested.

If this option is validated, it implies that the combined therapies of insulin and vanadium might be highly beneficial in human diabetes. Following receptor activation by insulin, the presence of vanadate may preserve the receptor in an activated state after insulin is removed or when its activating potency is decreased under certain pathophysiological states. In genetically obese fa/fa rats, oral vanadate therapy showed marked antidiabetic effects by increasing the low sensitivity of peripheral tissues (particularly muscle) to insulin (Brichard et al., 1989, 1992b).

5.2. Action of Other PTPase Inhibitors

5.2.1. Tungstate and Molybdate

Tungstate and molybdate mimic the biological effects of insulin in rat adipocytes. Also administration of tungstate and molybdate to streptozotocin-treated rats reduced blood glucose levels toward normal values. These compounds activate glucose transport and glucose oxidation, stimulate lipogenesis, and inhibit lipolysis. However, these effects are obtained at higher concentration as compared to vanadate (Li et al., 1995a). Similarly, higher concentrations of tungstate and molybdate were required to activate cytPTK (Elberg et al., 1994). These quantitative differences may be attributed to reduced capacity of tungstate and molybdate to inhibit cellular PTPases. It may result from a different mechanism of inhibition (Section 5.1) that seems to be less efficient.

5.2.2. Vanadyl

Exogenously added vanadate and vanadyl similarly activate intracellular protein tyrosine phosphorylation (Fig. 3), mimic the biological actions of insulin, and show a similar capacity to act as an antidiabetic agent in experimental animals (Venkatesan et al., 1991; Becker et al., 1994). Several studies also indicate that the active intracellular form of vanadium is vanadyl, and internalized vanadate is reduced intracellularly to vanadyl (Cantley and Aisen, 1979; Degani et al., 1981; Willsky et al., 1984). To determine whether vanadyl or vanadate is the active species, we have established a cell-free experimental system that makes it possible determine the activation of cytPTK by vanadium, under conditions in which vanadium interconversion is minimal (Elberg et al., 1995; Li et al., 1996a). The overall conclusions from our cell-free studies are as follows: a. When the cell-free system is constituted solely of the cytosolic fraction (40,000g supernatant fraction) vanadate (+5), but not vanadyl (+4), activates cytPTK. b. Similarly, vanadate is the active species when the cell-free system is constituted of both the cytosolic fraction and broken plasma membrane fragments. c. Solubilization of the broken plasma membrane fragments with Triton X-100 (1%) enables vanadyl (+4) to activate cytPTK. On the basis of these studies we have concluded that rat adipocytes possess two distinct vanadate- and vanadyl-dependent activating pathways. The latter is dependent on membranal protein phosphotyrosine phosphatases. It should be determined, however, whether in this case the cell-free experimental system represents the true situation at the intact cell level. Note that Triton-extractable PTPases (and not "intact" broken plasma membrane fragments) supported the activation of cytPTK by vanadyl (+4).

5.2.3. *Phenylarsine Oxide: Apparent Paradox of Inhibition of Protein Phosphotyrosine Phosphatase in Rat Adipocytes*

In general, inhibition of PTPases correlates with the manifestation of insulin-like effects in rat adipocytes (previous sections). We therefore put special effort into studying a notable exception. As for vanadate, phenylarsine oxide (PAO) is a documented inhibitor of PTPases as well (Liao et al., 1991; Shekels et al., 1992). It is a trivalent arsenic compound that can form a covalent adduct with two closely spaced protein cysteinyl residues. Low concentrations of PAO blocked the stimulating effects of insulin and of vanadate on hexose uptake and glucose metabolism, but not the antilipolytic effects of insulin or vanadate (Li et al., 1996b). The appearance of several phosphotyrosine-containing proteins in PAO-treated adipocytes suggested to us that, indeed, PAO does inhibit certain PTPase(s) in this cell type. This in turn led to a working hypothesis that assumes the presence of both "stimulatory" and "inhibitory" PTPases; inhibiting the latter arrests the activating effects of insulin or vanadate on hexose uptake and glucose metabolism. Attempts to identify the PAO-sensitive, "inhibitory" PTPase in cell-free experiments were unsuccesfull. Using indirect experimental approaches, we concluded that the "inhibitory" PTPase constitutes a minute fraction of the total adipocytic PTPase activity, which is exclusively associated with the membrane fraction. Several phosphotyrosine-containing proteins emerge in PAO-treated (but not in vanadate-treated) adipocytes. These include a 33-kDa protein, whose phosphorylation on tyrosine residue may produce a negative feedback mechanism and prohibit the activation of glucose uptake and its metabolism by either insulin or vanadate.

5.2.4. *Peroxovanadium (pV); The Basic Differences between Vanadium Salts and Peroxovanadium as Insulinomimetic Agents*

Although both peroxovanadium (pervanadate, pV) and vanadium salts mimic the actions of insulin, our long-term studies in rat adipocytes revealed that they operate through two distinct mechanisms.

Vanadate interacts with H_2O_2 to form peroxovanadium complexes, whereas peroxide ion (O_2^{2-}) enters the coordination sphere of vanadium. One or two such peroxo ligands can theoretically accumulate (Posner et al., 1994). Peroxovanadium (pV) is about 100 times more potent ($ED_{50} = 0.8-1$ μM) than vanadate in facilitating the rapid metabolic effects of insulin (Li et al., 1995b). In contrast with vanadate, in pV-treated adipocytes the insulin receptor undergoes rapid autophosphorylation and activation followed by the phosphorylation of receptor substrates such as the IRS-1 (Kadota et al., 1987; Fantus et al., 1987; Shisheva and Shechter, 1993b). These studies further confirmed that pV (but not vanadate) facilitates its insulin-like effects exclusively through receptor activation and IRS-1 phosphorylating pathways.

What are the unique features of pV that account for the above? Peroxovanadium differs from vanadium in being an oxidizing agent relative to glutathione. It oxidizes a stoichiometric amount of GSH to GSSG (Li et al., 1995b). In cell-free experiments, pV is only a slightly more potent inhibitor of adipose PTPases than is vanadate, but at the intact cell level pV is highly potent in this respect and substantially inhibits (or possibly inactivates) adipose PTPases (Shisheva and Shechter, 1993a,b). Membranal (rather than cytosolic) PTPases are more affected. About 75–85% of the total membranal PTPase is irreversibly inhibited prior to the phosphorylation and activation of the insulin receptor (submitted manuscript). Huyer et al. (1997) have recently demonstrated that prolonged activation of PTP1B with pV inactivates the enzyme by irreversibly oxidizing the catalytic cysteine moiety to cysteic acid. Such a process appears to be more efficient at the intact cell level following exposure to peroxovanadium.

6. PUTATIVE ROLE FOR THE INTRACELLULAR VANADIUM POOL IN HIGHER ANIMALS

Although we have focused in our studies on "enforcing" insulin-like effecs by enriching adipocytes with exogenously added vanadium, the data accumulated may also support a feasible physiological role for these minute quantities of intracellularly located vanadium. Vanadium is a dietary trace element that it has been suggested is essential for higher animals (Macara, 1980). Its intracellular concentration is approximately 20–100 nM. The bulk of the intracellular vanadium is probably in the vanadyl (+4) form. The large amount of cytPTK activity in the mammalian cytosolic compartment makes one wonder, whether it constitutes a reservoir for tackling physiological needs not directly controlled by external stimuli. Our cell-free studies, which exclusively represent the adipose cytosolic system, showed that vanadyl (+4) does not activate cytPTK but vanadate does (Elberg et al., 1994). It was believed for nearly two decades that vanadyl would not be oxidized to vanadate at the reducing intracellular atmosphere maintained by millimolar GSH concentrations. Recently, however, we found that at physiological pH and temperatures GSH is a very ineffectual reductant of vanadate to vanadyl (Li et al., 1995b). Also, free vanadyl undergoes spontaneous oxidation to vanadate in vitro at physiological pH and temperature and in the presence of GSH (Li et al., 1996a). Thus the above dogma appears to be discredited.

Vanadyl is readily oxidized to vanadate by one equivalent of hydrogen peroxide, at neutral pH values. The affinity of this reaction is enormous, and vanadyl would efficiently compete with GSH for endogenously formed H_2O_2 (manuscript in preparation). Any physiological conditions that activate NADPH oxidase and lead to the formation of H_2O_2 are expected to oxidize a fraction of the endogenous vanadyl pool to vanadate. This in turn will inhibit vanadate-sensitive PTPases, and correspondingly increase the steady states of phosphorylation and activation of cytPTK. Increased H_2O_2 levels occur upon treating noninflammatory cells such as fibroblasts and adipocytes with cytokines (IL-1 and TNFα), hormones (angiotensin II), or growth factors (PDGF, FGF, EGF and TGFβ) (Meier et al., 1989, 1991; Sundaresan et al., 1995; Tannickal and Fanburg, 1995; Krieger-Brauer and Kather, 1995). Most of the protein tyrosine kinases known to date seem to be activated as a result of autophosphorylation on tyrosine moieties located near or at their active sites. That vanadyl can be converted to vanadate via an NADPH-oxidative pathway has been previously demonstrated (Liochev and Fridovich, 1987) and a link between vanadate, NADPH, and activation of tyrosine phosphorylation in cells was frequently observed (Trudel et al., 1990, 1991; Grinstein et al., 1990).

ABBREVIATIONS

cytPTK: cytosolic protein tyrosine kinase

EGF: epidermal growth factor

FGF: fibroblast growth factor

GAP: G protein-activating protein

GLUT: glucose transporter

Grb2: growth factor-binding protein-2

GSH: reduced glutathione

IL: interleukin

insRTK: insulin receptor tyrosine kinase

IRS-1: insulin receptor substrate-1

JAK: janus kinase

MAP: mitogen-activated protein

MEK: MAP kinase kinase

membPTK: membranal protein tyrosine kinase

NGF: nerve growth factor

PAO: phenylarsine oxide

PDGF: platelet-derived growth factor
PI: phosphatidylinositol
PTK: protein tyrosine kinase
PTPase: protein phosphotyrosine phosphatase
pV: peroxovanadium
SOS: Son-of-Sevenless
TGF: transforming growth factors
TNF: tumor necrotic factor
STZ rats: streptozocin-treated diabetic rats
GSH: reduced glutathione
GSSG: oxidized glutathione
PTP1B: protein phosphotyrosine phosphatase 1B

ACKNOWLEDGMENTS

This study was supported in part by grants from the Minerva Foundation (Germany), the Israel Ministry of Health, the Israel Academy of Sciences and Humanities, the Levine Fund, Teva Pharmaceutical Company Fund, and the Rowland and Sylvia Shaefer Program in Diabetes Research, and by a postdoctoral fellowship received by G. E. from the Ministry of Science and Technology in Israel. Y. S. is the incumbent of the C. H. Hollenberg Chair in Metabolic and Diabetes Research, established by the Friends and Associates of Dr. C. H. Hollenberg of Toronto, Canada. We thank Mrs. Elana Friedman for typing the manuscript and Dr. Sandra Moshonov for editing it.

REFERENCES

Auger, K. R., Carpenter, C. L., Shoelson, S. E., Piwinica-Worms, H., and Cantley, L. C. (1992). Polyoma virus middle T antigen-pp60c-src complex associates with purified phosphatidylinositol 3-kinase in vitro. *J. Biol. Chem.* **267**, 5408–5415.

Barford, D., Flint, A. J., and Tonks, N. K. (1994). Crystal structure of human protein tyrosine phosphatase 1B. *Science* **263**, 1397–1404.

Becker, D. J., Ongemba, L. N., and Henquin, J. C. (1994). Comparison of the effects of various vanadium salts on glucose homeostasis in streptozotocin-diabetic rats. *Eur. J. Pharmacol.* **260**, 169–175.

Bosch, F., Hatzoglou, M., Park, E. A., and Hanson, R. W. (1990). Vanadate inhibits expression of the gene for phosphoenolpyruvate carboxykinase in rat hepatoma cells. *J. Biol. Chem.* **265**, 13677–13682.

Brichard, S. M., Bailey, C. J., and Henquin, J. C. (1990). Marked improvement of glucose homeostasis in diabetic ob/ob mice given oral vanadate. *Diabetes* **39**, 1326–1332.

Brichard, S. M, and Henquin, J. C. (1995). The role of vanadium in the management of diabetes. *Trends Pharmacol. Sci.* **16**, 265–270.

Brichard, S. M., Okitolonda, W., and Henquin, J. C. (1988). Long term improvement of glucose homeostasis by vanadate treatment in diabetic rats. *Endocrinology* **123**, 2048–2053.

Brichard, S. M., Ongemba, L. N., Girard, J., and Henquin, J. C. (1994). Tissue-specific correction of lipogenic enzyme gene expression in diabetic rats given vanadate. *Diabetologia* **37**, 1065–1072.

Brichard, S. M., Ongemba, L. N., and Henquin, J. C. (1992a). Oral vanadate decreases muscle insulin resistance in obese fa/fa rats. *Diabetologia* **35**, 522–527.

Brichard, S. M., Assimacopoulos-Jeannet, F., and Jeanrenaud, B. (1992b). Vanadate treatment markedly increases glucose utilization in muscle of insulin resistant fa/fa rats without modifying glucose transporter expression. *Endocrinology* **131**, 311–317.

Brichard, S. M., Poltier, A. M., and Henquin, J. C. (1989). Long term improvement of glucose homeostasis by vanadate in obese hyperinsulinemic fa/fa rats. *Endocrinology* **125**, 2510–2516.

Cantley, L. C., and Aisen, P. (1979). The fate of cytoplasmic vanadium. *J. Biol. Chem.* **254**, 1781–1784.

Cantley, L. C., Resch, M. D., and Guidotti, G. (1978). Vanadate inhibits the red cell (Na+, K+) ATPase from the cytoplasmatic side. *Nature (Lond.)* **272**, 552–554.

Carey, J. O., Azevedo, J. L., Morris, P. G., Pories, W. J., and Dohm, G. L. (1995). Okadaic acid, vanadate, and phenylarsine oxide stimulate 2-deoxyglucose transport in insulin-resistant human skeletal muscle. *Diabetes* **44**, 682–688.

Cobb, M. H., Boulton, T. G., and Robbins, D. J. (1991). Extracellular signal-regulated kinases: ERKs in progress. *Cell Regul.* **2**, 965–978.

Cross, D. A., Alessi, D. R., Cohen, P., Andjelkovitch, M., and Hemmings, B. A. (1995). Inhibition of glycogen synthase kinase-3 by insulin mediated by protein kinase B. *Nature (Lond.)* **378**, 785–789.

Cross, D. A. E., Alessi, D. R., Vandenheede, J. R., McDowell, H. E., Hundal, H. S., and Cohen, P. (1994). The inhibition of glycogen synthase kinase-3 by insulin or insulin-like growth factor 1 in the rat skeletal muscle cell line L6 is blocked by wortmannin, but not rapamycin: Evidence that wortmannin blocks activation of the mitogen-activated protein kinase pathway in L6 cells between Ras and Raf. *Biochem. J.* **303**, 21–26.

Czech, M. P. (1985). The nature and the regulation of the insulin receptor structure and function. *Annu. Rev. Physiol.* **47**, 357–381.

Degani, H., Gochin, M., Karlish, S. J. D., and Shechter, Y. (1981). Electron paramagnetic studies and insulin-like effects of vanadium in rat adipocytes. *Biochemistry* **20**, 5795–5799.

Denu J. M., Stuckey, J. A., Saper, M. A., and Dixon, J. E. (1996). Form and function in protein dephosphorylation. *Cell* **87**, 361–364.

D'Onofrio, F., Le, M. Q., Chiasson, J. L., and Srivastava, A. K. (1994). Activation of mitogen activated protein (MAP) kinases by vanadate is independent of insulin receptor autophosphorylation. *FEBS Lett* **340**, 269–275.

Dorrestijn J., Ouwens, D. M., Van den Berghe, N., Bos, J. L., and Maassen, J. A. (1996). Expression of a dominant-negative Ras mutant does not affect stimulation of glucose uptake and glycogen synthesis by insulin. *Diabetologia* **39**, 558–563.

Dubyak, G. R., and Kleinzeller, A. (1980). The insulin-mimetic effects of vanadate in isolated adipocytes. *J. Biol. Chem.* **255**, 5306–5312.

Elberg, G., Li, J., Leibovitch, A., and Shechter, Y. (1995). Non-receptor cytosolic protein tyrosine kinases from various rat tissues. *Biochim. Biophys. Acta* **1269**, 299–306.

Elberg, G., Li, J., and Shechter, Y. (1994). Vanadium activates or inhibits receptor and non-receptor protein tyrosine kinases in cell-free experiments, depending on its oxidation state. *J. Biol. Chem.* **269**, 9521–9527.

Elberg, G., He, Z. B., Li, J., Sekar, N., and Shechter, Y. (1997). Vanadate activates membranous non receptor protein tyrosine kinase in rat adipocytes. *Diabetes* **45**, in press.

Fantus, I. G., Ahmad, F., and Deragon, G. (1994). Vanadate augments insulin-stimulated insulin receptor tyrosine kinase activity and prolongs insulin action in rat adipocytes. *Diabetes* **43,** 375–383.

Fantus, G. I., Kadota, S., Deragon, G., Foster, B., and Posner, B. I. (1987). Pervanadate [peroxide(s) of vanadate] mimics insulin action via activation of the insulin-receptor tyrosine kinase. *Biochemistry* **28,** 8864–8871.

Ferber, S., Meyerovitch, J., Kriauciunas, K. M., and Khan, C. R. (1994). Vanadate normalizes hyperglycemia and phosphoenolpyruvate carboxykinase mRNA levels in ob/ob mice. *Metab. Clin. Exp.* **43,** 1346–1354.

Fingar, D. C., and Birnbaum, M. J., (1994). Characterization of the mitogen-activated protein kinase/90-kilodalton ribosomal protein S6 kinase signaling pathway in 3T3-1 adipocytes and its role in insulin-stimulated glucose transport. *Endocrinology* **134,** 728–735.

Folli, F., Saad, M. J. A., Backer, J. M., and Khan, C. R. (1992). Insulin stimulation of phosphatidylinositol 3-kinase and association with IRS-I in liver and muscle of intact rat. *J. Biol. Chem.* **267,** 22171–22177.

Gabbay, R. A., Sutherland, C., Gnudi, L., Kahn, B. B., O'Brien, R. M. Granner, D. K., and Flier, J. S. (1996). Insulin regulation of phosphoenolpyruvate carboxykinase gene expression does not require activation of the Ras/mitogen-activated protein kinase signaling pathway. *J. Biol. Chem.* **271,** 1890–1897.

Grinstein, S., Furuya, W., Lu, D. J., and Mills, G. B. (1990). Vanadate stimulates oxygen consumption and tyrosine phosphorylation in electropermeabilized human neutrophils. *J. Biol. Chem.* **265,** 318–327.

Hausdorff, S. F., Bennett, A. M., Neel, B. G., and Birnbaum, M. J. (1995). Different signaling roles of SHPTP2 in insulin-induced GLUT1 expression and GLUT4 translocation. *J. Biol. Chem.* **270,** 12965–12968.

Heyliger, C. E., Tahiliani, A. G., and McNeill, J. H. (1985). Effect of vanadate on elevated blood glucose and depressed cardiac performance of diabetic rats. *Science* **227,** 1474–1476.

Hosomi, Y., Shii, K., Ogawa, W., Matsuba, H., Yoshida, M., Okada, Y., Yokono, K., Kasuga, M., Baba, S., and Roth, R. A. (1994). Characterization of a 60-kilodalton substrate of insulin receptor kinase. *J. Biol. Chem.* **269,** 11498–11502.

Huyer, G., Liu, S., Kelley, J., Moffat, J., Payette, P., Kennedy, B., Tsaprailis, G., Gresser, M. J., and Ramachandran, C. (1997). Mechanism of inhibition of protein-tyrosine phosphatases by vanadate and pervanadate. *J. Biol. Chem.* **272,** 843–851.

Kadota, S., Fantus, G., Deragon, G., Guyda, J. H., Hersh, B., and Posner, B. I. (1987). Peroxide(s) of vanadium: A novel and potent insulin-mimetic agent which activates the insulin receptor kinase. *Biochem. Biophys. Res. Commun.* **147,** 259–266.

Kahn, C. R. (1994). Insulin action, diabetogenes, and cause of type II diabetes. *Diabetes* **43,** 1066–1084.

Klippel, A., Escobedo, J. A., Hirano, M., and Williams, L. T. (1994). The interaction of small domains between the subunits of phosphatidylinositol 3-kinase determines enzyme activity. *Mol. Cell. Biol.* **14,** 2675–2685.

Klippel, A., Reinhard, C., Kavanaugh, W. M., Appell, G., Escobedo, M. A., and Williams, L. T. (1996). Membrane localization of phosphatidylinositol 3-kinase is sufficient to activate multiple signal-transducing kinase pathways. *Mol. Cell. Biol.* **16,** 4117–4127.

Kozma, L., Baltensperger, K., Klarlund, J., Porras, A., Santos, E., and Czech, M. P. (1993). The Ras signaling pathway mimics insulin action on glucose transporter translocation. *Proc. Natl. Acad. Sci. USA* **90,** 4460–4464.

Krieger-Brauer, H. I., and Kather, H. (1995). Antagonistic effects of different members of the fibroblast and platelet-derived growth factor families on adipose conversion and NADPH-dependent H_2O_2 generation in 3T3-L1 cells. *Biochem. J.* **307,** 549–556.

Lavan, B. E., and Liendhard, G. E. (1993). The insulin-elicited 60-kDa phosphotyrosine protein in rat adipocytes is associated with phosphatidylinositol 3-kinase. *J. Biol. Chem.* **268,** 5921–5928.

Lazar, D. F., Wiese, R. J., Brady, M. J., Mastick, C. C., Waters, S. B., Yamauchi, K., Pessin, J. E., Cuatrecasas, P., and Saltiel, A. R. (1995). Mitogen-activated protein kinase inhibition does not block the stimulation of glucose utilization by insulin. *J. Biol. Chem.* **270,** 20801–20807.

Li, J., Elberg, G., Crans, D. C., and Shechter, Y. (1996a). Evidence for the distinct vanadyl (+4)–dependent activating system for manifesting insulin-like effects. *Biochemistry* **35,** 8314–8318.

Li, J., Elberg, G., and Shechter, Y. (1996b). Phenylarsine oxide and vanadate: Apparent paradox of inhibition of protein phosphotyrosine phosphatases in rat adipocytes. *Biochim. Biophys. Acta* **1312,** 233–230.

Li, J., Elberg, G., Libman, J., Shanzer, A., Gefel, D., and Shechter, Y. (1995a). Insulin-like effects of tungstate and molybdate: Mediation through insulin receptor independent pathways. *Endocrine* **3,** 631–637.

Li, J., Elberg, G., Gefel, D., and Shechter Y. (1995b). Permolybdate and pertungstate-potent stimulators of insulin effects in rat adipocytes: Mechanism of action. *Biochemistry* **34,** 6218–6225.

Li, J., Elberg, G., Sekar, N., He, Z. B., and Shechter, Y. (1997). Antilipolytic actions of vanadate and insulin in rat adipocytes mediated by distinctly different mechanisms. *Endocrinology* **138,** 2274–2279.

Liao, K., Hoffman, R. D., and Lane, M. D. (1991). Phospotyrosyl turnover in insulin signaling: Characterization of two membrane-bound ppl5 protein tyrosine phosphatase from 3T3-L1 adipocytes. *J. Biol. Chem.* **266,** 6544–6553.

Liochev, S., and Fridovich, I. (1987). The oxidation of NADPH by tetravalent vanadium. *Arch. Biochem. Biophys. Acta* **255,** 274–278.

Macara, I. G. (1980). Vanadium—an element in search for a role. *Trends Biochem. Sci.* **5,** 92–94.

Meier, B., Radeke, H. H., Selle, S., Younes, M., Sies, H., Resch, K., and Habermehl, G. (1989). Human fibroblasts release oxygen species in response to interleukin-1 or tumor necrosis factor α. *Biochem. J.* **263,** 539–545.

Meier B., Cross, A. R., Hancock, J. T., Kaup, F. J., and Jones, O. T. (1991). Identification of a superoxide generating NADPH oxidase system in human fibroblasts. *Biochem. J.* **275,** 241–245.

Meyerovitch, J., Farfel, Z., Sack, J., and Shechter, Y. (1987). Oral administration of vanadate normalizes glucose levels in streptozotocin-treated rats. *J. Biol. Chem.* **262,** 6658–6662.

Milarski K. L., Lazar, D. F., Wiese, R. J., and Saltiel, A. R. (1995). Detection of a 60 kDa tyrosine-phosphorylated protein in insulin-stimulated hepatoma cells that associates with the SH2 domain of phosphatidylinositol 3-kinase. *Biochem. J.* **308,** 579–583.

Milrapeix, M., Decaux, J. F., Kahn, A., and Bartrons, R. (1991). Vanadate induction of L-type pyruvate kinase mRNA in adult rat hepatocytes in primary culture. *Diabetes* **40,** 462–464.

Mooney, R. A., Bordwell, K. L., Luhowskyj, S., and Casnellie, J. E. (1989). The insulin-like effect of vanadate on lipolysis in rat adipocytes is not accompanied by insulin-like effect on tyrosine phosphorylation. *Endocrinology* **124,** 422–429.

Nguyen, T. T., Scimeca, J. C., Filloux, C., Peraldi, P., Carpentier, J. L., and Van Obberghen, E. (1993). Co-regulation of the mitogen-activated protein kinase, extracellular signal-regulated kinase 1, and the 90-kDa ribosomal S6 kinase in PC12 cells. Distinct effects of the neurotrophic factor, nerve growth factor, and the mitogenic factor, epidermal growth factor. *J. Biol. Chem.* **268,** 9803–9810.

Okada, T., Kawano, Y., Sakakibara, T., Hazeki, O., and Ui, M. (1994). Essential role of phosphatidylinositol 3-kinase in insulin-induced glucose transport and antilipolysis in rat adipocytes. *J. Biol. Chem.* **269,** 3568–3573.

Orvig C., Thompson, K. H., Battel, M., and McNeill, J. H. (1995). Vanadium compounds as insulin mimics. *Metal Ions Biol. Systems* **31,** 575–594.

Peraldi, P., Scimeca, J. C., Filloux, C., and Van Obberghen, E. (1993). Regulation of extracellular signal-regulated protein kinase-1 (ERK-1; pp44/mitogen-activated protein kinase) by epidermal growth factor and nerve growth factor in PC12 cells: Implication of ERK1 inhibitory activities. *Endocrinology* **132**, 2578–2585.

Piccione, E., Case, R. D., Domchek, S. M., Hu, P., Chaudhuri, M., Backer, J. M., Schlessinger, J., and Shoelson, S. E. (1993). Phosphatidylinositol 3-kinase p85 SH2 domain specificity defined by direct phosphopeptide/SH2 domain binding. *Biochemistry* **32**, 3197–3202.

Posner, B. I., Faure, R., Burgess, J. W, Bevan, A. P., Lachance, D., Zhang-Sun, G. Fantus, I. G., Ng, J. B., Hall, D. A., Lum, B. S., and Shaver, A. (1994). Peroxovanadium compounds. A new class of potent phosphotyrosine phosphatase inhibitors which are insulin mimetics. *J. Biol. Chem.* **269**, 4596–4604.

Saad, M. J. A., Carvalho, C. R. O., Thirone, A. C. P., and Velloso, L. A. (1996). Insulin induces tyrosine phosphorylation of JAK2 in insulin-sensitive tissues of intact rat. *J. Biol. Chem.* **271**, 22100–22104.

Saxena, A. K., Srivastava, P., and Baquer, N. Z. (1992). Effects of vanadate on glycolytic enzymes and malic enzyme in insulin-dependent and independent tissues of diabetic rats *Eur. J. Pharmacol.* **216**, 123–126.

Shechter, Y., and Karlish, S. J. D. (1980). Insulin-like stimulation of glucose oxidation in rat adipocytes by vanadyl(IV) ions. *Nature (Lond.)* **284**, 556–558.

Shechter Y., Li, J., Meyerovitch, J., Gefel, D., Bruck, R., Elberg, G., Miller, D. S., and Shisheva, A. (1995). Insulin-like actions of vanadate are mediated in an insulin-receptor-independent manner via non-receptor protein tyrosine kinases and protein phosphotyrosine phosphatases. *Mol. Cell. Biochem.* **153**, 39–47.

Shechter, Y., and Ron, A. (1986). Effect of depletion of phosphate and bicarbonate on insulin action in rat adipocytes. *J. Biol. Chem.* **261**, 14945–14950.

Shechter, Y., and Shisheva, A. (1993). Vanadium salts and the future treatment of diabetes. *Endeavour* **117**, 27–31.

Shekels, L. L., Smith, A. J., Van Etten, R. L., and Bernlohr, D. A. (1992). Identification of the adipocyte acid phosphatase as a PAO-sensitive tyrosyl phosphatase. *Protein Sci.* **1**, 710–721.

Shisheva, A., and Shechter, Y. (1991). A cytosolic protein tyrosine kinase in rat adipocytes. *FEBS Lett.* **300**, 93–96.

Shisheva, A., and Shechter, Y. (1992). Quercetin selectively inhibits insulin receptor function in vitro and the bioresponses of insulin and insulinomimetic agents in rat adipocytes. *Biochemistry* **31**, 8059–8063.

Shisheva, A., and Shechter, Y. (1993a). Role of cytosolic tyrosine kinase in mediating insulin-like actions of vanadate in rat adipocytes. *J. Biol. Chem.* **268**, 6463–6469.

Shisheva, A., and Shechter, Y. (1993b). Mechanism of pervanadate stimulation and potentiation of insulin-activated glucose transport in rat adipocytes: Dissociation from vanadate effect. *Endocrinology* **133**, 1562–1568.

Simons, T. J. B. (1979). Vanadate—A new tool for biologists. *Nature* **281**, 337–338.

Strout, H. V., Vicario, P. P., Superstein, R., and Slater, E. E. (1989). The insulin mimetic effect of vanadate is not correlated with insulin receptor tyrosine kinase activity nor phosphorylation in mouse diaphragm in vivo. *Endocrinology* **124**, 1918–1924.

Sun, X. J., Pons, S., Asano T., Myers M. G., Glasheen, E., and White, M. F. (1996). The Fyn tyrosine kinase binds IRS-1 and forms a distinct signaling complex during insulin stimulation. *J. Biol. Chem.* **271**, 10583–10587.

Sundaresan, M., Yu, Z.-X., Ferrans, V. J., Irani, K., and Finkel, T. (1995). Requirement for generation of H_2O_2 for platelet-derived growth factor signal transduction. *Science* **270**, 296–299.

Swarup, G., Cohen, S., and Garbers, D. L. (1982). Inhibition of membrane phosphotyrosyl-protein phosphatase activity by vanadate. *Biochem. Biophys. Res. Commun.* **107**, 1104–1109.

Tamura S., Brown T. A., Whipple, J. H., Yamaguchi, Y. F., Dubler, R. E., Cheng, K., and Larner, J. (1984). A novel mechanism for the insulin-like effects of vanadate on glycogen synthase in rat adipocytes. *J. Biol. Chem.* **259,** 6650–6658.

Tannickal, V. J., and Fanburg, B. L. (1995). Activation of an H_2O_2 NADH oxidase in human lung fibroblasts by transforming growth factor $\beta1$. *J. Biol. Chem.* **270,** 330334–330338.

Tolman, E. L., Barris, E., Burns, M., Pansini, A., and Partridge, R. (1979). Effect of vanadium on glucose metabolism in vitro. *Life Sci.* **25,** 1159–1164.

Tonks, N. K., and Neel, B. G. (1996). From form to function: Signaling by protein tyrosine phosphatases. *Cell* **87,** 365–368.

Trudel, S., Downey, G. P., Grinstein, S., and Paquet, R. M. (1990). Activation of permeabilized HL60 cells by vanadate. Evidence for divergent signalling pathways. *Biochem. J.* **269,** 127–131.

Trudel, S., Paquet, M., and Grinstein, S. (1991). Mechanism of vanadate-induced activation of tyrosine phosphorylation and of the respiratory burst in HL60 cells. Role of reduced oxygen metabolites. *Biochem. J.* **276,** 611–619.

Ui M., Okada, T., Hazeki, K., and Aseki, O. (1995). Wortmannin as a unique probe for a intracellular signalling protein, phosphoinositide 3-kinase. *Trends Biochem. Sci.* **20,** 303–307.

Venkatesan, N., Avidan, A., and Davidson, M. B. (1991). Antidiabetic action of vanadyl in rats independent of in vivo insulin-receptor kinase activity. *Diabetes* **40,** 492–498.

Waters, S. B., and Pessin, J. E. (1996). Insulin receptor substrate 1 and 2 (IRS1 and IRS2): What a tangled web we weave. *Trends Cell Biol.* **6,** 1–4.

Wilden, P. A., and Broadway, D. (1995). Combination of insulinomimetic agents H_2O_2 and vanadate enhances insulin receptor mediated tyrosine phosphorylation of IRS-1 association with phosphatidylinositol 3-kinase. *J. Cell. Biochem.* **58,** 279–291.

Williams, K. P., and Shoelson, S. E. (1993). A photoaffinity scan maps regions of the p85 SH2 domain involved in phosphoproptein binding. *J. Biol. Chem.* **268,** 5361–5364.

Willsky, G. R., White, D. A., and McCabe, B. C. (1984). Metabolism of added orthovanadate to vanadyl and high-molecular weight vanadates by *Saccharomyces cerevisiae.* *J. Biol. Chem.* **259,** 13273–13281.

Yamamoto-Honda, R., Tobe, K., Kaburagi, Y., Ueki, K., Asai, S., Yachi, M., Shirouzu, M., Yodoi, J., Akanuma, Y., Yokoyama, S., Yasaki, Y., and Kadowaki, T. (1995). Upsteam mechanisms of glycogen synthase activation by insulin and insulin-like growth factor-1. *J. Biol. Chem.* **270,** 2729–2734.

Yamauchi K., Milarski, K. L., Saltiel, A. R., and Pessin, J. E. (1995). Protein-tyrosine-phosphatase SHPTP2 is a required positive effector for insulin downstream signaling. *Proc. Natl. Acad. Sci. USA* **92,** 664–668.

Zhang-Sun G., Yang, C., Viallet, J., Feng, G., Bergeron, J. J., and Posner, B. I. (1996). A 60-kilodalton protein in rat hepatoma cells overexpressing insulin receptor was tyrosine phosphorylated and associated with Syp, phosphatidyl inositol 3-kinase, and Grb2 in an insulin-dependent manner. *Endocrinology* **137,** 2649–2658.

15

ANTIDIABETIC ACTION OF VANADIUM COMPLEXES IN ANIMALS: BLOOD GLUCOSE NORMALIZING EFFECT, ORGAN DISTRIBUTION OF VANADIUM, AND MECHANISM FOR INSULIN-MIMETIC ACTION

Hiromu Sakurai and Akihiro Tsuji

Department of Analytical and Bioinorganic Chemistry, Kyoto Pharmaceutical University, Nakauchi-cho 5, Misasagi, Yamashina-ku, Kyoto 607, Japan

Vanadium in the Environment. Part 2: Health Effects, Edited by Jerome O. Nriagu.
ISBN 0-471-17776-8. © 1998 John Wiley & Sons, Inc.

---

1. INTRODUCTION

Just before the twenty-first century, travels to vast and boundless space have been realized and medical treatment has made great progress. Nevertheless, we live every day feeling uneasy for the time when we might suffer from diseases such as cancer, myocardial infarction, cerebral infarct, Alzheimer's disease, rheumatism, hypertension, and diabetes mellitus. Among them, the diabetes mellitus is the most demographically and geographically widespread. The number of patients suffering from diabetes mellitus is increasing day by day.

A terrible aspect of suffering from diabetes mellitus is the development of many serious secondary complications, such as atherosclerosis, microangiopathy, renal dysfunction and failure, cardiac abnormality, diabetes retinopathy, and ocular disorders.

Diabetes mellitus is defined as a disease that results in chronic hyperglycemia due to an absolute or relative lack of insulin and/or insulin resistance, that in turn impairs glucose, protein, and lipid metabolism, and finally entrains the characteristic secondary complications (Atkinson and Maclaren, 1990).

According to the definition of WHO, diabetes mellitus (DM) is generally classified as either insulin-dependent (IDDM, type I) or non-insulin-dependent (NIDDM, type II) (WHO, 1985).

To treat NIDDM, several synthetic therapeutics have already been clinically used, involving sulfonylureas (Malaisse and Lebrun, 1990), sulfonamides, biguanides (Hermann and Melander, 1993), and triglydazone (CS-045) (Fujiwara et al., 1988), which has recently been developed. However, IDDM can be controlled only by daily injections of insulin. Thus, the development of compounds that cause insulin replacement or are insulin-mimetic on oral administration is an urgent task. In fact, oral vanadyl sulfate ($VOSO_4$) has been tested and reported to improve hepatic and peripheral insulin sensitivity in patients with NIDDM, but not with IDDM (Cohen et al., 1995; Halberstam et al., 1996). These results indicate the need for investigation to establish vanadium compounds of the safest and most long-lasting effectiveness in treating diabetes mellitus. For this purpose, several synthetic vanadyl complexes that are active on oral administration have been proposed in animal experiments.

This review describes recent progress in our investigations of orally active and long-acting vanadium complexes, as well as their possible action mechanisms.

2. FINDINGS OF INSULIN-MIMETIC VANADIUM COMPLEXES

Vanadium is an endogenous constituent of most mammalian tissues; but vanadium is not generally accepted as an essential trace element in humans. There has been as yet no demonstration of nutritional vanadium deficiency nor occurrences of vanadium-containing proteins or enzymes in either animals or humans. However, vanadium is proposed to be essential for development and growth of rats and chicks (Nechay, 1984; Nechay et al., 1986; Chasteen, 1983, 1990; French and Jones, 1992).

Several reviews of the biochemistry of vanadium have been published (Chasteen, 1983; Boyd and Kustin, 1984; Nechay et al., 1986), relating to redox chemistry (Liochev and Fridovich, 1990). Biologically relevant states of vanadium are vanadium(V), vanadate, in the pentavalent state (HVO_4^{2-} or $H_2VO_4^-$), vanadium(IV), vanadyl, in the tetravalent state (VO^{2+}), and vanadium(III), vanadic, in the trivalent state.

Vanadate was reported to be transported into erythrocytes via nonspecific anion channels and reduced there to vanadyl ion by endogenous reducing compounds such as a thiol-containing peptide, glutathione (Cantley et al., 1978; Heinz et al., 1982; Sakurai et al., 1981; Goda et al., 1988; Legrum, 1986). We have shown that most vanadium in organs of normal rats treated with vanadate is present exclusively in a vanadyl form (Sakurai et al., 1980a; Tsuchiya et al., 1990).

The vanadyl state was found to be less toxic than vanadate to rats (Hudson, 1964).

The occurrence of the vanadic state is exceptional; vanadium-rich blood cells, vanadocytes, of a limited species of tunicates or ascidians contain a high amount of vanadic ions (Lee et al., 1988).

During studies on the enzymatic mechanism of Na^+,K^+-ATPase, it was found that vanadate strongly inhibited ATPase activity (Cantley et al., 1977). Since this finding, many investigations of the biochemistry and physiology of vanadium have been reported (Chasteen, 1983; Boyd and Kustin, 1984; Nechay et al., 1986; Stern et al., 1993; Sigel and Sigel, 1995).

Among the findings, the insulin-like actions of vanadium ion are the most striking, as follows. (1) Vanadium increased intracellular K^+ but decreased intracellular Na^+ (Erdmann et al., 1984), and mobilized intracellular Ca^{2+} (Jamilson et al., 1988), as does insulin. (2) Like insulin, both vanadate and vanadyl (Bernier et al., 1988; Sakurai et al., 1990a) increased glucose uptake in adipocytes and increased the oxidation of glucose (Shechter and Karlish, 1980) and glycogen synthesis in adipocytes, liver cells, and human erythrocytes (Tolman et al., 1979; Dubyak and Kleinzeller, 1980; Tamura et al., 1983; Nanfali et al., 1983). (3) Vanadate stimulated autophosphorylation of the tyrosine kinase of insulin receptor (Tamura et al., 1983, 1984; Bernier et al., 1988). The actions of insulin and vanadate on the insulin receptor are similar, but the mechanisms have been reported not to be identical (Tamura et al., 1983, 1984; Bernier et al., 1988).

On the basis of these observations, Heyliger et al. (1985) and Meyerovitch et al. (1987) demonstrated that the blood glucose levels, cardiac performance, and fluid intake of rats with streptozotocin-induced diabetes (STZ rats) were normalized without any increase in the plasma insulin level when sodium orthovanadate was added to their drinking water containing sodium chloride.

After these observations, a great many investigations on in vivo vanadium-dependent insulin action were reported (Ramanadham et al., 1989a,b, 1990, 1991; Brichard et al., 1988; Bendayan and Gingras, 1989; Blondel et al., 1989, 1990; Pugazhenthin and Khandelwal, 1990; Shechter, 1990; Heffetz et al., 1990; Sekar et al., 1990; Cam et al., 1993; Thompson et al., 1993; Saxena et al., 1992; Thompson and McNeill, 1993; McNeill et al., 1992; Yen et al., 1993a,b; Caravan et al., 1995; Bevan et al., 1995).

Among these results, Pederson et al. (1989) reported an interesting observation, in which STZ rats that had had blood glucose levels normalized by 3 weeks of vanadyl sulfate treatment remained normoglycemic after 13 weeks of withdrawal from the treatment, suggesting that permanent changes might have occurred in glucose metabolism in the absence of raised vanadium levels in the tissues.

In addition, we studied the insulin-mimetic action of vanadyl sulfate, which is less toxic to rats than vanadate, in respect to the action on adipocytes, as well as the organ distribution of vanadium (Sakurai et al., 1990a; Nakai et al., 1995).

3. ORGAN DISTRIBUTION OF VANADIUM AND THE IN VIVO COORDINATION MODE AROUND THE VANADYL ION

When STZ rats were given daily ip injections of vanadyl sulfate, their serum glucose levels dropped from hyperglycemic to normal levels within 2 days (Sakurai et al., 1990a) and serum free fatty acid (FFA) levels also dropped to normal (Nakai et al., 1995). However, the plasma insulin levels remained low, suggesting that the vanadyl action is peripheral. Vanadium was incorporated into most organs as well as adipose tissues, as determined by neutron activation analysis (NAA) (Nakai et al., 1995).

Vanadium was incorporated into the organs in the following order (μg V/g wet weight): kidney > liver > bone > pancreas; it was also incorporated into the supernatant of the kidney and mitochondria of the liver. Vanadium was thus assumed to act on the islet of the pancreas, on mineralization of bone, on electron transport systems or induction of metallothionein in the kidney, and on the liver.

It is interesting that no significant differences in vanadium uptake of normal and STZ rats was observed. Both vanadyl and total vanadium in several organs of STZ rats treated with vanadyl sulfate could be determined by electron spin resonance (ESR) spectrometry at $-196\,°C$. In almost all organs, approximately 90% of the vanadium was found to be present in the vanadyl form.

Possible ligands to the vanadyl in organs were also estimated by ESR spectrometry for freshly isolated organs such as liver, kidney, and serum of STZ rats treated with vanadyl sulfate, in which a characteristic eight-line signal due to vanadyl ion was observed. To know the coordination mode around the detected vanadyl species, ESR parameters ($A_\parallel$-value as a hyperfine coupling constant and $g_\parallel$-factor as a universal constant, characteristic of the paramagnetic species) were compared with those of several model vanadyl complexes with various coordination modes around the vanadyl ion. The relationship between two ESR parameters ($A_\parallel$ vs. $g_\parallel$) for vanadyl complexes suggested that the vanadyl species in tissues was predominantly in an oxovanadium(VO^{2+}) form with a square pyramidal structure, in which VO^{2+} was coordinated with four oxygen ligands of either water or oxyamino acid residues in proteins (Sakurai et al., 1990a).

Recently, ESEEM (electron spin echo envelope modulation) spectrometry was applied to reveal a more detailed in vivo coordination structure of the vanadyl state in the organs of rats treated with vanadyl sulfate (Fukui et al., 1995). The ESEEM spectra of the kidney and liver measured at $-196\,°C$ demonstrated the occurrence of nitrogen coordination to a certain percentage of vanadyl ion, when they were compared with several model vanadyl complexes and proteins. The ratios of nitrogen-coordinating vanadyl ion were estimated as 70–80% in the liver, and 50–55% in the kidney. Isotopic portions of the ^{14}N hyperfine coupling were estimated as $|A_{iso}| = \sim5.0$ MHz for the liver, and ~5.2 MHz for the kidney, indicating that the coordinating nitrogen was an amino nitrogen. Thus, in vivo coordination of Lys ε-amine or N-terminal α-amine of a protein (or a peptide) to vanadyl ion was suggested.

4. VANADIUM OXIDATION STATE FOR INSULIN-MIMETIC ACTION

The pharmacologically active form of vanadium was investigated in respect to the interaction of isolated rat adipocytes and vanadium ions, and the following results were obtained (Nakai et al., 1995). (1) Vanadyl ion enhanced glucose uptake; (2) vanadyl ion inhibited FFA release in the absence of glucose and suppressed FFA release in the presence of glucose; (3) glucose inhibited FFA release and the effect was suppressed by cytochalasin B (Cyt B), which is an inhibitor of glucose transporter; (4) the suppressed FFA release by vanadyl ion was restored by Cyt B; (5) vanadyl ion was taken up into adipocytes; and (6) vanadate was not incorporated into the adipocytes and hence it was not activated in suppressing FFA release from adipocytes. In addition, ESR study indicated that vanadate was reduced extracellularly in the presence of glucose and the reduced vanadyl was then incorporated into the cells. On the basis of these results, the vanadyl state has been proposed to be a possible active form of vanadium for insulin-mimetic action and to act on the glucose transporter.

During in vitro experiments using isolated rat adipocytes, we noticed that vanadyl sulfate potentiated glucose incorporation into the adipocytes and suppressed release of FFA from adipocytes stimulated with epinephrine, mimicking the effects of insulin. Therefore, we used an in vitro test system using isolated rat adipocytes treated with epinephrine for evaluating the insulin-mimetic action of a vanadyl complex, in which the effect of the vanadyl complex could be compared with that of insulin by measuring the effect on FFA release. Since an addition of insulin to adipocytes inhibited the release of FFA dose-dependently, a complex that caused dose-dependent inhibition of FFA release was expected to have an insulin-mimetic action in vivo.

5. ORALLY ACTIVE AND LONG-ACTING
VANADYL COMPLEXES

Myerovitch et al. (1987) reported that hyperglycemia of STZ rats was normalized with free access to drinking water containing vanadate, the vanadate treatment having been initiated 1 week after the administration of STZ. Pederson et al. (1989) indicated that hyperglycemia of STZ rats was normalized by 3 weeks of free access to drinking water containing vanadyl ion and the normoglycemic state remained after 13 weeks of withdrawal from the treatment. This discrepancy was explained by the fact that the vanadate treatment was initiated 1 week after the induction of STZ in the rats compared with 3 days in the vanadyl treatment experiment, suggesting progressive destruction of B-cells with time for 1 week after STZ administration (Pederson et al., 1989).

To overcome this discrepancy as well as in consideration of the results for vanadyl sulfate given by ip injection to STZ rats, we used STZ rats after 1 week of STZ administration and aimed to develop therapeutic compounds that are effective by daily and single oral administrations after the end of long-term administration of the complex.

Among many synthetic vanadyl complexes, we first found a dose-dependent hypoglycemic effect of bis(methyl cysteinato) oxovanadium(IV) (V-CYS) complex (VS_2N_2 type coordination complex) given by oral administration (Sakurai et al., 1990b).

In general, the coordination bond between VO^{2+} as a hard Lewis acid and thiolate as a soft Lewis base is not expected to be stronger than that due to the combinations of hard acid/hard base or soft acid/soft base, according to Pearson's HSAB (hard and soft acids and bases) rule (Pearson, 1963). Nevertheless, V-CYS complex has been found to have a monomeric and trans-configuration with strong coordination bonds (Sakurai et al., 1980b, 1988).

Based on the results, we prepared several types of vanadyl complexes with V-S coordination mode and evaluated them in vitro using isolated rat adipocytes treated with epinephrine. The following in vitro results were obtained. A gray-green bis(pyrrolidine-N-carbodithioato)oxovanadium(IV) (V-P) complex (VS$_4$ type coordination complex) was found to be the most effective among six complexes examined (Watanabe et al., 1994), the effect being dose-dependent by in vitro evaluation as observed by inhibition of FFA release from isolated rat adipocytes treated with epinephrine in the absence of glucose.

Thus, V-P complex was given to STZ rats in daily ip injections or oral administrations. When the complex was administered at a dose of 10 mg (0.195 mmol) of vanadium equivalent per kilogram of body weight for the first 2 days, the serum glucose levels decreased to the normal range within 2 days, and it was maintained in the normal range by daily administrations of 5 mg (0.098 mmol) of vanadium per kilogram. In STZ rats whose serum glucose level was normalized within 1 week on administration of V-P complex the glucose level remained normal for 1 week after the end of treatment and then gradually became hyperglycemic.

In addition, orally active insulin-mimetic vanadyl complexes with other coordination modes have been developed. McNeill et al. (1992) reported the usefulness of bis(maltolato) oxovanadium(IV) (V-MAL) (VO$_4$ type coordination complex) which was activated by vanadium in drinking water. V-MAL at a dose of 0.4 mmol/kg body weight per day was effective in normalizing blood glucose and lipid levels during a 6-month study. We proposed new complexes with a VN$_2$O$_2$ coordination mode, bis(picolinato)oxovanadium(IV) (V-PA) complex having been demonstrated to have a strong insulin-mimetic effect as evaluated by in vitro experiment using isolated rat hepatocytes treated with epinephrine (Sakurai et al., 1995). V-PA was effective in normalizing the serum glucose level of STZ rats when given intraperitoneally or orally. The serum glucose level was maintained in the normal range for about 30 days with a gain of body weight after the end of oral administration of V-PA for 14 days, confirming it to be a possible orally active and long-acting insulin-mimetic vanadyl complex for treating IDDM in rats. Using V-PA complex as a leading complex, several analogs have been prepared and evaluated for insulin-mimetic activity by both in vitro and in vivo tests, in which a bis(methyl-picolinato)oxovanadium(IV) (V-MPA) complex was found to be the best for treating IDDM in STZ rats so far (Fujimoto et al., 1996).

The orally active insulin-mimetic vanadyl complexes proposed are thus summarized in Figure 1 (Sakurai, 1996). However, finding a clear relationship between the structure of the complex and insulin-mimetic activity is yet very difficult, given the interrelationship of many important factors involving physicochemical and physiological characters such as stability and electronic charge at physiological pH values, hydrophilicity or lipophilicity, availability for gastrointestinal absorption, organ and subcellular distributions of vanadium, and toxicity and safety of the complex.

On the other hand, the potentiation of vanadate complexes by addition of hydrogen peroxide, termed peroxovanadium complex, has been demonstrated (Kadota et al., 1987; Fantus et al., 1989). The prepared complexes stimulated

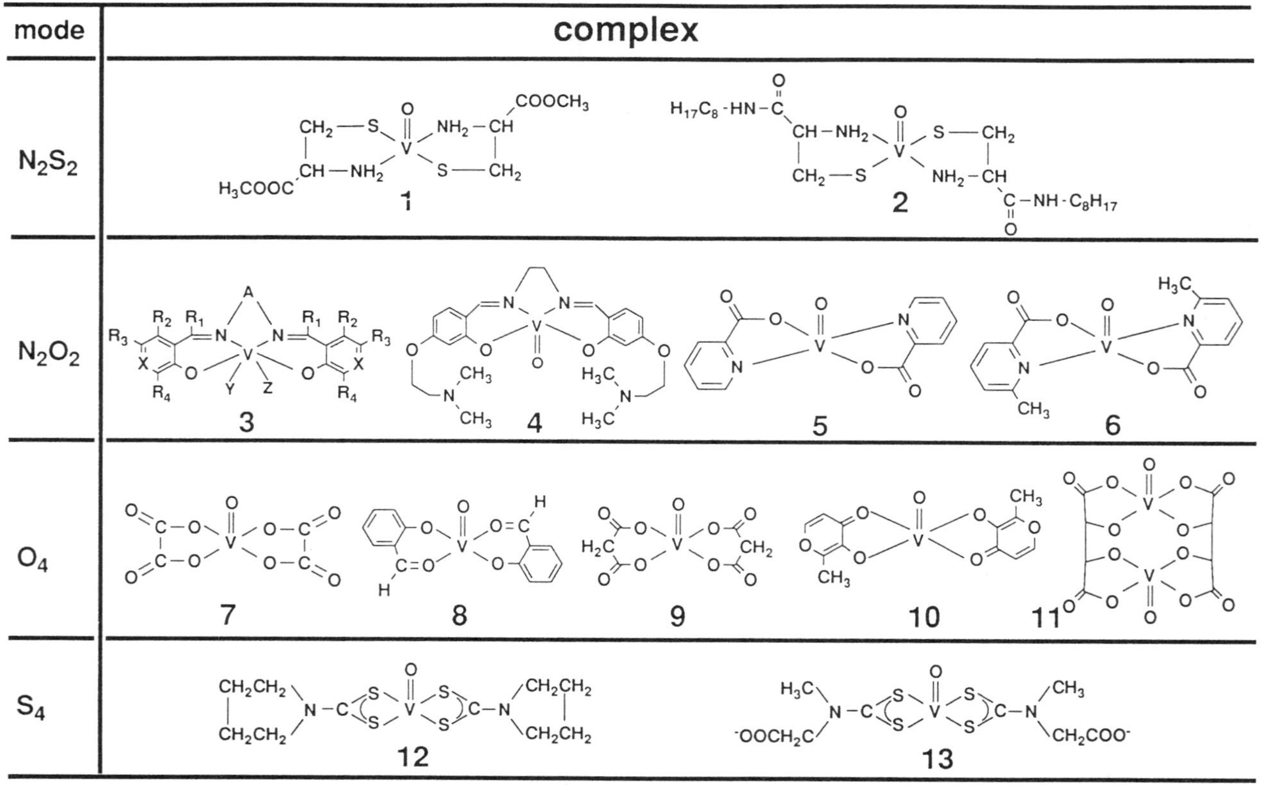

Figure 1. Orally active insulin mimetic vanadyl complexes. Reproduced from Sakurai (1996).

1 Sakurai et al. 1990, 2 EP patent 305246, 3, 4 EP patent 521787A1, 5, 6 Sakurai et al. 1994, 7~9 Sakurai et al. 1990, 10 McNeill et al. 1992, 11 Sakurai et al. 1990, 12 Watanabe et al. 1994 13 Sakurai et al. 1995

lipogenesis, inhibited lipolysis, and promoted protein synthesis in rat adipocytes. The iv administration of $K_2[VO(O_2)(picolinate)]$, $VO(O_2)(picolinate)$, and $K[VO(O_2)(4,7-dimethyl-1,10-phenanthlorine)]$ produced the maximal hypoglycemic effects, in which ancillary ligand of the peroxovanadium complex might target one tissue because of differing capacity to act on skeletal muscle (Bevan et al., 1995).

6. MECHANISM OF INSULIN-MIMETIC ACTION OF VANADIUM

Since vanadate behaves like phosphate, the effect of vanadium in biochemistry has been understood to inhibit protein phosphotyrosine phosphatase, which in turn stimulates protein tyrosine phosphorylation. Thus, vanadate was reported to activate autophosphorylation of solubilized insulin receptors (Tamura et al., 1983, 1984; Bernier et al., 1988; Gherji et al., 1988), similarly to the action of insulin. Vanadate also stimulated the tyrosine kinase activity of the insulin receptor β subunit (Ueno et al., 1987; Smith and Sale, 1988). In addition, both vanadate and vanadyl were found to be effective in stimulating glucose metabolism in rat adipocytes (Dubyak and Kleinzeller, 1980; Shechter and Karlish, 1980; Tamura et al., 1983, 1984; Bernier et al., 1988).

We have proposed that the vanadyl state is a possible active form of vanadium for insulin-mimetic action and for acting on glucose transporter (Nakai et al., 1995). Evidence for the proposal comes from the observation that vanadate, which was in turn reduced to vanadyl, restored expression of the insulin-sensitive glucose transporter of skeletal muscle in rats (Okumura and Shimizu, 1992) and induced the recruitment of GLUT4 glucose transporter to the plasma membrane of adipocytes (Paquet et al., 1992).

In addition, the effect of vanadium on lipid metabolism has been examined. The adenosine $3',5'$-cyclic monophosphate (c-AMP)-mediated protein phosphorylation cascade in adipocytes was activated during diabetes (in vivo) or in the presence of epinephrine (in vitro), and both glucose and vanadyl ion, which were taken up in adipocytes by vanadyl treatment, led to restored regulation of this cascade in peripheral cells (Boyd and Kustin, 1984; Erdmann, et al., 1984; Ramasarma and Crane, 1981). Thus, it was proposed that FFA release from adipocytes is inhibited by vanadyl ion. The suppressed FFA release by vanadate ion depended on the enhancement of glucose uptake by the metal ion, which was reduced to vanadyl by added glucose. Therefore, vanadate was concluded not to inhibit FFA release in adipocytes in a dose-dependent manner, and the effect in the presence of 1 mg/ml glucose was completely reversed by 10^{-5} M Cyt B. These effects were not seen in the absence of glucose. Therefore, it was suggested that glucose, which was taken up in adipocytes by either insulin or vanadyl ion, suppressed FFA release (Nakai et al., 1995). On the basis of these observations, we have proposed a tentative mechanism by which vanadyl-dependent insulin-mimetic action in the peripheral cells normalizes both glucose and FFA levels in STZ rats, as shown in Figure 2.

7.   PREVENTION OF THE ONSET OF DIABETES MELLITUS BY VANADIUM

IDDM, which is characterized by hyperglycemia due to an absolute deficiency of insulin, is initiated by destruction of the islet B-cells in autoimmune disease (Eisenbarth, 1986). A novel hypothesis has recently been proposed, in which nitric oxide (NO) production from macrophages (mø) mediates autoimmune destruction of islet B-cells of IDDM (Kroncke et al., 1991). An immunosuppressant, cyclosporin A, and a poly(ADP-ribose)polymerase inhibitor, nicotinamide, were both found to extend the remission phase and preserve islet B-cell function in patients (Elliot and Chase, 1991). In recent years, the relationship

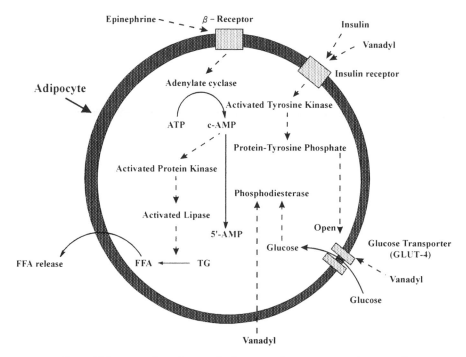

Possible mechanism of glucose incorporation and FFA release from rat adipocytes

– – – – ▶   Activation   ───────▶   Transport or metabolism of the compound

Figure 2. Possible mechanism of glucose incorporation and FFA release from isolated rat adipocytes. Reproduced from Sakurai (1996).

between mø and NO in relation to diabetes was extensively investigated. In experimental animals, the following many observations have been reported. Burkart et al. (1992) reported that cyclosporin A inhibited NO synthesis in murine mø and prevented the toxic action of NO on islet B-cells in vitro. Paul et al. (1995) reported that nicotinamide inhibited inducible NO synthase (iNOS) in murine mø. In addition, Lukic et al. (1991) and Kolb et al. (1991) reported that administration of NO synthase inhibitors, N^G-monomethyl-L-arginine (L-NMMA), and N-nitro-L-arginine-methylester (L-NAME) prevented the induction of diabetes by administration of low doses of STZ. Recently, Andrade et al. (1993) reported that isolated peritoneal mø increased NO production in STZ mice. Wu and Flyan (1993) and Kasuga et al. (1993) also reported that peritoneal mø increased NO production in experimental animals such as BB rats and NOD mice who spontaneously developed diabetes.

These observations suggested that NO release from mø might play an important role in B-cell destruction in IDDM. It appears that suppressing NO production from mø in the prediabetic phase prevents the onset of diabetes.

To clarify the mechanism of antidiabetic activity of vanadyl sulfate as it related to the suppression of NO production from mø, we examined the NO production from the peritoneal mø of diabetic mice in which diabetes had been induced with low doses of STZ, after which they had received vanadyl sulfate injection. Diabetic mice treated with low doses of STZ were seen as animals relevant to this study, because they resemble human IDDM subjects in many aspects. In our experiments, changes in the basal serum glucose levels of BALB/c mice with STZ-induced diabetes (STZ mice) were monitored following daily ip injections of vanadyl sulfate. To avoid a direct chemical reaction of STZ and vanadyl ion in mice, administration of vanadyl sulfate was started 48 h after the final STZ treatment. BALB/c mice that had received daily STZ injections at a dose of 40 mg/kg for the first 5 days (from day 0 to day 4) developed hyperglycemia (glucose about 200 mg/dL serum) on the sixth day following the discontinuation of STZ administration (day 10 of the experiment). However, STZ mice that had received vanadyl sulfate at a dose of 10 mg/kg of body weight for the first 2 days (from day 6 to day 7), followed by 5 mg/kg for the next 5 days (from day 8 to day 15), maintained serum glucose levels in the normal range (glucose about 150 mg/dL). And mice that had received vanadyl sulfate alone had no significant changes in their serum glucose levels. Furthermore, the administration of vanadyl sulfate for 6 days to STZ mice partially improved the decrease in serum insulin. These observations indicated that daily administration of vanadyl sulfate was effective in preventing the onset of STZ-induced diabetes in mice.

On the basis of these observations, it was speculated that the effect of vanadyl sulfate against the onset of diabetes related to the function of mø, and to the possibility that mediator molecules such as NO played an important role in the process of pancreatic B-cell destruction. To examine the effect of vanadyl sulfate on the peritoneal mø of STZ mice, the NO production of the peritoneal mø was determined in the presence of vanadyl sulfate. Vanadyl sulfate dose-dependently inhibited NO production from isolated peritoneal mø activated with interferon-γ (IFNγ) plus lipopolysaccharide (LPS).

We expected that administration of vanadyl sulfate during the prediabetic phase might inhibit NO production from peritoneal mø. Therefore, the in vivo effect of vanadyl sulfate on NO production in the mø of STZ mice was examined. NO production was enhanced in peritoneal mø of STZ mice compared with that of normal mice, but NO production was significantly suppressed in the peritoneal mø of STZ mice that had received vanadyl sulfate. However, the control normal mice, which had received vanadyl sulfate alone, had enhanced NO production. Vanadyl sulfate given during the prediabetic phase of STZ mice suppressed NO production in peritoneal mø to the normal level, indicating that the functions of vanadyl ion and peritoneal mø were closely related to the vanadium-dependent inhibition of the onset of diabetes.

The difference in NO production in the mø of STZ mice that had received vanadyl sulfate and of control normal mice, that had received vanadyl sulfate was then examined by evaluation of vanadyl uptake into mø. Total vanadium levels in mø were determined by ESR spectrometry. Vanadyl uptake in mø of STZ mice receiving vanadyl sulfate was found to be significantly greater than that of normal mice receiving vanadyl sulfate. These results indicated an interaction between vanadyl ion and the NO-producing system in mø.

From these results (Tsuji and Sakurai, 1996), a possible mechanism for vanadium-dependent prevention of the onset of diabetes was proposed as follows (Fig. 3): Mø (indicated as ND-mø) of normal mice treated with vanadyl sulfate is relatively low in uptake of vanadium and in enhanced NO production activity. On the other hand, in the prediabetic phase of mice treated with low doses of STZ, activated mø is exuded through the islets, and the cytotoxic mediator, such as extensive NO production from the activated mø, destroys normal islet B-cells. In addition, Nukatsuka et al. (1988) previously proposed a mechanism for the onset of diabetes by STZ administration in terms of the enhancement of superoxide anion ($\cdot O_2^-$) generation in islet B-cells. Thus, it was assumed that NO reacts with the generated $\cdot O_2^-$ to form an unstable intermediate peroxynitrite, $ONOO^-$. Since one of the degradation products of peroxynitrite, it has been proposed, is a hydroxyl radical ($\cdot OH$) with a powerful cytotoxic activity (Beckman et al., 1990), free radicals such as $\cdot O_2^-$ and $\cdot OH$ have been thought to destroy the normal islet B-cells. However, mø (indicated as D-mø) of low-dosed STZ mice treated with vanadyl sulfate suppresses NO production, because of the relatively enhanced vanadium uptake in the low-dose STZ mice. Suppression of the cytotoxic mediator, such as NO and $\cdot OH$, preserves the damage to the islet B-cells in the prediabetic phase. However, the biochemical mechanism for the suppression of NO production is still unknown. In this regard, Cohen et al. (1996) reported that vanadium might affect mø IFNγ-binding and -inducible responses. Thus,

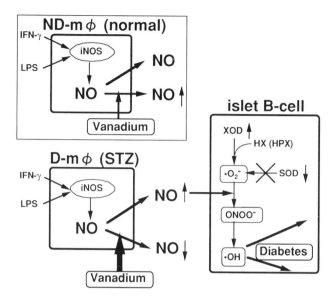

Figure 3. A possible mechanism for vanadium-dependent prevention of the onset of STZ-induced diabetes in terms of nitric oxide released from macrophages.

vanadium-dependent modulation of immune responses may be responsible for suppression of NO production.

In conclusion, vanadyl ion suppresses the excess NO production from mø induced during the prediabetic phase by STZ treatment in low doses. Thus, the islet B-cells destruction by cytotoxic NO produced from mø is partially suppressed, and hence the onset of diabetes is prevented in the mice model. Further, it would be important to determine not only the role of cytotoxic free radicals like NO but also cytokines such as tumor necrosis factor-α and interleukin-1β, which activate inducible NO synthase in islet B-cells and mediate islet B-cell destruction.

8.   SUMMARY

Both inorganic and chelated vanadium compounds have been demonstrated by a great many in vitro and in vivo observations to have an insulin-mimetic activity. On the basis of these results, in recent years, vanadyl sulfate has been clinically used to improve hepatic and peripheral insulin sensitivity in patients with NIDDM (Cohen et al., 1995; Halberstam et al., 1996).

Although vanadium compounds have been used to improve STZ-induced IDDM in experimental animals, vanadyl sulfate has been shown to improve human NIDDM. The information encourages researchers in the field of vanadium chemistry, biochemistry, and physiology to investigate this important problem.

Elucidation of the action mechanism of the insulin-mimetic activity of vanadium, in which many factors are involved, should be very difficult; however, investigation of the mechanism is essential to developing better orally active vanadium compounds or complexes.

Recently, the possible contribution of nitric oxide (NO) to the action of vanadium has been proposed, in an effort not only to improve the condition of diabetics but to prevent the onset of diabetes mellitus (Tsuji and Sakurai, 1996). This observation might open a new insight into understanding the action mechanism of vanadium in vivo.

Furthermore, establishment of a structure-activity relationship for the development of orally active and safe vanadyl complexes should be urgently prosecuted work.

The finding of unique properties of vanadium and its complexes may contribute to maintaining human health and to developing therapy for and prevention of important diseases like diabetes mellitus.

ACKNOWLEDGMENTS

H. S. thanks the Ministry of Education, Science, and Culture of Japan and the Kyoto Pharmaceutical University Foundation for funds. H. S. also thanks Drs. J. Kawada, M. Nukatsuka, H. Kakegawa, T. Satoh, J. Takada, R. Matsushita, K. Fukui, H. Ohya-Nishiguchi, and H. Kamada; Misses H. Watanabe, C. Fujiwara, K. Komazawa, and K. Fujii; and Messrs. M. Nakai, H. Tamura, and S. Fujimoto for their scientific work and discussion for this project.

REFERENCES

Andrade J., Conde, M., Sobrino, F., and Bedoya, F. J. (1993). Activation of peritoneal macrophages during the prediabetic phase in low-dose streptozotocin-treated mice. *FEBS Lett.* **327**, 32–34.

Atkinson, M. A., and Maclaren, N. K. (1990) What causes diabetes?. *Sci. Am.* **260**, 42–49.

Beckman, J. S., Beckman, T. W., Chen, J., Marshall, P. A., and Freeman, B. A. (1990). Apparent hydroxyl radical production by peroxynitrite: Implications for endothelial injury from nitric oxide and superoxide. *Proc. Natl. Acad. Sci. USA* **87**, 1620–1624.

Bendayan, M., and Gingras, D. (1989). Effect of vanadate administration on blood glucose and insulin levels as well as in the exocrine pancreatic function in streptozotocin-diabetic rats. *Diabetologia* **32**, 561–567.

Bernier, M., Laind, D. M., and Lane, M. D. (1988). Effects of vanadate on the cellular accumulation of pp15, an apparent product of insulin receptor tyrosine kinase action. *J. Biol. Chem.* **263**, 13626–13634.

Bevan, A. P., Burgess, J. W., Yale, J. F., Drake, P. G., Lachance, D., Baquiran, G., Shaver, A., and Posner, B. I. (1995). In vivo insulin mimetic effects of pV compounds: Role for tissue targeting in determining potency. *Am. J. Physiol.* **268**, E60–E66.

Blondel, O., Bailbe, D., and Portha, B. (1989). In vivo insulin resistance in streptozotocin-diabetic rats—Evidence for reversal following oral vanadate treatment. *Diabetologia* **32**, 185–190.

Blondel, O., Simon, J., Chevalier, B., and Portha, B. (1990). Impaired insulin action but normal insulin receptor activity in diabetic rat liver: Effect of vanadate. *Am. J. Physiol.* **258,** E459–E467.

Boyd, D. W., and Kustin, K. (1984). Vanadium: A versatile biochemical effector with an elusive biological function. *Adv. Inorg. Biochem.* **6,** 311–365.

Brichard, S. M., Okitolonda, W., and Henquin, J. C. (1988). Long-term improvement of glucose homeostasis by vanadate treatment in diabetic rats. *Endocrinology* **123,** 2048–2953.

Burkart, V., Imai, Y., Kallmann, B., and Kolb, H. (1992). Cyclosporin A protects pancreatic islet cells from nitric oxide-dependent macrophages cytotoxicity. *FEBS Lett.* **313,** 56–58.

Cam, M. C., Pederson, R. A., Brownsey, R. W., and McNeill, J. H. (1993). Long-term effectiveness of oral vanadyl sulfate in streptozotocin-diabetic rats. *Diabetologia* **36,** 218–224.

Cantley, L. C. Jr., Josephson, L., Warner, R., Yanagisawa, M., Lechene, C., and Guidotti, G. (1977). Vanadate is a potent (Na$^+$-K$^+$)-ATPase inhibitor found in ATP derived from muscle. *J. Biol. Chem.* **242,** 7421–7423.

Cantley, L. C. Jr., Tresh, M. D., and Guidotti, G. (1978). Vanadate inhibits the red cell Na$^+$-K$^+$-ATPase from the cytoplasmic side. *Nature* (*Lond.*) **272,** 552–554.

Caravan, P., Gelmini, L., Glover, N., Herring, F. G., Li, H., McNeill, J. H., Retting, S. J., Setyawati, I. A., Shuter, E., Sun, Y., Tracey, A. S., Yuen, V. G., and Orvig, C. (1995). Reaction chemistry of BMOV, bis(maltolato)oxovanadium(IV)—a potent insulin mimetic agent. *J. Am. Chem. Soc.* **117,** 12759–12770.

Chasteen, N. D. (1983). The biochemistry of vanadium. *Struct. Bonding* (Berlin) **53,** 105–138.

Chasteen, N. D. (1990). Vanadium in biological systems. Kluwer Academic Publishers, Norwell, MA.

Cohen, M. D., McManus, T. P., Yang, Z., Qu, Q., Schlesinger, R. B., and Zelikoff, J. T. (1996). Vanadium affects macrophage interferon-γ-binding and -inducible responses. *Toxicol. Appl. Pharmacol.* **138,** 110–120.

Cohen, N., Halberstam, M., Shlimovich, P., Chang, C. J., Shamoon, H., and Rosseti, L. (1995). Oral vanadyl sulfate improves hepatic and peripheral insulin sensitivity in patients with non-insulin-dependent diabetes mellitus. *J. Clin. Invest.* **95,** 2501–2509.

Dubyak, G. R., and Kleinzeller, A. (1980). The insulin-mimetic effects of vanadate in isolated rat adipocytes. *J. Biol. Chem.* **255,** 5306–5312.

Eisenbarth, G. S. (1986). Type I diabetes mellitus: A chronic autoimmune disease. *N. Eng. J. Med.* **314,** 1360–1368.

Elliot, R. B., and Chase, H. P. (1991). Prevention or delay of type I (insulin-dependent) diabetes mellitus in children using nicotinamide. *Diabetologia* **34,** 362–365.

Erdmann, E., Wardan, K., Drawietz, W., Schmitz, W., and Scholz, H. (1984). Vanadate and its significance in biochemistry and pharmacology. *Biochem. Pharmacol.* **33,** 945–950.

Fantus, I. G., Kadota, S., Deragon, G., Fosner, B., and Posner, B. I. (1989). Pervanadate [peroxide(s) of vanadate] mimics insulin action in rat adipocytes via activation of the insulin receptor tyrosine kinase. *Biochemistry* **28,** 8864–8871.

French, R. J., and Jones, P. J. H. (1992). Role of vanadium in nutrition: Metabolism, essentiality and dietary considerations. *Life Sci.* **52,** 339–346.

Fujimoto S., Tamura, H., and Sakurai, H. (1996). Vanadyl-picolinates: Orally active insulin-mimetic vanadyl complexes. *31st Int. Conf. Coord. Chem. Vancouver, Canada.* Abstract, p. 21.

Fujiwara, T., Yoshida, S., Yoshida, T., Ushiyama, I., and Horikoshi, H. (1988). Characterization of new oral antidiabetic agent CS-045. Studies in KK and ob/ob mice and Zucker fatty rats. *Diabetes* **37,** 1549–1558.

Fukui, K, Ohya-Nishiguchi, H., Nakai, M., Sakurai H., and Kamada, H. (1995). Detection of vanadyl–nitrogen interaction in organs of the vanadyl-treated rat: Electron spin echo envelope modulation study. *FEBS Lett.* **368,** 31–35.

Gherji, R., Caratti, C., Andraghetti, G., Bertolini, S., Montemurro, A., Sesti, G., and Cordera, R. (1988). Direct modulation of insulin receptor protein tyrosine kinase by vanadate and anti-insulin receptor monoclonal antibodies. *Biochem. Biophys. Res. Commun.* **152**, 1474–1480.

Goda, T., Sakurai, H., and Yoshimura, T. (1988). Structure of oxovanadium(IV)–glutathione complexes and reductive complex formation between glutathione and vanadate (+5 oxidation state) (in Japanese). *J. Chem. Soc. Jap., Chem. Ind. Chem.*, 1988, pp. 654–661.

Halberstam, M., Cohen, N., Shlimovich, P., Rossetti, L., and Shamoon, H. (1996). Oral vanadyl sulfate improves insulin sensitivity in NIDDM but not in obese nondiabetic subjects. *Diabetes* **45**, 659–666.

Heffetz, D., Bushkin, I., Dror, R., and Zick Y. (1990). The insulinomimetic agents H_2O_2 and vanadate stimulate protein tyrosine phosphorylation in intact cells. *J. Biol. Chem.* **265**, 2896–2902.

Heinz, A., Rubinson, K. A., and Grantham, J. T. (1982). The transport and accumulation of oxyvanadium compounds in human erythrocytes *in vitro. J. Lab. Clin. Med.* **100**, 593–612.

Hermann, L. S., and Melander, A. (1993). Biguanides: Basic aspects and clinical uses. In K. G. M. M. Alberti, R. A. Defrozo, H. Keen, and P. Zimmet (Eds.), *International Textbook of Diabetes Mellitus.* John Wiley & Sons, Chichester, pp. 733–795.

Heyliger, C. E., Tahiliani, A. G., and McNeill, J. H. (1985). Effect of vanadate on elevated blood glucose and depressed cardiac performance of diabetic rats. *Science* **227**, 1474–1477.

Hudson, F. T. G. (1964). *Toxicology and Biological Significance.* Elsevier, New York, p. 140.

Jamilson, G. A. Jr., Etscheid, B. G., Muldoon, L. L., and Villereal, M. L. (1988). Effects of phorbol ester on mitogen and orthovanadate stimulated responses of cultured human fibroblasts. *J. Cell. Physiol.* **134**, 220–228.

Kadota, S., Fantus, I. G., Deragon, G., Guyda, H. J., Hersh, B., and Posner, B. I. (1987). Peroxida(s) of vanadium: A novel and potent insulin-mimetic agent which activates the insulin receptor kinase. *Biochem. Biophys. Res. Commun.* **147**, 259–266.

Kasuga, A., Maruyama, T., Takei, I., Shimada, A., Kasatani, T., Watanabe, K., Saruta, T., Nakaki T., Habu, S., and Miyazaki, J. (1993). The role of cytotoxic macrophages in non-obese diabetic mice: Cytotoxicity against murine mastocytoma and beta-cell lines. *Diabetologia* **36**, 1252–1257.

Kolb, H., Kiesel, U., Kroncke, K. D., and Kolb-Bachofen, V. (1991). Suppression of low dose streptozotocin induced diabetes in mice by administration of a nitric oxide synthase inhibitor. *Life Sci.* **49**, 213–217.

Kroncke, K. D., Kolb-Bachofen, V., Berschick, B., Burkart, V., and Kolb, H. (1991). Activated macrophages kill pancreatic syngeneic islet cells via arginine-dependent nitric oxide generation. *Biochem. Biophys. Res. Commun.* **175**, 752–758.

Lee, S., Kustin, K., Robinson, W. E., Frankel, R. B., and Spartalian, K. (1988). Magnetic properties of tunicate blood cells I. *Ascidia nigna. J. Inorg. Biochem.* **33**, 183–192.

Legrum, W. (1986). The mode of reduction of vanadate(+V) to oxovanadium(+IV) by glutathione and cysteine. *Toxicology* **42**, 281–289.

Liochev, S. I., and Fridovich, I. (1990). Vanadate-stimulated oxidation of NAD(P)H in the presence of biological membrans and other sources of O_2^-. *Arch. Biochem. Biophys.* **279**, 1–7.

Lukic, M. L., Stosic-Grujicic, S., Ostojic, N., Chan, W. L., and Liew, F. Y. (1991). Inhibition of nitric oxide generation affects the induction of diabetes by streptozotocin in mice. *Biochem. Biophys. Res. Commun.* **178**, 913–920.

Malaisse, W. J., and Lebrun, P. (1990). Mechanisms of sulfonylurea-induced insulin release. *Diabetes Care* **13** (Suppl. 3), 9–17.

McNeill, J. H., Yuen, V. G., Hoveyda, H. R., and Orvig, C. (1992). Bis(maltolato)-oxovanadium(IV) is a potent insulin mimetic. *J. Med. Chem.* **35**, 1489–1491.

Meyerovitch, J., Farfel, Z., Sack, J., and Shechter, Y. (1987). Oral administration of vanadate normalizes blood glucose levels in streptozotocin-treated rats. *J. Biol. Chem.* **262**, 6658–6662.

Nakai, M., Watanabe, H., Fujiwara, C., Kakegawa, H., Satoh, T., Takada, J., Matsushita, R., and Sakurai, H. (1995). Mechanism on insulin-like action of vanadyl sulfate: Studies on interaction between rat adipocytes and vanadium compounds. *Biol. Pharm. Bull.* **18**, 719–725.

Nanfali, P., Accorsi, A., Faji, A., Palma, F., and Fornaini, G. (1983). Vanadate affects glucose metabolism of human erythrocytes. *Arch. Biochem. Biophys.* **226**, 441–447.

Nechay, B. R. (1984). Mechanisms of action of vanadium. *Annu. Rev. Pharmacol. Toxicol.* **24**, 501–524.

Nechay, B. R., Nanninga, L. B., Nechay, P. S. E., Post, R. L., Grantham, J. J., Macara, I. G., Kubena, L. F., Phillips, T. D., and Nielsen, F. H. (1986). Role of vanadium in biology. *Fed. Proc.* **45**, 123–132.

Nukatsuka, M., Sakurai, H., Yoshimura, Y., Nishida, M., and Kawada, M. (1988). Enhancement by streptozotocin of O_2^- radical generation by the xanthine oxidase system of pancreatic β-cells. *FEBS Lett.* **239**, 295–298.

Okumura, N., and Shimazu, T. (1992). Vanadate stimulates D-glucose transport into sarcolemmal vesicles from rat skeletal muscles. *J. Biochem.* (Tokyo) **112**, 107–111.

Paquet, M. R., Romanek, R. J., and Sargeant, R. J. (1992). Vanadate induces the recruitment of glut-4 glucose transporter to the plasma membrane of rat adipocytes. *Mol. Cell. Biochem.* **109**, 149–155.

Paul, A., Pendreigh, R. H., and Plevin, R. (1995). Protein kinase C and tyrosine kinase pathways regulate lipopolysaccharide-induced nitric oxide synthase activity in RAW 264.7 murine macrophages. *Br. J. Pharmacol.* **114**, 482–488.

Pearson, R. G. (1963). Hard and soft acids and bases. *J. Am. Chem. Soc.* **85**, 3533–3539.

Pederson, R. A., Ramanadham, S., Buchan, A. M. J., and McNeill, J. H. (1989). Long-term effects of vanadyl treatment on streptozotocin-induced diabetes in rats. *Diabetes* **38**, 1390–1395.

Pugazhenthin, S., and Khandelwal, R. L. (1990). Insulin-like effects on vanadate on hepatic glycogen metabolism in nondiabetic and streptozotocin-induced diabetic rats. *Diabetes* **39**, 821–827.

Ramanadham, S., Cros, G. H., Mongold, J. J., Serrano, J. J., and McNeill, J. H. (1990). Enhanced in vivo sensitivity of vanadyl-treated diabetic rats to insulin. *Can. J. Physiol. Pharmacol.* **68**, 486–491.

Ramanadham, S., Heyligner C., Gresser, M. J., Tracey, A. S., and McNeill, J. H. (1991). The distribution and half-life for retention of vanadium in the organs of normal and diabetic rats orally fed vanadium(IV) and vanadium(V). *Biol. Trace Elem. Res.* **30**, 119–124.

Ramanadham, S., Mongold, J. J., Brownsey, R. W., Cros, G. H., and McNeill, J. H. (1989a). Oral vanadyl sulfate in the treatment of diabetes mellitus in rats. *Am J. Physiol.* **257**, H904–H911.

Ramanadham, S., Brownsey, R. W., Cros, G. H., Mongold, J. J., and McNeill, J. H. (1989b). Sustained prevention of myocardial and metabolic abnormalities in diabetic rats following withdrawal from oral vanadyl treatment. *Metabolism* **38**, 1022–1028.

Ramasarma, T., and Crane, F. L. (1981). Does vanadium play a role in cellular regulation? *Curr. Top. Cell. Regul.* **20**, 249–301.

Sakurai, H. (1996). Vanadium complexes as a possible therapeutic of diabetes mellitus (in Japanese). *Chemistry Today*, no. 304, pp. 14–20.

Sakurai, H., Fujii, K., Watanabe, H., and Tamura, H. (1995). Orally active and long-term acting insulin-mimetic vanadyl complex: Bis(picolinato) oxovanadium(IV). *Biochem. Biophys. Res. Commun.* **214**, 1095–1101.

Sakurai, H., Shimomura, S., Fukuzawa, K., and Ishizu, K. (1980a). Detection of oxovanadium(IV) and characterization of its ligand environment in subcellular fractions of the liver of rats treated with pentavalent vanadium(V). *Biochem. Biophys. Res. Commun.* **96**, 293–298.

Sakurai, H., Hamada, Y., Shimomura, S., Yamashita, S., and Ishizu, K. (1980b). Cysteine-methyl ester-oxo-vanadium(IV) complex, preparation and characterization. *Inorg. Chim. Acta* **46**, L119–L120.

Sakurai H., Shimomura, S., and Ishizu, K. (1981). Reduction of vanadate(V) to oxovanadium(IV) by cysteine, and mechanism and structure of the oxovanadium(IV)–cysteine complex subsequently formed. *Inorg. Chim. Acta.* **57,** L67–L69.

Sakurai, H., Taira, Z., and Sakai, N. (1988). Crystal structure of an L-cysteine methyl ester–vanadyl(IV) complex. *Inorg. Chim. Acta* **151,** 85–86.

Sakurai, H., Tsuchiya, K, Nukatsuka, M., Sofue, M., and Kawada, J. (1990a). Insulin-like effect of vanadyl ion on streptozotocin-induced diabetic rats. *J. Endocrinol.* **126,** 451–459.

Sakurai, H., Tsuchiya, K, Nukatsuka, M., Kawada, J., Ishikawa, S., Yoshida, H., and Komatsu, M. (1990b). Insulin-mimetic action of vanadyl complexes. *J. Clin. Biochem. Nutr.* **8,** 193–200.

Saxena, A. K. Srivastava, P., and Baquer, N. Z. (1992). Effects of vanadate on glycolytic enzymes and malic enzyme in insulin-dependent and -independent tissues of diabetic rats. *Eur. J. Pharmacol.* **216,** 123–126.

Sekar, N., Kanthasamy, A., Williams, S., Subramanian, S., and Govindasamy, S. (1990). Insulinic actions of vanadate in diabetic rats. *Pharmacol. Res.* **22,** 207–217.

Shechter, Y. (1990). Insulin-mimetic effects of vanadate. *Diabetes* **39,** 1–5.

Shechter, Y., and Karlish, S. J. D. (1980). Insulin-like stimulation of glucose oxidation in rat adipocytes by vanadyl(IV) ions. *Nature* (Lond.) **286,** 556–558.

Sigel, H., and Sigel, A. (Eds). (1995). *Vanadium and Its Role in Life.* Vol. 31: *Metal Ion in Biological Systems.* Marcel Dekker, New York.

Smith, D. M., and Sale, G. J. (1988). Evidence that a novel serine kinase catalyses phosphorylation of the insulin receptor in an insulin-dependent and tyrosine kinase-dependent manner. *Biochem. J.* **256,** 903–909.

Stern, A., Yin, X., Tsang, S.-S., Davison, A., and Moon, J. (1993). Vanadium as a modulator of cellular regulatory cascades and oncogene expression. *Biochem. Cell. Biol.* **71,** 103–112.

Tamura, S., Brown, T. A., Dubler, R. F., and Lerner, J. (1983). Insulin-like effect of vanadate on adipocyte glycogen synthase and on phosphorylation of 95,000 dalton subunit of insulin receptor. *Biochem. Biophys. Res. Commun.* **113,** 80–86.

Tamura, S., Brown, T. A., Whipple, J. H., Fujita-Yamaguchi, Y., Dubler, R. F., Cheng, K., and Farner, J. (1984). A novel mechanism for the insulin-like effect of vanadate on glycogen synthase in rat adipocytes. *J. Biol. Chem.* **259,** 6650–6658.

Thompson, K. H., Leitchter J., and McNeill, J. H. (1993). Studies of vanadyl sulfate as a glucose-lowering agent in STZ-diabetic rats. *Biochem. Biophys. Res. Commun.* **197,** 1549–1555.

Thompson, K. H., and McNeill, J. H. (1993). Effect of vanadyl sulfate feeding on susceptibility to peroxidative change in diabetic rats. *Res. Commun. Chem. Pathol. Pharmacol.* **80,** 187–200.

Tolman, E. L., Barris, E., Burns, M., Prasini, R., and Partridge, R. (1979). Effects of vanadium on glucose metabolism *in vitro. Life Sci.* **25,** 1159–1164.

Tsuchiya, K., Sakurai, H., Nishida, M., Takada, J., and Koyama M. (1990). Selective determination of vanadium in organs of rats treated with vanadium complexes (in Japanese). *Trace Elem. Res.* **7,** 59–63.

Tsuji, A., and Sakurai, H. (1996). Vanadyl ion suppresses nitric oxide production from peritoneal macrophages of streptozotocin-induced diabetic mice. *Biochem. Biophys. Res. Commun.* **226,** 506–511.

Ueno, A., Arakaki, N., Takeda, Y., and Fujio, H. (1987). Inhibition of tyrosine autophosphorylation of the solubilized insulin receptor by an insulin-stimulating peptide derived from bovine serum albumin. *Biochem. Biophys. Res. Commun.* **144,** 11–18.

Watanabe, H., Nakai, M., Komazawa, K., and Sakurai, H. (1994). A new orally active insulin mimetic vanadyl complex: Bis(pyrrolidine-N-carbodithioato) oxovanadium(IV). *J. Med. Chem.* **37,** 876–877.

WHO (1985). Diabetes mellitus: Reports of a WHO study group. *WHO Technical Report Series* **727,** p. 10.

Wu, G., and Flynn, N. E. (1993). The activation of the arginine–citrulline cycle in macrophages from the spontaneously diabetic BB rat. *Biochem. J.* **294**, 113–118.

Yen, V. G. Orvig, C., and McNeill, J. H. (1993a). Glucose-lowering effects of a new organic vanadium complex, bis(maltolato)oxovanadium(IV). *Can. J. Physiol. Pharmacol.* **71,** 263–269.

Yen, V. G., Orvig, C., Thompson, K. H., and McNeill, J. H. (1993b). Improvement in cardiac dysfunction in streptozotocin-induced diabetic rats following chronic oral administration of bis(maltolato)oxovanadium(IV). *Can. J. Physiol. Pharmacol.* **71,** 270–276.

16

VANADIUM DETOXIFICATION

Enrique J. Baran

*Centro de Química Inorgánica (CEQUINOR), Facultad de
Ciencias Exactas, Universidad Nacional de La Plata,
C. Correo 962, 1900-La Plata, Argentina*

Vanadium in the Environment. Part 2: Health Effects, Edited by Jerome O. Nriagu.
ISBN 0-471-17776-8. © 1998 John Wiley & Sons, Inc.

1. INTRODUCTION

Vanadium presents a wealthy and fascinating chemistry (Clark, 1968; Cotton and Wilkinson, 1980; Greenwood and Earnshaw, 1984). Relevant and relatively singular aspects of its behavior are the very rich structural chemistry of the oxoanions (vanadates), the high stability of VO^{2+}, considered the most stable diatomic ion known (Selbin, 1966), and the well-developed coordination chemistry of its most usual oxidation states, which range from $+2$ to $+5$ (Vilas Boas and Costa Pessoa, 1987; Butler and Carrano, 1991).

The essentiality, distribution, and toxicology of vanadium, like its biological and pharmacological activity, are areas of increasing research that are interesting for chemists and biochemists.

Because vanadium compounds are very active in vitro and pharmacologically, numerous biochemical and physiological functions have been suggested for it (Kustin and Macara, 1982; Chasteen, 1983). Notwithstanding, and despite the magnitude of the knowledge so far accumulated, vanadium still lacks a clearly defined role in higher organisms (Chasteen, 1990; Sigel and Sigel, 1995). Nowadays, the best evidence of a biological role of vanadium comes from bacteria [vanadium-containing nitrogenase in *Azotobacter* species (Erfkamp and Müller, 1990; Eady and Leigh 1994)] and from plants [vanadium-dependent haloperoxidases in algae and lichens (Wever and Kustin, 1990; Baran, 1995a)]. On the other hand, although the accumulation of relatively high concentrations of vanadium in tunicates and in the toadstool *Amanita muscaria* has been well established, the possible function of the systems containing the element (hemovanadine and amavadin, respectively) remains obscure (Baran, 1995a).

Like molybdenum, vanadium assumes an exceptional position among the biometals in that both its anionic and cationic forms can participate in biological processes (Rehder, 1991, 1992). In its anionic forms, vanadates, or vanadium (V), it strongly resembles phosphates, but in its cationic forms—mainly as VO^{2+}, but in certain cases also as vanadium(III)—it behaves like a typical transition metal ion, which competes with other metal cations in coordination to biogenic ligands and compounds. This duality, which allows vanadium to behave in its highest oxidation state like the representative element phosphorus and in all the lowest states as a typical transition metal cation, together with the facility with which it changes oxidation state, may be responsible for the very peculiar and unparalleled behavior of this new bioelement, the characteristics of which have just begun emerging (Baran, 1994, 1995a).

Owing to the increase in vanadium levels in the environment as a result of its widespread use in different industrial processes and of the increased atmospheric pollution with vanadium species generated by the combustion of vanadium-containing fuels, and owing to the emerging interest in the pharmacological effects of some of its compounds, the toxicology of vanadium has become an important area of research.

The older literature about vanadium toxicology has been reviewed in the book by Faulkner-Hudson (1964); brief general and more recent accounts are also found in the books edited by Merian (1984), Frieden (1984), Seiler et al. (1988), Chasteen (1990), and Siegel and Sigel (1995). The two chapters by Lener et al. (Ch. 1) and Thompson et al. (Ch. 2) in the present volume also cover the most relevant aspects of this subject matter.

In relation to the toxicological problems generated by vanadium, it also seems interesting to analyze the different possible biological detoxification mechanisms, as well as those induced by chelating agents and similar drugs. These two aspects constitute the main goal of the present chapter.

2. GENERAL ASPECTS OF VANADIUM METABOLISM AND TOXICITY

In order to facilitate the reading and comprehension of the main subject of this chapter, it is necessary to have in mind the most relevant aspects of vanadium metabolism and toxicity.

2.1. Metabolism

Existing information about the metabolism of physiological amounts of vanadium in animals is very scarce. Nevertheless, it is apparent that most ingested vanadium remains unabsorbed and is excreted via the feces. Although apparently absorption is less than 5%, some studies indicate it could be higher (Nielsen and Uthus, 1990).

Dietary vanadium probably occurs mainly as $H_2VO_4^-$ and in vitro studies suggest that anionic vanadium(V) can enter cells through the phosphate transport mechanism (Neilsen, 1995). Most or all of the vanadium(V) undergoes a one-electron reduction to form VO^{2+}, in the gastrointestinal tract (Chasteen et al., 1986b). Much evidence suggests that the binding of the generated vanadyl(IV) to iron-containing proteins (i.e., VO^{2+}/transferrin and VO^{2+}/ferritin) is very important in vanadium metabolism.

Very little vanadium is retained under normal conditions in the body. Animal studies suggest that accumulation can be related directly to level of administered dose. At high doses, vanadium is rapidly accumulated in bone, kidney, and liver. Overall, experience with ^{4B}V has indicated that distribution occurs in the following order: bone > kidney > liver > spleen > intestine > stomach > muscle > testis > lung > brain (Sharma et al., 1980). Bone apparently is a major sink for retained vanadium (Talvitie and Wagner, 1954), an aspect that is of great relevance in relation to detoxification mechanisms, as will be discussed in Section 4.2.1. Excretion of absorbed vanadium occurs mainly through the urine as small-molecular-weight complexes (Chasteen et al., 1986b).

The dietary requirement for vanadium is very small. Vanadium deficiency has not been identified in humans. Most diets supply between 15 and 30 μg daily, suggesting that a dietary intake of 10 μg daily probably meets any postulated vanadium requirement. Most fats, oils, fruits, and vegetables possess vanadium levels of 1–5 ng/g, whereas cereals, liver, and fish present levels of about 5–40 ng/g, and only a few items such as spinach, oysters, shellfish, black pepper, and parsley contain relatively higher levels (Nielsen and Uthus, 1990; Nielsen, 1995).

2.2. Toxicity

Vanadium toxicity has been reported in experimental animals and humans. The degree of toxicity depends on the route of administration, valence, and chemical form and is also to some extent species-dependent (Stoecker and Hopkins, 1984). Small experimental animals, such as the rat and mouse, tolerate the metal relatively well; the rabbit and horse, and also humans, are apparently more sensitive (Faulkner-Hudson, 1964).

In general, the toxicity of vanadium is high when given by injection, low by the oral route, and intermediate by the respiratory tract. Toxicity also varies considerably with the nature of the compound, but in general it increases as valence increases, pentavalent vanadium being the most toxic (Nechay et al., 1986). Toxic effects in humans and animals under natural conditions do not occur frequently, but at high doses or as a consequence of chronic exposure it is a relatively toxic element for humans (Nielsen, 1995).

The upper respiratory tract is the main target in occupational exposure. Vanadium compounds, especially V_2O_5, are strong irritants of the eyes and the airways. Acute and chronic exposure gives rise to conjunctivitis, rhinitis, reversible irritation of the respiratory tract, and to bronchitis, bronchospasms, and asthma-like diseases in more severe cases (Faulkner-Hudson, 1964; Chiriatti, 1971; Wennig and Kirsch, 1988). Likewise, vanadium has shown to produce gastrointestinal distress, fatigue, cardiac palpitation, and kidney damage, as well as other physiological effects such as cardiovascular changes, disturbances of the central nervous system, and metabolic alterations in laboratory animals (Nechay et al., 1986; Elfant and Keen, 1987). In human beings, acute vanadium toxicity has been observed in vanadium miners, as well as other industrial workers exposed to high concentrations of vanadium. The classic symptoms of this malady, referred to as "green tongue" syndrome, are a green discoloration of the tongue, accompanied by some of the above-mentioned disorders.

Determination of vanadium by radiochemical neutron activation analysis in urine appears to be one of the most suitable methods for the control of vanadium levels in exposed humans (Kucera et al., 1994).

Taking into account the low contents of the element in practically all types of food and beverage, toxic effects due to the intake of high amounts of vanadium in the diet are unlikely. Toxicity usually occurs only as a result of industrial exposure to high amounts of airborne vanadium. On the other hand, the high pharmacological activity of certain vanadium compounds suggests that beneficial pharmaceutical roles for this element may be found and exploited currently. Furthermore, nutritional supplements containing vanadium are now being marketed. All these facts point to the possibility of oral toxicity of vanadium in the near future (Nielsen, 1995); therefore, a sound knowledge of the possible detoxification mechanisms and processes seems highly desirable.

3. DETOXIFICATION MECHANISMS

All living systems have developed defense mechanisms to deal with the reactive and potentially harmful by-products that arise from cellular metabolism and to control the effects of exogenous substances that eventually invade the organism (biological detoxification).

On the other hand, a series of drugs have been developed that are capable of chelating metal ions in vivo in order to eliminate excesses of essential metals and to prevent possible damage caused by nonessential, toxic elements (chemical detoxification).

In this section a brief description of the most usual and well-known mechanisms of these two types is given.

3.1. Biological Detoxification

Typical examples of biological detoxification mechanisms, related to the control of by-products of normal metabolism, are the action of enzymatic systems such as catalase, peroxidases, and superoxide dismutase, which eliminate reactive species generated by incomplete reduction from O_2 to H_2O (H_2O_2 and O_2^-). Other well-known examples include the correct balance of essential metals through participation of specific binding proteins (for example, the iron storage protein ferritin or the copper storage system ceruloplasmin) (Baran, 1995a).

More interesting in the context of the present chapter are the natural defense mechanisms against exogenous metal ions, specially heavy metals, which are toxic when they interfere with normal cell metabolism. These toxic effects may be caused by the following processes: (a) blocking of an essential functional group of a biomolecule; (b) displacement of an essential metal; (c) modification of the active conformation of a biomolecule; (d) disruption of the integrity of biomembranes (Baran, 1995a; Occhiai, 1995). All these toxicity mechanisms are based on the usually strong binding abilities of these metal ions.

Resistance mechanisms include chemical redox processes, transmembrane ion pumps to export the toxic ion out of the cell, methylation to a volatile form or, simply, binding of the ion by ligands, proteins, or cell membranes.

Some general, nonspecific, and limited tolerance mechanisms include the confination of toxic elements to places that have little influence on general biological activities, including cell walls, hard tissues such as bone, or extra body tissues such as hair. The so-called inclusion bodies, granules of variable composition often occurring in the form of "amorphous" concentrically layered spherules, which are present in virtually every phylum of animals in a wide variety of tissues, appear also as a cellular route for the detoxification of heavy metals. These granules consist mainly of Ca^{2+}, Mg^{2+}, PO_4^{3-}, and CO_3^{2-} ions in widely different ratios and usually contain small amounts of other metals (Ag, Al, Cd, Co, Cr, Fe, Pb, Sn) (Krampitz and Witt, 1979; Baran, 1992; Baran, 1995a).

Yeasts, fungi, higher plants, and animals contain low-molecular-weight cysteine-rich proteins called metallothioneins—or phytochelatins, for those of vegetal origin (Grill and Zenk, 1989)—that bind heavy metals. Although metallothioneins are important proteins in zinc and copper metabolism, they function as detoxification proteins for other nonessential metals, as suggested by the fact that their synthesis can be induced by heavy-metal ions such as Cd(II), Hg(II), and Ag(I) (Vasák and Kägi, 1983; Vasák, 1984; Baran, 1995a; Occhiai, 1995; Stillman, 1995).

The tripeptide thiol glutathione (GSH) has also an essential role in the cellular processes controlling the level of heavy metals. Direct interaction between GSH and metal ions can take place by two different mechanisms: The first is the one-electron reduction of the metal with concomitant oxidation of GSH to GSSG; the second is chelation of metals by either GSH or GSSG, with the formation of a complex that allows transport of the metal in a controlled manner (Rabenstein, 1989).

Biomethylation constitutes another interesting biological detoxification mechanism. The main natural methylating agents are methylcobalamin, S-adenosyl methionine, and methyl iodide. For example, Hg(II) can be converted into volatile $(CH_3)_2Hg$ by certain bacteria. Methylation of arsenic compounds to $(CH_3)_4As^+$ or $(CH_3)_3As$ may serve as a defense mechanism against the toxic effects of this element. Also, in the case of the essential micronutrient selenium, the generation of gaseous $Se(CH_3)_2$, is probably a form to eliminate an excess of this element (Krishnamurthy, 1992; Baran, 1995a).

3.2. Chemical Detoxification

The toxic effects caused by the presence of excess quantities of an essential metal, as well as those arising from the entry of nonessential metals into the cells, can be controlled or suppressed in different chemical ways. The most usual one involves chelation therapy, in which a metal-specific chelating agent is administered to complex, and facilitate excretion of, the unwanted excess element (Baran, 1995a; Taylor and Williams, 1995).

Some typical examples of chelation therapies are the use of 2,3-dimercapto-propanol (BAL) or D-penicillamine for the removal of copper in Wilson's disease or the use of desferrioxamine in cases of hemochromatosis (accumulation of excess iron).

In the case of nonessential metals, the use of EDTA and similar chelating ligands (CDTA, DTPA, or TTHA) are well known in the cases of lead and (radioactive) strontium detoxification therapies. The above-mentioned BAL has been used for arsenic detoxification.

4. VANADIUM DETOXIFICATION

4.1. General Aspects

It has usually been accepted that vanadium toxicity increases with increasing valence, pentavalent vanadium being the most toxic (Llobet and Domingo, 1984; Nechay et al., 1986). Nevertheless, a number of recent animal experiments suggest that, in vivo, vanadium is converted to a common form, since the organ distribution of vanadium is essentially independent of the oxidation state of the originally administered vanadium species (Sabbioni et al., 1978; Chasteen, 1983; Harris et al., 1984; Chasteen et al., 1986a).

Taking into account that the standard potential for the couple $H_2VO_4^-$ + $4H^+$ + e^- $\rightleftharpoons$ VO^{2+} + $3H_2O$ is 1.31 V (Rehder, 1992), it is evident that vanadylI(IV) undergoes autoxidation to vanadate in the presence of oxygen, and vanadate in turn can be reduced, as previously discussed, by reductants such as glutathione or ascorbic acid. Kinetic measurements with serum in vitro indicate that the interconversion time between the two vanadium oxidation states is short relative to the residence time of most of the metals in circulation. Endogenous reducing agents and dissolved oxygen ensure that both the +4 and the +5 oxidation states are present in serum (Chasteen et al., 1986a). However, in the case of chronic inhalation exposures, where the entry of vanadium in the lungs occurs gradually and the oxygen content of the blood is at its highest, one might expect a very efficient oxidation of all the vanadium species to vanadium(V) (Harris et al., 1984).

On the other hand, it must be clearly emphasized that an important number of the most relevant aspects of vanadium detoxification are based on model and/or in vitro studies as well as on animal experiments (mainly with rats, mice, rabbits, and chicks and least frequently with dogs and horses). Therefore, the direct extension of all the available information to humans should be treated with caution.

4.2. Biological Detoxification

4.2.1. Detoxification by Accumulation in Bone and Connective Tissues

As stated in Section 2.1. of this chapter, bone seems to be a very active and important vanadium accumulator. Experiments by Talvitie and Wagner (1954) showed that the greater part (up to 84%) of the vanadium retained by rabbits could be detected in the skeleton. These investigators believed that this high skeletal retention was probably due to its rapid exchange with the bone phosphate. This behavior may be explained by the fact that both the PO_4^{3-} and VO_4^{3-} anions can be easily incorporated into lattices of apatites, a situation that is known for natural minerals (McConnell, 1973; Elliott, 1994).

In order to investigate the above-mentioned exchange, we have made different studies using calcium hydroxylapatite, $Ca_{10}(PO_4)_6(OH)_2$, as a model for the inorganic matrix of bone (Etcheverry et al., 1984). Under physiological conditions the exchange could be observed only with the amorphous material, suggesting that vanadium incorporation into the hard tissues should be especially important in the case of younger tissues, in which the inorganic phase is known to be especially amorphous (Montel et al., 1981).

On the other hand, our studies also demonstrate that the incorporation of moderate or low concentrations of vanadate into phosphate sites produces only weak distortions at the macroscopic (crystallographic parameters and crystal ordering) and microscopic (local distortions, weakening of chemical bonds) levels of the apatite lattice (Etcheverry et al., 1984; Baran, 1994).

The results of these model studies suggest that the incorporation of vanadate into bone may constitute a useful biological detoxification mechanism, favored by the great chemical and structural similarities between PO_4^{3-} and VO_4^{3-}.

Another interesting point to explore was to find out if VO^{2+}, the other relevant vanadium species present in biological systems, could also interact with hydroxylapatite, entering in competition with the Ca(II) ions of the material. Precipitation of calcium hydroxylapatite in the presence of the VO^{2+} cation (Oniki and Doi, 1983) as well as interaction of apatite suspensions with the cation (Narda et al., 1992) demonstrated that VO^{2+} is not incorporated into the apatite lattice, but is strongly absorbed onto the surface of the material, suggesting that this process may be considered a second possible detoxification mechanisms along with the participation of bone.

The ESR spectra of materials obtained from apatite suspensions show a typical axial symmetry and suggest an inhomogeneous distribution of the surface-adsorbed VO^{2+} ions. Interestingly, the generated $O=V(O)_4$ moieties show a high stability towards oxidation (Narda et al., 1992).

Finally, it was also interesting to verify that the vanadyl(IV) cation interacts with the organic matrix of hard tissues. Initial studies on this subject were recently performed by investigating the behavior of VO^{2+} in relation to chondroitin sulfate A (CSA), a well-known acid mucopolysaccharide present in connective tissues and other mineralized systems that contain alternate units of D-glucuronic acid and N-acetyl D-galactosamine. The interaction was investigated in solution by electron absorption and infrared spectroscopy, and the generation of a $VO(CSA)_2$ species was demonstrated, in which the oxocation coordinates through the carboxylate group and the glycosidic oxygen of the D-glucuronate moieties (Etcheverry et al., 1994).

More recently, we have also shown that the two isolated component moieties of CSA behave towards VO^{2+} in a way similar to that in the mucopolysaccharide (Etcheverry et al., 1996a), and it also becomes possible to isolate a solid VO^{2+}/glucuronate complex in alkaline media (Etcheverry et al., 1996b).

Finally, it is also interesting that, at nearly physiological pH values, the VO^{2+} cation interacts with tropocollagen at room temperature to give nitrogen monocoordinated complexes in aqueous solution and chelate complexes of the type $VO(N_2O_2)$ when the protein is in a rigid matrix state. Apparently, vanadyl(IV) can occupy one of the potential sites for cross-link formation in mature collagen (Ferrari, 1990).

Interestingly, the formation of "inclusion bodies" has also been reported in bivalve mollusks, as a possible detoxification mechanism for vanadium and other heavy metals (Kustin et al., 1983) but such a possibility has so far not been postulated for other organisms.

4.2.2. Detoxification by Reductive Processes and Complexation

As previously stated, reduction of vanadium(V) to vanadium(IV) plays an important role in the first steps of vanadium metabolism and may also be related, indirectly, to vanadium detoxification mechanisms.

Vanadate in blood is apparently quickly converted to VO^{2+} in the erythrocytes by glutathione or in plasma by reductants such as ascorbate, cysteine, and catecholamines (Nielsen, 1995). Most of these and similar reducing agents, or eventually their oxidation products, also interact with the generated oxocation, consequently constituting potentially important detoxification systems. The incorporation of vanadium species into serum albumin or into proteins such as transferrin or ferritin probably plays a similar role. Some of these systems will be discussed in the next sections.

4.2.2.1. Interactions with Glutathione. Glutathione (GSH), the tripeptide γ-L-glutamyl-L-cysteinyl-glycine (**1a;** figures in bold type identify organic compounds presented in schemes) is the major nonprotein thiol present in most animal cells. It is an important source of reducing equivalents, has a number of regulatory functions, and is involved in detoxification processes of exogenous

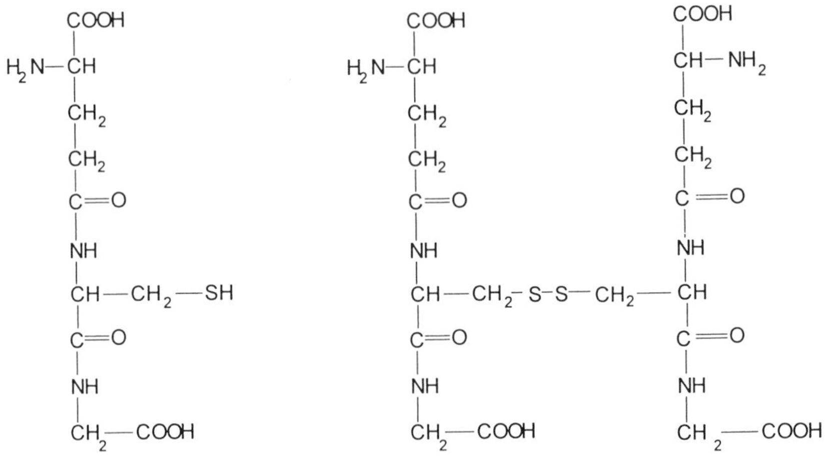

```
        COOH                    COOH                    COOH
         |                       |                       |
H₂N—CH                   H₂N—CH                   CH—NH₂
         |                       |                       |
        CH₂                     CH₂                     CH₂
         |                       |                       |
        CH₂                     CH₂                     CH₂
         |                       |                       |
        C=O                     C=O                     C =O
         |                       |                       |
        NH                      NH                      NH
         |                       |                       |
        CH—CH₂—SH        CH—CH₂-S-S—CH₂—CH
         |                       |                       |
        C=O                     C=O                     C =O
         |                       |                       |
        NH                      NH                      NH
         |                       |                       |
        CH —COOH            CH —COOH            CH  —COOH
          2                       2                      2
```

1 a   1 b

Scheme 1

materials (Rabenstein, 1989). Glutathione, as well as its oxidation product (GSSG) (**1b**) possesses very interesting complexing properties (Rabenstein et al., 1979; Rabenstein, 1989).

It has been shown that red cells, in vitro, reduced vanadate almost quantitatively to VO^{2+} (Macara et al., 1980). GSH was found to be responsible for the reduction and to act as a ligand for the VO^{2+} cations formed within adipocytes (Macara et al., 1980; Degani et al., 1981).

The system VO^{2+}/GSH has been investigated in detail by electronic absorption spectroscopy at different metal-to-ligand ratios and pH values (Ferrer et al., 1991). The interaction strongly depends on the initial metal-to-ligand ratio. Starting with a tenfold GSH excess, coordination takes place through the two carboxylate groups of the ligand, generating at pH = 7 a blue 2:1 GSH/VO^{2+} complex, which has also been characterized previously by NMR studies (Delfini et al., 1985).

Higher GSH concentrations produce a violet complex, which can also be generated by addition of the ligand to the blue species, and which apparently is similar to that obtained in the VO_3^-/GSH system—that is, a 1:1 complex in which the peptide coordinates with the VO^{2+} cation through the cysteinyl-thio group, the two amide nitrogens, and the amino group, in equatorial position (Degani et al., 1981). This violet complex seems to be the most stable species in the VO^{2+}/GSH system (Ferrer et al., 1991).

Oxidized glutathione shows a similar behavior, as two different VO^{2+}/GSSG complexes can be obtained, depending on the initial metal-to-ligand ratio. But in this case, these two complexes can be easily transformed into each other by simply changing the metal-to-ligand relations. This different behavior is probably due to the fact that in GSG the SH group participates in the formation and stabilization of the violet complex, whereas in the case of GSSG, this group is not available because of the formation of the disulfide bridge (Ferrer et al., 1993). At low GSSG concentrations, coordination takes place through carboxylate groups, whereas at higher concentrations, mainly N-donors are involved in coordination.

Interestingly, despite the fact that it has also blocked the sulfhydril group, S-methyl-glutathione (GSMe) only generates a single $2:1$ GSMe:VO^{2+} complex at pH $= 7$ and at all investigated metal-to-ligand ratios (Williams and Baran, 1994).

The formation of different complexes of the vanadyl(IV) cation with both the reduced and the oxidized glutathione suggests that both molecules may participate in the stabilization and transport of VO^{2+} immediately after the GSH-mediated reduction of vanadate(V) in living systems.

4.2.2.2. Interactions with L-Ascorbic Acid. L-Ascorbic acid (vitamin C (**2**)) is widely distributed in nature. Although its biochemical role is not completely understood, its essentiality for man has been clearly established (Davies et al., 1991; Davies, 1992), and it has often been considered one of the possible natural reducing agents of vanadate(V) to vanadyl(IV) in biological systems (Kustin and Toppen, 1973; Ding et al., 1994; Baran, 1995a; Nielsen, 1995).

The kinetics of the reduction has been investigated by stopped-flow techniques and an inner-sphere one-electron reduction mechanism has been proposed (Kustin and Toppen, 1973).

It has also been established that the reduced species, VO^{2+}, is able to interact with the acid and with some of its oxidation products (dehydroascorbic acid, 2,3-diketogulonic acid) (Kriss et al., 1975; Kriss and Kurbatova, 1976; Baran et al., 1995; Ferrer et al., 1997a), and the generation of free radicals with strong oxidation potential, during the reduction process, has also been postulated (Ding et al., 1994).

Different VO^{2+}/ascorbic acid complexes may be generated in solution at different pH values and some of them have been obtained as microcrystalline solids. Dehydroascorbic acid begins to interact with the oxocation at pH $= 4$, but these solutions are highly unstable towards oxidation of the ligand at increasing pH values (Ferrer et al., 1997a).

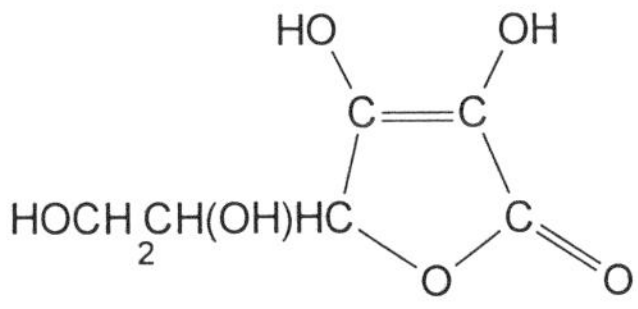

2

Scheme 2

4.2.2.3.   Interactions with Other Systems. It is possible that the essential amino acid cysteine (**3a**) also plays a role in the reduction process of vanadates to VO^{2+}, as its concentration in plasma has been estimated to lie around 10 μM(Rabenstein et al., 1979).

Model studies with the VO_3^-/cysteine system show that vanadate is reduced by cysteine irrespective of the pH value and that at pH = 6.8, reduction is followed by the formation of a purple complex (Sakurai et al., 1981). In this purple complex, the VO^{2+} cation apparently interacts with the amino nitrogen atom and the deprotonated –SH group of two amino acid molecules, generating a 2:1 (ligand-to-metal) species. This complex seems to be very similar to the vanadyl(IV) complexes of cysteine-methyl ester (Sakurai et al., 1980) and cysteine-ethyl ester (Ferrer and Baran, 1992), of the same stoichiometry, which have been investigated in the solid state. Recently, we were able to demonstrate that the VO^{2+} cation can also interact with cystine (**3b**), the oxidation product of cysteine, apparently through its carboxylate and amino groups (Ferrer et al., 1997b).

Taking into account that catechol and its derivatives react rapidly with VO_2^+ at high pH values and with vanadate at physiological pH, it has been postulated that catecholamines also may be able to reduce and complex vanadate(V) under certain circumstances (Cantley et al., 1978).

As carbohydrates constitute the most abundant class of compounds by weight in the biosphere (Whitfield et al., 1993), the interaction of cationic species of vanadium with polysaccharides may also be a detoxification route. In fact, a number of macromolecular polysaccharides are known to participate in different regulatory and metabolic processes and also appear to be very useful in the development of new pharmaceutical agents (Whitfield et al., 1993).

It is also well known that many sugars are able to reduce vanadates to VO^{2+} and to complex this cation. Owing to its strong hydrolytic tendency, this cation usually needs the presence of additional donor groups (e.g., carboxylates) in the sugar moiety, but, once bound to the ligand, it can easily deprotonate the hydroxyl groups and strongly coordinate up to four of them (Branca

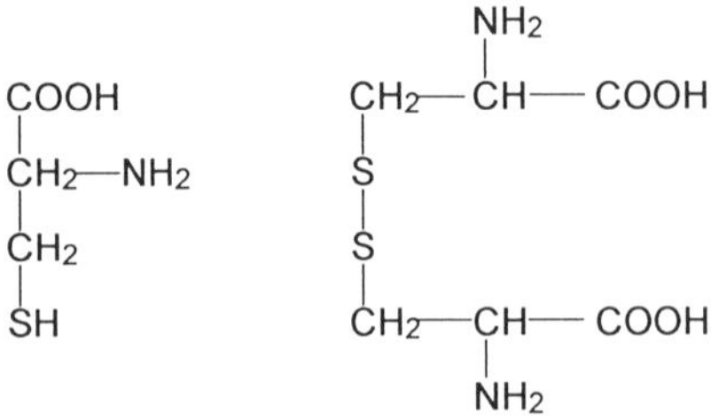

COOH
|
CH₂—NH₂
|
CH₂
|
SH

NH₂
|
CH₂—CH—COOH
|
S
|
S
|
CH₂—CH—COOH
|
NH₂

3 a   3 b

Scheme 3

et al., 1989, 1992; Williams and Baran, 1993a). Different studies with simple sugars have demonstrated that complexation is possible only with ligands provided with cis couples of adjacent OH groups (Branca et al., 1992). Recently, it was also possible to isolate some solid vanadyl(IV) complexes with different sugars, allowing a wider understanding of the main characteristics of the VO^{2+}/sugar interactions (Sreedhara et al., 1994; Etcheverry et al., 1996b, 1997; Williams et al., 1996).

Vanadate esterification of hydroxyl groups in simple saccharides and nucleosides has also been reported (Geraldes and Castro, 1989; Tracey et al., 1990; Angus-Dunne et al., 1995; Crans, 1995), and it is well known that the VO^{2+} cation interacts strongly with nucleotides and its constituents under different experimental conditions (Baran, 1995b).

Moreover, the interaction of the vanadyl(IV) cation with phosphate groups, which are also common ligands in biological systems, must be briefly brought up for comment. A number of phosphate complexes of different stoichiometries are known. Some of them are very stable and have been well characterized (Baran, 1995b; Buglyó et al., 1995b). From the nutritional point of view, phytic acid (mio-inositol hexaphosphate) appears to be especially important, because of its possible effects on the bioavailability of various essential metals. It has been shown that the VO^{2+} cation interacts with phytate to form both soluble and insoluble complexes. A soluble 1:1 complex is formed at pH < 1, whereas at higher pH values insoluble species are generated (Williams and Baran, 1993b).

Finally, as was stated in Section 2.1, evident relations exist between iron and vanadium metabolisms. This fact suggests that iron transport and/or reserve proteins also may be involved to some extent in vanadium detoxification. It has been shown that vanadyl(IV) transferrin complexes are probably physiologically important species because VO^{2+} is bound in serum to transferrin following injection of the metal ion (Harris et al., 1984). Also a vanadium(V) transferrin complex is formed either by addition of vanadate to apotransferrin or by air-oxidation of the vanadyl(IV)/transferrin complex. It has been suggested that in this case vanadium is bound to the protein as the VO_2^+ cation and has no requirement for a synergistic anion, in contrast to di- and trivalent metal cations (Harris and Carrano, 1984; Bertini et al., 1985; Butler and Carrano, 1991; Saponja and Vogel, 1996). Both the vanadyl(IV) and the vanadium(V) species can be displaced from the protein by Fe(III) and some other cations.

Furthermore conalbumin, from egg white, interacts with VO^{2+}, but the iron-saturated protein does not bind vanadyl(IV) ions, suggesting common binding sites for iron and vanadium. The apparent stability constants indicate that vanadyl(IV) ions bind to conalbumin approximately 12 orders of magnitude more weakly than iron to human serotransferrin but still sufficiently strongly to overcome hydrolysis (Casey and Chasteen, 1980).

The accumulation of vanadium in ferritin is also well established (Chasteen et al., 1986b,c; Gerfen et al., 1991), and some aspects of the equilibrium between vanadium in transferrin and ferritin have also been investigated (Sabbioni et al., 1978).

4.3. Chemical Detoxification

Different chelating agents have so far been assayed for vanadium detoxification, with varying success. To be useful, such an agent must be highly selective and, although we do not know with certainty all the factors necessary for designing a specific chelating species, the search must be oriented to systems whose action will be confined to the alteration of a particular biological site or equilibrium (Albert, 1961). Many excellent chelating agents are unsuited to therapeutic use in humans simply because they are not selective enough.

Thus a number of important criteria are to be considered in the selection of a therapeutic chelating agent (Taylor and Williams, 1995): (a) the formation constants for the complexes with the metal to be removed in comparison with those for H^+, calcium, and other essential metals; (b) the rate of reaction with the toxic metal deposits; (c) the total concentration of the ligand achievable at the desired site of action; (d) the chemical and biochemical stability and the toxicity of the ligand and its metal complexes; (e) the route of excretion of the complexes; and (f) the solubility of the ligand and its lipophilicity.

Another simple starting point in the selection of an adequate chelator may be the hard and soft acid and base approach (HSAB) (Porterfield, 1984). In this frame of reference, the biologically most relevant vanadium species (i.e., V^{3+}, VO^{2+}, VO^{3+}, VO_2^+) can be classified as hard acids (Porterfield, 1984; Taylor and Williams, 1995). Therefore, one may expect that the best chelating agents for these species are ligands that offer oxygen or nitrogen donors (hard bases in the HSAB classification).

On the other hand, in the particular case of the VO^{2+} cation, affinity for oxygen donors is particularly remarkable; the presence of one or two nitrogens in its coordination sphere lowers its stability appreciably (Chasteen, 1981).

4.3.1. Animal and in Vitro Studies

In the following sections we present a brief summary of relevant literature reporting in vitro and animal studies with different chelating agents and other chemical systems assayed for vanadium detoxification.

Chelating agents usually negate metal toxicity by complexation of the toxic species and subsequent excretion of the generated complex or by prevention of the absorption of the toxic species. Most of the systems investigated so far apparently act following the second route.

4.3.1.1. EDTA and Related Systems. Different well-known chelating agents have been tested, beginning with ethylenediaminetetraacetic acid (EDTA, **4**) and some of its salts. The use of EDTA (Mitchell, 1953a; Jones and Basinger, 1983) and its comparison with ascorbic acid (**2**) (Mitchell and Floyd, 1954) showed that both of these were effective detoxification agents

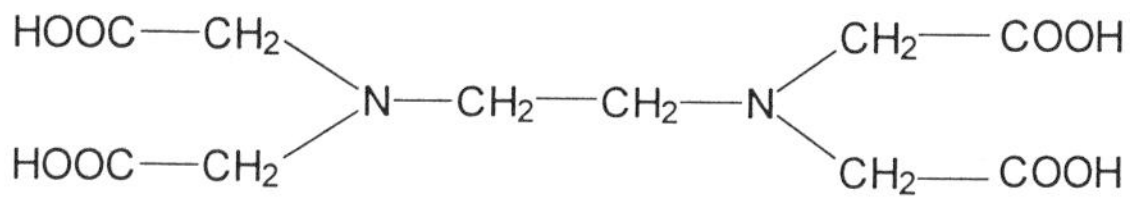

$$\text{HOOC—CH}_2\diagdown\underset{\text{HOOC—CH}_2\diagup}{N}\text{—CH}_2\text{—CH}_2\text{—}\underset{\diagdown\text{CH}_2\text{—COOH}}{\overset{\diagup\text{CH}_2\text{—COOH}}{N}}$$

4

Scheme 4

and that ascorbic acid apparently reacted more rapidly. Signs of vanadium toxicity were present for a longer period of time in vanadium-poisoned mice and rats after CaNa$_2$EDTA than after ascorbic acid administration. The effectiveness of EDTA as an antidote to both oxidation states of vanadium has also been demonstrated in mice experiments (Jones and Basinger, 1983), although its effectiveness at high vanadium doses has been questioned (Domingo et al., 1986). Mitchell (1953a) showed the convenience using the calcium disodium salt CaNa$_2$EDTA instead of the pure sodium salt. This reduces the toxicity of the latter salt, which, because of its tendency to combine with calcium, causes hypocalcemic tetany in laboratory animals.

Later experiments, performed with chicks and rats, also suggested that in the presence of EDTA less vanadium is absorbed from the intestinal tract (Hathcock et al., 1964).

Especially interesting in this context is the fact that the stability constant for the equilibrium:

$$VO^{2+} + [EDTA]^{4-} \rightleftharpoons [VOEDTA]^{2-}$$

at 20°C (log K = 18.35) (Schwarzenbach and Sandera, 1953) is markedly higher than the values determined for Mg^{2+} (log K = 8.79) and Ca^{2+} (log K = 10.69) (Martell and Smith, 1974b), but similar as those found for Ni^{2+}, Cu^{2+}, and Pb^{2+} (18.4, 18.3, and 18.2, respectively) (Schwarzenbach and Freitag, 1951). Also, values for the VO$_2^+$ cation are relatively high (Przyborowski et al., 1965):

$$VO_2^+ + [EDTA]^{4-} \rightleftharpoons [VO_2EDTA]^{3-} \text{ (log K = 15.55)}$$
$$VO_2^+ + [HEDTA]^{3-} \rightleftharpoons [VO_2HEDTA]^{2-} \text{ (log K = 9.60)}$$

Other polyaminopolycarboxylic acids known as good chelators, such as cyclohexane-1,2-diaminetetraacetic acid (CDTA), triethylene-tetraminehexaacetic acid (TTHA), and diethylenetriaminepentaacetic acid (DTPA), have not apparently been explored in relation to vanadium detoxification. Only for

DTPA (**5**) are some data available; they show that the CaNa$_3$DTPA salt was very effective in removing vanadium injected as VOSO$_4$, but less effective when the vanadium was given as vanadate (Hansen et al., 1982). More recently, it has been claimed that it is only effective at low vanadium doses (Domingo et al., 1985, 1986).

In regard to these types of chelating agents, it is interesting to comment that EDTA forms very stable complexes not only with the VO^{2+} cation (Chasteen, 1981) but also with vanadate, even though both are anions. This 1:1 complex, which is very stable and will be formed even at micromolecular concentrations of both ligand and vanadate, is recognized as a complex between the anion EDTA and the cation VO$_2^+$ although no observable concentration of this oxocation exists in the neutral pH range (Crans, 1994).

Finally, it should be mentioned that nitrilotriacetic acid (NTA, **6**) also has been assayed, and in spite of the fact that it is mentioned as an adequate detoxification agent, no detailed information about this system has been published (Hansen et al., 1982). But it should be emphasized that also in this case the 1:1 NTA:VO^{2+} complex has a higher stability constant (log K = 12.30) (Napoli, 1977) than the respective Ca^{2+} and Mg^{2+} cations (log K = 6.41 and 5.41, respectively) (Martell and Smith, 1974a).

4.3.1.2.   Sulfur-Containing Ligands.   The compound 2,3-dimercaptopropanol (BAL, British antilewisite, **7**), one of the best-known chelating agents used in medicine, was found to be ineffective as a therapeutic agent in the case of acute vanadium intoxications (Lusky et al., 1949; Sjöberg, 1951). This fact is not totally unexpected, given the above-noted HSAB principles.

Nevertheless, some other sulfur-containing systems present a certain degree of effectiveness. However, it is related essentially to the reducing power of these ligands rather than to its chelating ability (Jones and Basinger, 1983). This is surely the case in regard to glutathione (**1a**) and D-penicillamine (**8**). GSH is possibly one of the biological reductants of vanadate to VO^{2+} and, as discussed in Section 4.2.2.1., both reduced and oxidized glutathione are good chelators for the vanadyl(IV) cation.

As is known, D-penicillamine forms a variety of complexes with the VO^{2+} cation, depending on the metal-to-ligand ratio and the pH value (Costa Pessoa et al., 1990), despite the contradictory reports on its pharmacological effectivity. Jones and Basinger (1983) stated that it is an effective antidote for both vanadium oxidation states, a fact that suggests that probably both its reducing

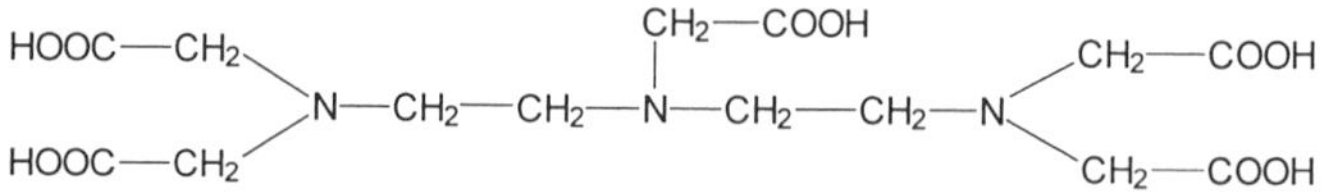

5

Scheme 5

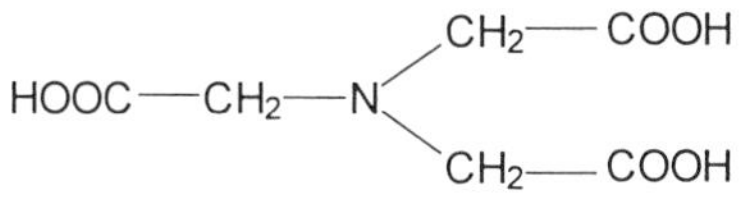

$$HOOC-CH_2-N \Big\langle \begin{array}{l} CH_2-COOH \\ CH_2-COOH \end{array}$$

6

Scheme 6

power and its complexation capability play a role in its action. On the other hand, Domingo et al. (1985) showed that it has no significant antidotal effect, even at low vanadate concentrations.

Also, L-cysteine (**3a**) showed only a very poor protective action in animal studies (Domingo et al., 1986).

4.3.1.3. Phosphonic Acids and Related Systems. Phosphonic acids are also known to form very stable chelates with different metal ions and a number of them were introduced recently in medical praxis for various purposes (Baran, 1995a). But they were not investigated practically in relation to vanadium detoxification, although these systems may be potentially useful, taking into account the stability of vanadyl(IV)/phosphate complexes as well as the affinity that vanadium oxocations have for this type of ligand, as predicted by the HSAB approach.

The only system of this type investigated so far seems to be the calcium salt of ethylenediaminetetramethylenephosphonic acid (EDTMP, **9**), which is of considerable interest because it shows a very high activity towards both vanadate(V) and vanadyl(IV), even though it is not a reducing agent and its complexes appear to be more stable than those of EDTA (Jones and Basinger, 1983).

Recently, the formation of very stable complexes between VO^{2+} and various phosphonic acids was investigated by pH potentiometry and spectroscopic techniques (Sanna et al., 1996). As in the case of phosphate complexes (Buglyó et al., 1995b), polyprotic phosphonic ligands yield anions able to bind VO^{2+} in rather acidic solutions.

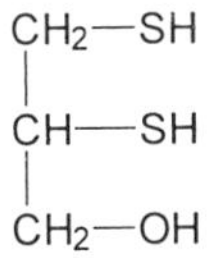

$$\begin{array}{l} CH_2-SH \\ | \\ CH-SH \\ | \\ CH_2-OH \end{array} \qquad\qquad CH_3-\underset{\underset{HS}{|}}{\overset{\overset{H_3C}{|}}{C}}-\underset{\underset{NH_2}{|}}{\overset{\overset{H}{|}}{C}}-COOH$$

7 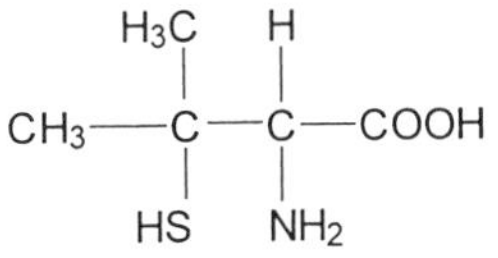   **8**

Schemes 7 and 8

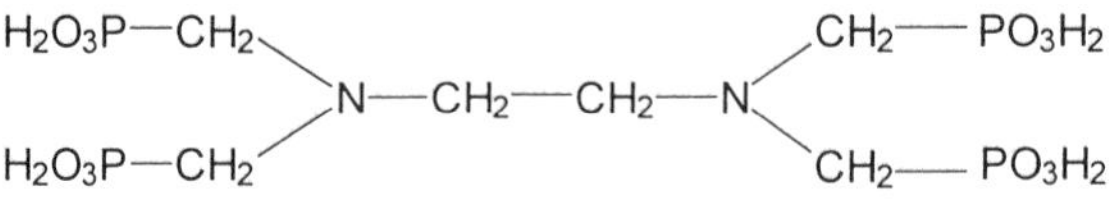

H₂O₃P—CH₂ ... N—CH₂—CH₂—N ... CH₂—PO₃H₂
H₂O₃P—CH₂ ... CH₂—PO₃H₂

9

Scheme 9

4.3.1.4.   Ascorbic Acid. From all the detoxification systems investigated so far, ascorbic acid (**2**) appears to be the most promising for human use, as shown by a systematic and comparative study of a great number of antidotes of very different chemical characteristics (Jones and Basinger, 1983; Domingo et al., 1985, 1986). It is probably the least toxic of all the examined drugs and can be administered orally in large doses.

In the case of chicks, it was shown that this vitamin reduced the growth retardation induced by the administration of vanadium and other toxic elements (Se, Co, Cd) and that its effect could not be mimicked by Fe(II) administration (Hill, 1979a).

4.3.1.5.   Other Systems. Among the other investigated detoxification agents for vanadium two others merit special comment. One of them is *Tiron* (the sodium salt of 4,5-dihydroxy-1,3-benzene-disulfonic acid, **10**); the other is *Desferrioxamine B* (**11**).

The action of Tiron was first investigated by Braun and Lusky (1959) on rats, rabbits, and pigeons intoxicated with NH_4VO_3. The ligand, which is relatively innocuous, reduces vanadate to VO^{2+} and apparently forms with that oxocation, a very stable complex which is rapidly excreted by urine. Lately, Jones and Basinger (1983) reinvestigated this system and confirmed that it is a very good antidote for either vanadyl(IV) sulfate or sodium vanadate, acting as an effective reducing agent for vanadium(V) and as a good chelator for vanadyl(IV).

Desferrioxamine B is the most widely used chelating agent for the treatment of iron overload conditions (Baran, 1988, 1995a; Taylor and Williams, 1995). In vitro studies on the interaction of this ligand with the vanadyl(IV) cation

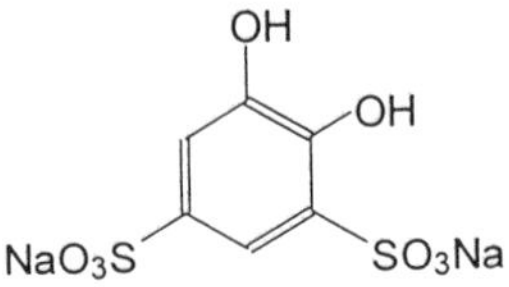

OH
OH
NaO₃S ... SO₃Na

10

Scheme 10

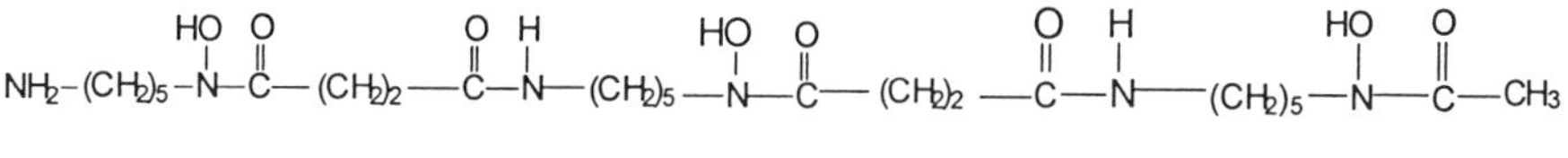

$$NH_2-(CH_2)_5-\overset{\overset{\displaystyle HO}{|}}{N}-\overset{\overset{\displaystyle O}{\|}}{C}-(CH_2)_2-\overset{\overset{\displaystyle O}{\|}}{C}-\overset{\overset{\displaystyle H}{|}}{N}-(CH_2)_5-\overset{\overset{\displaystyle HO}{|}}{N}-\overset{\overset{\displaystyle O}{\|}}{C}-(CH_2)_2-\overset{\overset{\displaystyle O}{\|}}{C}-\overset{\overset{\displaystyle H}{|}}{N}-(CH_2)_5-\overset{\overset{\displaystyle HO}{|}}{N}-\overset{\overset{\displaystyle O}{\|}}{C}-CH_3$$

11

Scheme 11

have shown that in very acidic media it is capable of displacing the oxygen ligand of the VO^{2+} moiety (Keller et al., 1991; Buglyó et al., 1995a). Also in the case of the vanadium(V) VO_2^+ oxocation, the ligand is able to displace one or both oxo groups (Buglyó et al., 1995a). In these non-oxo species, the bare vanadium(IV) and vanadium(V) cations are coordinated to the three hydroxamato functions of the ligand. With increasing pH, oxo coordination is restored and VO^{2+} and VO_2^+ complexes are generated with metal bonding to one or two hydroxamato functions (Buglyó et al., 1995a).

Desferrioxamine B appears to be a very effective antidote for vanadium poisoning, as shown by rat experiments (Hansen et al., 1982). It raises urinary and fecal vanadium excretion and is effective in the removal of both vanadate and vanadyl(IV) species. These findings were also confirmed by later experiments (Jones and Basinger, 1983; Domingo et al., 1985, 1986).

4.3.1.6. Probable Mechanisms of Action of These Chemical Systems. Although a discussion about the modes of action of the described chemical detoxification agents remains rather speculative, some possible mechanisms become evident from the results of the animal studies discussed above, as well as from general knowledge of the chemical characteristics of the ligands and chelators employed.

As vanadate is absorbed three to five times more effectively than VO^{2+} (Nielsen and Uthus, 1990; Nielsen, 1995), it is evident that one possible mechanism that prevents absorption may be the rapid reduction of vanadate, followed by the chelation of the generated vanadyl(IV) cation, or the direct binding of any cationic species derived from vanadium(V).

The known behavior of EDTA suggests that the formation of very stable complexes with both the vanadyl(IV) cation and cationic species derived from vanadates(V) probably has a direct impact on the vanadium absorption. The same is probably true for Desferrioxamine B, which also binds strongly to cationic species in both oxidation states.

In the case of Tiron, the ligand acts simultaneously as a reducing agent for vanadium(V) and as a chelator for VO^{2+}. But the fact that the finally generated vanadyl(IV) complex is excreted in the urine suggests that this ligand probably possesses activity on previously absorbed vanadium.

GSH is apparently more effective for the lower oxidation state, an indication that its chelate with VO^{2+} is involved in the detoxification process.

In spite of the fact that the activity of EDTMP against VO^{2+} is obviously related to its strong chelating ability, the origin of its activity against vanadate remains obscure.

The strong detoxification activity observed in the case of ascorbic acid is especially interesting. Evidently, one mode of action is its efficient reducing power against vanadium(V), transforming all the vanadium into the form of VO^{2+}. This cation could be then complexed by an excess of ligand but, as is known from the general behavior of ascorbate complexes (Davies, 1992), and was confirmed by our own model studies with the vanadyl(IV)/ascorbate system (Baran et al., 1995; Ferrer et al., 1997a), the stability constants of ascorbato complexes are relatively low, indicating that this ligand does not participate in the formation of chelates. This fact suggests that these complexation mechanisms would not be useful for the stabilization and excretion of the vanadium(IV) species. Another way of eliminating the generated vanadyl(IV) may be its complexation with the oxidation products of the vitamin.

As is known, dehydroascorbic acid, generated as the primary oxidation product, is also a very unstable species and undergoes a rapid series of transformations. As shown schematically in Figure 1, dehydroascorbic acid degrades first to 2,3-diketogulonic acid, which can further be degraded to a mixture of

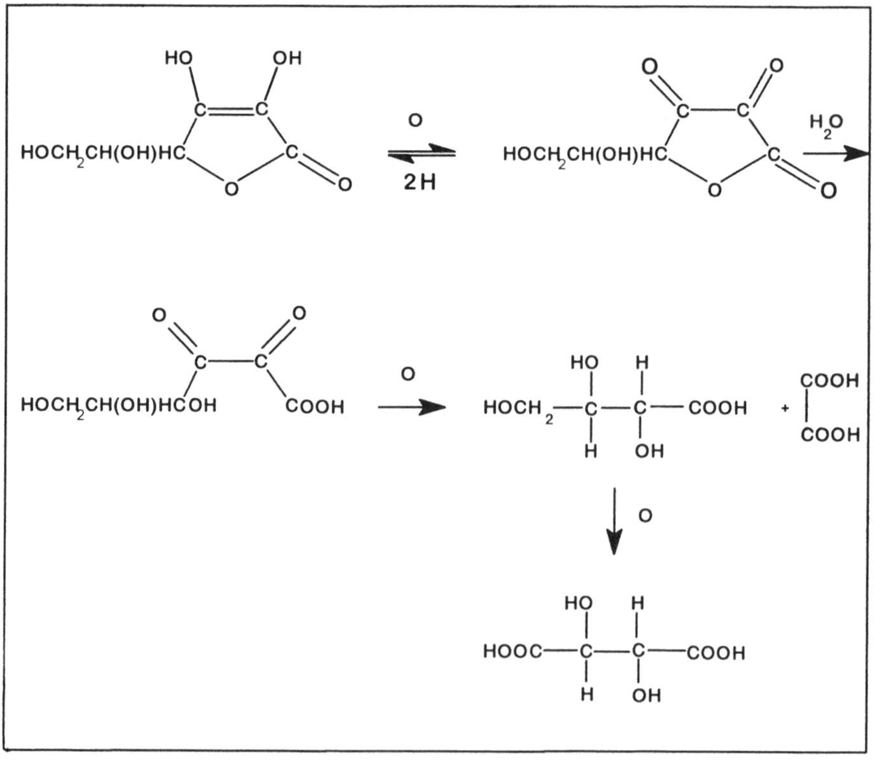

Figure 1. Schematic representation of the stepwise oxidation of L-ascorbic acid.

oxalic and L-threonic acids. At higher pH values, the latter acid oxidizes to tartaric acid. All these species could, in principle, interact with the VO^{2+} cation but, as shown by model studies, the primary complex generated by interaction of VO^{2+} with dehydroascorbic acid is very unstable towards oxidation, hydrolyzes irreversibly with opening of the lactone ring, and generates 2,3-diketogulonic acid. Finally, a $2:1$ ligand-to-metal complex is produced, in which an enolized form of the cited acid acts as a bidentate chelator of the cation (Ferrer et al., 1997a).

Evidently, some or all of the useful reducing systems could also generate another detoxification route, namely, reduction of previously absorbed vanadium(V) and elimination of the vanadyl(IV) cation complexed with some biogenic chelators (the low-molecular-weight ligands suggested by Chasteen et al., 1986b).

Another interesting way that remains to be explored in detail is the one suggested by Hansen et al. (1982), that is, the consecutive application of an efficient reducing agent (for example, ascorbic acid) followed by a strong VO^{2+} chelator (desferrioxamine B, EDTA, EDTMP, etc.).

4.4. Interactions That Reduce Vanadium Toxicity and Enhance Vanadium Tolerance

Various factors may affect vanadium absorption, retention, and toxicity (Nielsen, 1995). Some typical examples will be discussed in this section.

In some earlier animal studies, it was shown that the toxicity of vanadium shows a strong pH dependence (Mitchell, 1953b): NH_4Cl administered prior to vanadium to mice decreased vanadium toxicity; also acidification of vanadium solutions prior to injection markedly reduced the toxicity. In contrast, injections of sodium bicarbonate did result in an increase in vanadium toxicity. As the alkalinization of vanadate solutions did not increase the toxicity, it appears that excessive blood bicarbonate may retard the process of converting vanadate to a less toxic form.

Also very interesting is the observation that, much like phosphates, $Al(OH)_3$ may prevent tissue accumulation of vanadate from dietary sources by reducing its intestinal absorption (Wiegmann et al., 1982).

In a number of recent studies the effect of different metal ions and other chemical species were investigated in relation to vanadium toxicity in chicks. An increase in the dietary supplementation of NaCl resulted in amelioration of vanadate toxicity, as measured by growth rate (Hill, 1990a). Other effects, such as a reduction of vanadium concentration in different organs, were also observed. Nevertheless, there was little correlation between these reductions and reversal of vanadium toxicity. Although the origin of the chemical and biochemical effects of the chloride anion on toxicity remains unclear, it is possible that higher chloride concentrations improve renal metabolism, possibly deranged by vanadate.

Unexpectedly, the growth-retarding effect of 30 mg vanadium/kg diet in chicks was completely overcome by the inclusion of 500 mg mercury/kg diet, added as HgO. The mechanism of this mercury–vanadate interaction is not clear, although one possible explanation is that the addition of Hg(II) to the diet results in the stimulation of a vanadium detoxification mechanism (Hill, 1990b). Similar beneficial effects could also be observed by copper supplementation (at levels of 200 mg/kg diet, added as $CuSO_4 \cdot 5H_2O$), whereas neither Zn(II) nor sulphate show any effect on the toxicity of vanadium (Hill, 1990c) and the effect of chromium is not totally clear (Nielsen et al., 1980).

The impact of iron on vanadium toxicity in chicks has also been investigated (Blalock and Hill, 1987). As expected, significant interactions between the two elements could be observed. The toxicity of vanadium was significantly greater in iron-deficient than in iron-supplemented animals. Radioisotope studies with ^{48}V revealed that the absorption of vanadium is not influenced by the iron concentration in the diet, but the iron-deficient chicks retained more vanadium in the blood and liver and less in the bone than did the iron-supplemented animals. These results may be explained by the degree of saturation of transferrin and ferritin, to which vanadium can bind.

To conclude, it should be noted that the binding of vanadium, probably in its cationic forms, to complex macromolecules such as proteins and carbohydrates also has an important effect on its absorption and probably on its toxic effects as well (Hill, 1979b; Nielsen, 1995).

5. SUMMARY AND CONCLUSIONS

In this chapter an introductory overview of the most relevant aspects of vanadium metabolism and toxicity was given. Furthermore, the most relevant and usual biological and chemical detoxification mechanisms are briefly discussed. On this basis, a detailed analysis of the different routes of vanadium detoxification, based essentially on model and animal studies, is presented.

Accumulation of vanadium in bone, favored by the great structural and chemical similarities of PO_4^{3-} and VO_4^{3-}, appears to be one of the fundamental biological detoxification mechanisms. As suggested by model studies, the VO^{2+} cation also could interact with bone, as it can be strongly adsorbed or to the surface of hydroxyapatite. Possible interactions of this cation with mucopolysaccharides present in the organic matrix of hard tissues, as well as with tropocollagen, should also be considered.

As reduction of vanadium(V) to vanadium(IV) is surely one important step in vanadium metabolism, this process could also be indirectly related to detoxification. In this context, different stable vanadyl(IV) complexes of reduced and oxidized glutathione have been identified, and the participation of ascorbic acid (vitamin C) and of some of its oxidation products, as well as cysteine and catecholamines, must also be considered.

The incorporation of vanadium species into serum albumin, transferrin, and ferritin also probably plays an important role. In addition, the possible generation of VO^{2+} complexes with carbohydrates, macromolecular polysaccharides, and phosphate groups should be considered as potentially useful detoxification routes.

In relation to chemical detoxification processes, different chelating agents have been shown to be very effective. EDTA, especially in the form of its calcium disodium salt, Tiron, and Desferrioxamine B may be considered as very useful antidotes for both vanadium(IV) and vanadate(V), whereas sulphur-containing ligands are less effective. Nevertheless, of all the systems investigated to date, ascorbic acid appears to be the most convenient detoxicant for human use because it is probably the least toxic of all the systems examined and can be orally administered.

The possible utilization of phosphonates and its derivatives should be further explored, as they also appear to be very promising detoxification systems.

ACKNOWLEDGMENTS

It is a great pleasure to acknowledge the contributions of the collaborators whose names appear in the references. Work from this laboratory reported herein has been supported by the Consejo Nacional de Investigaciones Científicas y Técnicas de la República Argentina and the Comisión de Investigaciones Científicas de la Provincia de Buenos Aires.

REFERENCES

Albert, A. (1961). Design of chelating agents for selected biological activity. *Fed. Proc.* **20,** 137–147.

Angus-Dunne, S. J., Batchelor, R. J., Tracey, A. S., and Einstein, F. W. B. (1995). The crystal and solution structures of the major products of the reaction of vanadate with adenosine. *J. Am. Chem. Soc.* **117,** 5292–5296.

Baran, E. J. (1988). La nueva farmacoterapia inorgánica. VII. Compuestos de hierro. *Acta Farm. Bonaerense* **7,** 33–39.

Baran, E. J. (1992). El fascinante mundo de los biominerales. *Ciencia Invest.* **45,** 110–118.

Baran, E. J. (1994). Algunas contribuciones a la química bioinorgánica del vanadio. *Anales Acad. Nac. Cienc. Exact. Fis. Nat.,* **46,** 35–43.

Baran, E. J. (1995a). *Química Bioinorgánica.* McGraw-Hill Interamericana de España, Madrid.

Baran, E. J. (1995b). Vanadyl(IV) complexes of nucleotides. In H. Sigel and A. Sigel. (Eds.), *Metal Ions in Biological Systems.* Marcel Dekker, New York, Vol. 31, pp. 129–146.

Baran, E. J., Ferrer, E. G., and Williams, P. A. M. (1995). Interaction of the vanadyl(IV) cation with ascorbic acid and related systems. *J. Inorg. Biochem.* **59,** 600.

Bertini, I., Luchinat, C., and Messori, L. (1985). Spectral characterization of vanadium–transferrin systems. *J. Inorg. Biochem.* **25,** 57–60.

Blalock, T. L., and Hill, C. H. (1987). Studies on the role of iron in the reversal of vanadium toxicity in chicks. *Biol. Trace Elem. Res.* **14,** 225–235.

Branca, M., Micera, G., Dessi, A., and Kozlowski, H. (1989). Proton electron double resonance study of oxovanadium(IV) complexes of D-galacturonic and polygalacturonic acids. *J. Chem. Soc. Dalton Trans.,* pp. 1283–1287.

Branca, M., Micera, G., Dessi, A., and Sanna, D. (1992). Oxovanadium(IV) complex formation by simple sugars in aqueous solution. *J. Inorg. Biochem.* **45,** 169–177.

Braun, H. A., and Lusky, L. M. (1959). The protective action of disodium catechol disulfonate in experimental vanadium poisoning. *Toxicology* **1,** 38–41.

Buglyó, P., Culeddu, N., Kiss, T., Micera, G., and Sanna, D. (1995a). Vanadium(IV) and vanadium(V) complexes of Deferoxamine B in aqueous solution. *J. Inorg. Biochem.* **60,** 45–59.

Buglyó, P., Kiss, T., Alberico, E., Micera, G., and Dewaelle, D. (1995b). Oxovanadium complexes of di- and triphosphate. *J. Coord. Chem.* **36,** 105–116.

Butler, A., and Carrano, C. J. (1991). Coordination chemistry of vanadium in biological systems. *Coord. Chem. Rev.,* **109,** 61–105.

Cantley, Jr., L. C., Ferguson, J. H., and Kustin, K. (1978). Norepinephrine complexes and reduces vanadium(V) to reverse vanadate inhibition of the (Na,K)-ATPases. *J. Am. Chem. Soc.* **100,** 5210–5212.

Casey, J. D., and Chasteen, N. D. (1980). Vanadyl(IV) conalbumin. I. Metal binding site configurations. *J. Inorg. Biochem.* **13,** 111–126.

Chasteen, N. D. (1981). Vanadyl(IV) EPR spin probes. Inorganic and biochemical aspects. In L. Berliner and J. Reuben (Eds.), *Biological Magnetic Resonance.* Plenum, New York, Vol. 3, pp. 53–119.

Chasteen, N. D. (1983). The biochemistry of vanadium. *Struct. Bonding,* **53,** 105–138.

Chasteen, N. D. (Ed.). (1990). *Vanadium in Biological Systems.* Kluwer, Dordrecht.

Chasteen, N. D., Grady, J. K., and Holloway, C. E. (1986a). Characterization of the binding, kinetics and redox stability of vanadium(IV) and vanadium(V) protein complexes in serum. *Inorg. Chem.* **25,** 2754–2760.

Chasteen, N. D., Lord, E. M., and Thompson, H. J. (1986b). Vanadium metabolism. Vanadyl(IV) electron paramagnetic resonance spectroscopy of selected tissues in the rat. In A. V. Xavier (Ed.), *Frontiers in Bioinorganic Chemistry.* Verlag Chemie, Weinheim, pp. 133–141.

Chasteen, N. D., Lord, E. M., Thompson, H. J., and Grady, J. K. (1986c). Vanadium complexes of transferrin and ferritin in the rat. *Biochim. Biophys. Acta* **884,** 84–92.

Chiriatti, G. N. (1971). Prevenzione della intossicazione professionale da vanadio. *Folia Med.* **54,** 57–76.

Clark, R. J. H. (1968). *The Chemistry of Titanium and Vanadium.* Elsevier, Amsterdam.

Costa Pessoa, J., Vilas Boas, L. F., and Gillard, R. D. (1990). Oxovanadium(IV) and aminoacids— IV. The systems L-cysteine or D-penicillamine + VO^{2+}; a potentiometric and spectroscopic study. *Polyhedron* **9,** 2101–2125.

Cotton, F. A., and Wilkinson G. (1980). *Advanced Inorganic Chemistry.* 4th ed. Wiley, New York.

Crans, D. C. (1994). Aqueous chemistry of labile oxovanadates: Relevance to biological studies. *Comments Inorg. Chem.* **16,** 1–33.

Crans, D. C. (1995). Interaction of vanadates with biogenic ligands. In H. Sigel and A. Sigel. (Eds.), *Metal Ions in Biological Systems.* Marcel Dekker, New York, Vol. 31, pp. 147–209.

Davies, M. B. (1992). Reactions of L-ascorbic acid with transition metal complexes. *Polyhedron* **11,** 285–321.

Davies, M. B., Austin, J., and Partridge, D. A. (1991). *Vitamin C: Its Chemistry and Biochemistry.* Royal Society of Chemistry, Cambridge.

Degani, H., Gochin, M., Karlish, S. J. D., and Schechter, Y. (1981). Electron paramagnetic resonance studies and insulin-like effects of vanadium in rat adipocytes. *Biochemistry* **20,** 5795–5799.

Delfini, M., Gaggelli, E., Lepri, A., and Valensin, G. (1985). Nuclear magnetic resonance study of the oxovanadium(IV)–(glutathione)₂ complex. *Inorg. Chim. Acta* **107,** 87–89.

Ding, M., Gannett, P. M., Rojanasakul, Y., Liu, K., and Shi, X. (1994). One-electron reduction of vanadate by ascorbate and related free radical generation at physiological pH. *J. Inorg. Biochem.* **55,** 101–112.

Domingo, J. L., Llobet, J. M., and Corbella, J. (1985). Protection of mice against the lethal effects of sodium metavanadate: A quantitative comparison of a number of chelating agents. *Toxicol. Lett.* **26,** 95–99.

Domingo, J. L., Llobet, J. M., Tomas, J. M., and Corbella, J. (1986). Influence of chelating agents on the toxicity, distribution and excretion of vanadium in mice. *J. Appl. Toxicol.* **6,** 337–341.

Eady, R. R., and Leigh, G. F. (1994). Metals in the nitrogenases. *J. Chem. Soc. Dalton Trans.,* pp. 2739–2747.

Elfant, M., and Keen, C. L. (1987). Sodium vanadate toxicity in adults and developing rats. *Biol. Trace Elem. Res.* **14,** 193–208.

Elliott, J. C. (1994). *Structure and Chemistry of the Apatites and Other Calcium Orthophosphates.* Elsevier, Amsterdam.

Erfkamp, J., and Müller, A. (1990). Die Stickstoff-Fixierung. *Chem. Unserer Zeit* **24,** 267–279.

Etcheverry, S. B., Apella, M. C., and Baran, E. J. (1984). A model study of the incorporation of vanadium in bone. *J. Inorg. Biochem.* **20,** 269–274.

Etcheverry, S. B., Williams, P. A. M., and Baran, E. J. (1994). The interaction of the vanadyl(IV) cation with chondroitin sulfate A. *Biol. Trace Element Res.* **42,** 43–52.

Etcheverry, S. B., Williams, P. A. M., and Baran, E. J. (1996a). A spectroscopic study of the interaction of the VO²⁺ cation with the two components of chondroitin sulfate. *Biol. Trace Element Res.* **51,** 169–176.

Etcheverry, S. B., Williams, P. A. M., and Baran, E. J. (1996b). Synthesis and characterization of a solid vanadyl(IV) complex of D-glucuronic acid. *J. Inorg. Biochem.* **63,** 285–289.

Etcheverry, S. B., Williams, P. A. M., and Baran, E. J. (1997). Synthesis and characterization of vanadyl(IV) complexes with saccharides. *Carbohydr. Res.* **302,** 131–138.

Faulkner-Hudson, T. G. (1964). *Vanadium: Toxicology and Biological Significance.* Elsevier, Amsterdam.

Ferrari, R. P. (1990). Metal binding sites of oxovanadium(IV) in native and modified soluble collagen. *Inorg. Chim. Acta* **176,** 83–86.

Ferrer, E. G., and Baran, E. J. (1992). A spectroscopic study of the VO²⁺ complexes of cysteinate esters. *An. Asoc. Quim. Argent.* **80,** 429–437.

Ferrer, E. G., Williams, P. A. M., and Baran, E. J. (1991). A spectrophotometric study of the VO²⁺–glutathione interactions. *Biol. Trace Element Res.* **30,** 175–183.

Ferrer, E. G., Williams , P. A. M., and Baran, E. J. (1993). The interaction of the VO²⁺ cation with oxidized glutathione. *J. Inorg. Biochem.* **50,** 253–262.

Ferrer, E. G., Williams, P. A. M., and Baran, E. J. (1997a). Interaction of the vanadyl(IV) cation with L-ascorbic acid and related systems. Submitted for publication.

Ferrer, E. G., Williams, P. A. M., and Baran, E. J. (1997b). On the interaction of the VO²⁺ cation with cystine. *J. Trace Elem. Med. Biol.* In press.

Frieden, E. (Ed.). (1984). *Biochemistry of the Essential Ultratrace Elements.* Plenum, New York.

Geraldes, C. F. G. C., and Castro, M. M. C. A. (1989). Interaction of vanadate with monosaccharides and nucleosides: A multinuclear NMR study. *J. Inorg. Biochem.* **35,** 79–93.

Gerfen, G. J., Hanna, P. M., Chasteen, N. D., and Singel, D. J. (1991). Characterization of the ligand environment of vanadyl complexes of apoferritin by multifrequency electron spin-echo envelope modulation. *J. Am. Chem. Soc.* **113,** 9513–9519.

Greenwood, N. N., and Earnshaw, A. (1984). *Chemistry of the Elements.* Pergamon, Oxford.

Grill, E., and Zenk, M. H. (1989). Wie shützen sich Pflanzen vor toxischen Schwermetallen? *Chem. Unserer Zeit* **23**, 193–199.

Hansen, T. V., Aaseth, J., and Alexander, J. (1982). The effect of chelating agents on vanadium distribution in the rat body and on uptake by human erythrocytes. *Arch. Toxicol.* **50**, 195–202.

Harris, W. R., and Carrano, C. (1984). Binding of vanadate to human serum transferrin. *J. Inorg. Biochem.* **22**, 201–218.

Harris, W. R., Friedman, S. B., and Silberman, D. (1984). Behavior of vanadate and vanadyl ion in canine blood. *J. Inorg. Biochem.* **20**, 157–169.

Hathcock, J. N., Hill, C. H., and Matrone, G. (1964). Vanadium toxicity and distribution in chicks and rats. *J. Nutr.* **82**, 106–110.

Hill, C. H. (1979a). Studies on the ameliorating effect of ascorbic acid on mineral toxicities in the chick. *J. Nutr.* **109**, 84–90.

Hill, C. H. (1979b). The effect of dietary protein levels on mineral toxicity in chicks. *J. Nutr.* **109**, 501–507.

Hill, C. H. (1990a). Interaction of vanadate and chloride in chicks. *Biol. Trace Elem. Res.* **23**, 1–10.

Hill, C. H. (1990b). The effect of dietary mercury on vanadate toxicity in the chick. *Biol. Trace Elem. Res.* **23**, 11–16.

Hill, C. H. (1990c). Effect of dietary copper on vanadate toxicity in chicks. *Biol. Trace Elem. Res.* **23**, 17–23.

Jones, M. M., and Basinger, M. A. (1983). Chelate antidotes for sodium vanadate and vanadyl sulfate intoxication in mice. *J. Toxicol. Environ. Health* **12**, 749–756.

Keller, R. J., Rush, J. D., and Grover, T. A. (1991). Spectrophotometric and ESR evidence for vanadium(IV) Deferoxamine complexes. *J. Inorg. Biochem.* **41**, 269–276.

Krampitz, G., and Witt, W. (1979). Biochemical aspects of biomineralization. *Top. Curr. Chem.* **78**, 59–144.

Krishnamurthy, S. (1992). Biomethylation and environmental transport of metals. *J. Chem. Ed.* **69**, 347–350.

Kriss, E. E., and Kurbatova, G. T. (1976). Complex formation by vanadium(IV) with ascorbic and dehydroascorbic acids. *Russ. J. Inorg. Chem.* **21**, 1302–1306.

Kriss, E. E., Yatsimirskii, K. B., Kurbatova, G. T., and Grigor'eva, A. S. (1975). An electron spin resonance study of the reduction of vanadate by ascorbic acid. *Russ. J. Inorg. Chem.* **20**, 55–59.

Kucera, J., Lener, J., and Mnuková J. (1994). Vanadium levels in urine and cystine levels in fingernails and hair of exposed and normal persons. *Biol. Trace Elem. Res.* **43/45**, 327–334.

Kustin, K., and Macara, I. G. (1982). The new biochemistry of vanadium. *Comments Inorg. Chem.* **2**, 1–22.

Kustin, K., and McLeod, G. C., Gilbert, T. R., and Briggs, Le B. R. (1983). Vanadium and other metal ions in the physiological ecology of marine organisms. *Struct. Bonding* **53**, 139–160.

Kustin, K., and Toppen, D. L. (1973). Reduction of vanadium(V) by L-ascorbic acid. *Inorg. Chem.* **12**, 1404–1407.

Llobet, J. M., and Domingo, J. L. (1984). Acute toxicity of vanadium compounds in rats and mice. *Toxicol. Lett.* **23**, 227–231.

Lusky, L. M., Braun, H. A., and Laug, E. P. (1949). The effect of BAL on experimental lead, tungsten, vanadium, uranium, copper, and copper-arsenic poisoning. *J. Ind. Hyg. Toxicol.* **31**, 301–305.

Macara, I. G., Kustin, K., and Cantley Jr., L. C. (1980). Glutathione reduces cytoplasmic vanadate. Mechanism and physiological implications. *Biochim. Biophys. Acta* **629**, 95–106.

McConnell, D. (1973). *Apatite. Its Crystal Chemistry, Mineralogy, Utilization, and Geologic and Biologic Occurrences.* Springer-Verlag, Wien.

Martell, A. E., and Smith, M. R. (1974a). *Critical Stability Constants.* Plenum, New York, Vol. 1, pp. 139–144.

Martell, A. E., and Smith, M. R. (1974b). *Critical Stability Constants.* Plenum, New York, Vol. 1, pp. 204–211.

Merian, E. (Ed.). (1984). *Metalle in der Umwelt.* Verlag Chemie, Weinheim.

Mitchell, W. G. (1953a). Antagonism of toxicity of vanadium by ethylenediamine tetra acetic acid in mice. *Proc. Soc. Exp. Biol. Med.* **83,** 346–348.

Mitchell, W. G. (1953b). Influence of pH on toxicity of vandium in mice. *Proc. Soc. Exp. Biol. Med.* **84,** 404–405.

Mitchell, W. G., and Floyd, E. P. (1954). Ascorbic acid and ethylene diamine tetraacetate as antidotes in experimental vanadium poisoning. *Proc. Soc. Exp. Biol. Med.* **85,** 206–208.

Montel, G., Bonel, G., Heughebaert, J. C., Trombe, J. C., and Rey, C. (1981). New concepts in the composition, crystallization and growth of the mineral component of calcified tissues. *J. Cryst. Growth* **53,** 74–99.

Napoli, A. (1977). Vanadyl complexes of some N-substituted iminodiacetic acids. *J. Inorg. Nucl. Chem.* **39,** 463–466.

Narda, G. E., Vega, E. D., Pedregosa, J. C., Etcheverry, S. B., and Baran, E. J. (1992). Ueber die Wechselwirkung des Vanadyl(IV)-Kations mit Calcium-Hydroxylapatit. *Z. Naturforsch.* **47b,** 395–398.

Nechay, B. R., Nanninga, L. B., Nechay, P. S. E., Post, R. L., Grantham, J. J., Macara, I. G., Kubena, L. F., Phillips, T. D., and Nielsen, F. H. (1986). Role of vanadium in biology. *Fed. Proc.* **45,** 123–132.

Nielsen, F. H. (1995). Vanadium in mammalian physiology and nutrition. In H. Sigel and A. Sigel (Eds.), *Metal Ions in Biological Systems.* Marcel Dekker, New York, Vol. 31, pp. 543–573.

Nielsen, F. H., Hunt, C. D., and Uthus, E. O. (1980). Interactions between essential trace and ultratrace elements. *Ann. N. Y. Acad. Sci.* **355,** 152–164.

Nielsen, F. H., and Uthus, E. O. (1990). The essentiality and metabolism of vanadium. In N. D. Chasteen (Ed.), *Vanadium in Biological Systems.* Kluwer, Dordrecht, pp. 51–62.

Occhiai, E. I. (1995). Toxicity of heavy metals and biological defense. *J. Chem Ed.* **72,** 479–484.

Oniki, T., and Doi, Y. (1983). ESR spectra of VO^{2+} ions adsorbed on calcium phosphates. Calc. *Tissue. Int.* **35,** 538–541.

Porterfield, W. W. (1984). *Inorganic Chemistry. A Unified Approach.* Addison-Wesley, Reading, MA.

Przyborowski, L., Schwarzenbach, G., and Zimmermann, T. (1965). Die EDTA-Komplexe des Vanadiums(V). *Helv. Chim. Acta* **48,** 1556–1565.

Rabenstein, D. L. (1989). Metal complexes of glutathione and their biological significance. In D. Dolphin, O. Avramovic, and R. Poulson (Eds.), *Glutathione.* Wiley, New York, Part A, pp. 147–186.

Rabenstein, D. L., Guevremont, R., and Evans, C. A. (1979). Glutathione and its metal complexes. In H. Sigel (Ed.), *Metal Ions in Biological Systems.* Marcel Dekker, New York, Vol. 9, pp. 103–141.

Rehder, D. (1991). The bioinorganic chemistry of vanadium. *Angew. Chem. Int. Ed. Engl.* **30,** 148–167.

Rehder, D. (1992). Structure and function of vanadium compounds in living organisms. *Biometals* **5,** 3–12.

Sabbioni, E., Marafante, E., Amantini, L., Ubertalli, L., and Birattari, C. (1978). Similarity in metabolic patterns of different chemical species of vanadium in the rat. *Bioinorg. Chem.* **8,** 503–515.

Sakurai, H., Hamada, Y., Shimomura, S., Yamashita, S., and Ishizu, K. (1980). Cysteine methylester oxovanadium(IV) complex, preparation and characterization. *Inorg. Chim. Acta* **46,** L119–L120.

Sakurai, H., Shimomura, S., and Ishizu, K. (1981). Reduction of vanadate(V) to oxovanadium(IV) by cysteine and mechanism and structure of the oxovanadium(VI)–cysteine complex subsequently formed. *Inorg. Chim. Acta* **55,** L67–L69.

Sanna, D., Micera, G., Buglyó, P., and Kiss, T. (1986). Oxovanadium(IV) complexes of ligands containing phosphonic acid moieties. *J. Chem. Soc. Dalton Trans,* pp. 87–92.

Saponja, J. A., and Vogel, H. J. (1996). Metal-ion binding properties of the transferrins: A vanadium-51 NMR study. *J. Inorg. Biochem.* **62,** 253–270.

Schwarzenbach, G., and Freitag, E. (1951). Stabilitätskonstanten von Schwermetallkomplexen mit Äthylendiamin-tetraessigsäure. *Helv. Chim. Acta* **34,** 1503–1508.

Schwarzenbach, G., and Sandera, I. (1953). Die Vanadin Komplexe der Äthylendiamintetraessigsäure. *Helv. Chimica Acta* **36,** 1089–1101.

Seiler, H. G., Sigel, H., and Sigel, A. (Eds.). (1988). *Handbook on Toxicity of Inorganic Compounds.* Marcel Dekker, New York.

Selbin, J. (1966). Oxovanadium(IV) complexes. *Coord. Chem. Rev.* **1,** 293–314.

Sharma, R. P., Oberg, S. G., and Parker, R. D. (1980). Vanadium retention in rat tissues following acute exposures to different dose levels. *J. Toxicol. Environ. Health* **6,** 45–54.

Sigel, H., and Sigel, A. (Eds.). (1995). Metal Ions in Biological Systems. Vol. 31: *Vanadium and Its Role in Life.* Marcel Dekker, New York.

Sjöberg, S. G. (1951). Health hazards in the production and handling of vanadium pentoxide. *Arch. Ind. Hyg.* **3,** 346–348.

Sreedhara, A., Srinivasa-Raghavan, M. S., and Rao, C. P. (1994). Transition metal–saccharide interactions: Synthesis and characterization of vanadyl saccharides. *Carbohydr. Res.* **264,** 227–235.

Stillman, M. J. (1995). Metallothioneins. *Coord. Chem. Rev.* **144,** 461–511.

Stoecker, B. J., and Hopkins, L. L. (1984). Vanadium. In E. Frieden (Ed.), *Biochemistry of the Essential Ultratrace Elements.* Plenum, New York, pp. 239–255.

Talvitie, N. A., and Wagner, W. D. (1954). Studies in vanadium toxicology: Distribution and excretion of vanadium in animals. *Arch. Industr. Hyg.* **9,** 414–419.

Taylor, D. M., and Williams, D. R. (1995). *Trace Element Medicine and Chelation Therapy.* Roy. Soc. Chem., Cambridge.

Tracey, A. S., Jaswal, J. S., Gresser, M. J., and Rehder, D. (1990). Condensation reactions of aqueous vanadate with the common nucleosides. *Inorg. Chem.* **29,** 4283–4288.

Vasák, M. (1984). Metallothionein: A novel class of diamagnetic metal–thiolate cluster proteins. *J. Mol. Struct.* **23,** 293–302.

Vasák, M., and Kägi, J. H. R. (1983). Spectroscopic properties of metallothionein. In H. Sigel (Ed.), *Metal Ions in Biological Systems.* Marcel Dekker, New York, Vol. 15, pp. 213–273.

Vilas-Boas, L., and Costa-Pessoa, J. (1987). Vanadium. In G. Wilkinson, R. D. Gillard, and J. A. McCleverty (Eds.), *Comprehensive Coordination Chemistry.* Pergamon, Oxford, Vol. 3, pp. 453–583.

Wennig, R., and Kirsch, N. (1988). Vanadium. In H. G. Seiler, H. Sigel, and A. Sigel (Eds.), *Handbook on Toxicity of Inorganic Compounds.* Marcel Dekker, New York, pp. 749–765.

Wever, R., and Kustin, K. (1990). Vanadium: A biologically relevant element. *Adv. Inorg. Chem.* **35,** 81–115.

Wiegmann, T. B., Day, H. D., and Patak, R. V. (1982). Intestinal absorption and secretion of radioactive vanadium ($^{48}VO_3^-$) in rats and effect of $Al(OH)_3$. *J. Toxicol. Environ. Health* **10,** 233–245.

Whitfield, D. M., Stojkovski, S., and Sarkar, B. (1993). Metal coordination to carbohydrates. Structure and function. *Coord. Chem. Rev.* **122,** 171–225.

Williams, P. A. M., and Baran, E. J. (1993a). A spectrophotometric study of the interaction of VO^{2+} with monophosphate nucleotides. *J. Inorg. Biochem.* **50,** 101–106.

William, P. A. M., and Baran, E. J. (1993b). The interaction of the vanadyl(IV) cation with phytic acid. *Biol. Trace Elem. Res.* **36,** 143–150.

Williams, P. A. M., and Baran, E. J. (1994). The interaction of the VO^{2+} cation with S-methyl-glutathione. *J. Inorg. Biochem.* **54,** 75–78.

Williams, P. A. M., Etcheverry, S. B., and Baran, E. J. (1997). Synthesis and characterization of solid vanadyl(IV) complexes of D-ribose and D-ribose-5-phosphate. *J. Inorg. Biochem.* **65,** 133–136.

17

VANADIUM—A NEW TOOL FOR CANCER PREVENTION

*Malay Chatterjee and Anupam Bishayee**

Division of Biochemistry, Department of Pharmaceutical Technology, Jadavpur University, Calcutta 700 032, India

1. **Introduction**
2. **Thinking about Vanadium: A Possible New Tool for Cancer Chemoprevention**
3. **Vanadium and Chemical Rat Hepatocarcinogenesis**
   3.1. Evaluation of Morphological and Morphometric Responses to Vanadium in Preneoplastic Liver Lesions
   3.2. Biochemical Significance of Vanadium in the Morphological and Morphometric Analysis of Hepatocellular Lesions in Two-Stage Hepatocarcinogenesis
   3.3. Hematological, Histological, Histochemical, and Biochemical Basis for the Anticarcinogenic Potential of Vanadium
4. **Probable Mechanism of the Antitumor Effect of Vanadium**
   4.1. Role of Hepatic Biotransformation Enzymes in Vanadium-Mediated Inhibition of Hepatocarcinogenesis
   4.2. Cytogenetic, Biochemical, and Molecular Aspects of the Anticancer Effect of Vanadium
5. **Conclusion**
   References

Present address: MSB F-451, Division of Radiation Research, Department of Radiology, New Jersey Medical School, University of Medicine and Dentistry of New Jersey, Newark, New Jersey 07103-2714.

Vanadium in the Environment. Part 2: Health Effects, Edited by Jerome O. Nriagu.
ISBN 0-471-17776-8.   © 1998 John Wiley & Sons, Inc.

1. INTRODUCTION

Like the biochemistry and functions of vanadium, its possible role in health as well as disease remains one of the most fascinating stories in biology. Studies undertaken in the last decade suggest that dietary micronutrient vanadium could be considered a representative of a new class of nonplatinum group metal antitumor agents (Köpf-Maier, 1987). Although Kingsnorth et al. (1986) observed that vanadate supplementation in diet or drinking water did not modify the incidence or type of dimethylhydrazine-induced colon tumors in mice, Djordjevic and Wampler (1985) reported a significant antitumor activity of vanadium complexes against L1210 murine leukemia. Vanadium compounds have also been shown to possess pronounced antineoplastic activity against mouse liver tumors (Köpf-Maier and Köpf, 1979), fluid and solid Ehrlich ascites tumor (Köpf-Maier and Köpf, 1988), and TA3Ha murine mammary adenocarcinoma (Murthy et al., 1988). They have also markedly inhibited the growth of human tumor colony formation (Hanauske et al., 1987), HEp-2 human epidermoid carcinoma cells (Murthy et al., 1988), as well as xenografted human carcinomas of the lung, breast, and gastrointestinal tract (Köpf-Maier and Köpf, 1988; Köpf-Maier, 1994). Furthermore, feeding a purified diet supplemented with vanadyl sulfate prevented the induction by N-methyl-N-nitrosourea of mammary cancers when given during the postinitiation stages of neoplastic process (Thompson et al., 1984). The observations of Toney et al. (1985) concerning the biodistribution and pharmacokinetics of vanadocene dichloride, an organometalic antitumor compound, are comparable to the findings of previous mammalian studies with cis-diamminedichloroplatinum (CDDP) and related "second generation" platinum derivatives. Recently, [48V]vanadyl-chlorine e6Na, a synthetic chlorophyll derivative, has been reported to be promising as a tumor-imaging agent in conjunction with photodynamic therapy of tumors (Iwai et al., 1990). More recently, Kachinskas et al. (1994) observed that treatment of cultured malignant human keratinocytes with vanadate suppressed the expression of involucrin, a specific marker of keratinocyte differentiation. Very recently, the effect of vanadate on tumor promoter 12-O-tetradecanoylphorbol-13-acetate-induced monocytic differentiation in human promyelocytic leukemia cell line HL-60 has been documented by Wei and Yung (1995). Djordjevic (1995) reported the antitumor activities of a number of vanadium compounds in animal model systems. Sakurai et al. (1995) have also found strong antitumor activities of vanadyl complexes of 1,10-phenanthroline [$VO(Phen)^{2+}$] and related derivatives. In the 50% inhibition concentration (IC_{50}) of cell growth test using the human nasopharyngeal carcinoma, KB cell line, the cytotoxic effects of the vanadium complexes (16–22 ng/ml) are superior to that of the therapeutic drug CDDP (30 ng/ml).

Recent studies in our laboratory have established vanadium as a novel biological regulator in assessing the physiological and biochemical state of animals in a dose-related manner. We have observed that vanadium is capable of exhibiting some unique beneficial effects under a very low dose in comparison with the toxic higher doses (Bishayee and Chatterjee, 1994a; Chakraborty et al., 1994, 1995a,b). Our studies have also confirmed preceding reports regarding the "biphasic" effect of this element, that is, essentiality at low concentrations and toxicity at high concentrations (Smith, 1983; Jones and Reid, 1984; Hanauske et al., 1987). The beneficial effects of low levels of vanadium on normal cellular functions prompted us to investigate its role on pathophysiological systems. Previously, a significant antitumor response of vanadium against mice bearing Dalton's ascitic lymphoma and the increased survival of tumor-bearing hosts have been reported from our laboratory (Sardar et al., 1993). A possible correlation between vanadium-mediated alterations in biochemical indices such as reduced glutathione (GSH), extent of lipid peroxidation, and enzymatic activity of glutathione peroxidase (EC 1.11.1.9) in hepatic tissue, and the prolongation of survival of tumor-bearing animals receiving vanadium treatment, has also been observed (Sardar and Chatterjee, 1993). More recently, our laboratory has confirmed the antitumorigenic potential of vanadium in the control of tumor progression in lymphoma-bearing mice through modulation of several factors involving erythropoiesis (Chakraborty and Chatterjee, 1994). In the light of its effects on general aspects of tumor biology, it is more worthwhile and highly rational to look at its possible antitumorigenic effect, particularly on a long-term basis, in contrast with a more sophisticated model system of experimentally induced carcinogenesis. In view of this, the work detailed in this chapter was aimed at deciphering the impact(s) of vanadium on chemical carcinogenesis. The objective was, first, to see whether the trace element vanadium could be effective as a chemopreventive agent against experimentally induced chemical rat hepatocarcinogenesis as it is known that an anticarcinogenic agent for the liver may reflect anticarcinogenicity for other organs (Popper, 1978) and, second, to gain an eventual understanding of the mechanistic basis of vanadium-mediated inhibition of rat hepatocarcinogenesis at the cellular, subcellular, biochemical, chromosomal, and molecular levels. This may entail a clear understanding of the relevant biological and biochemical basis of the cancer chemopreventive action of this dietary micronutrient.

2. THINKING ABOUT VANADIUM: A POSSIBLE NEW TOOL FOR CANCER CHEMOPREVENTION

Recently we have reported a short- and long-term enzyme assay for screening potential inhibitors of tumorigenesis based on the determination of the induction of the detoxifying enzyme systems in which glutathione S-transferase (GST, EC 2.5.1.18) is the predominant one (Bishayee and Chatterjee, 1994b,c). GSTs are a family of phase II detoxifying enzymes that catalyze the reaction of GSH, one of the major endogenous antioxidant tripeptides, with numerous electrophiles, including activated forms of chemical carcinogens, to yield less toxic conjugates that are easily eliminated by excretion (Chasseaud, 1979; Jakoby, 1980). Thus, an elevation of GST activity indicates an increase in the ability to detoxify carcinogens. It has also been established that any compounds that induce an increase in the activity of this detoxifying enzyme system may be potential inhibitors of chemically induced tumorigenesis (Zheng et al., 1992; Nijhoff et al., 1993). The correlation of the induction of increased GST activity with the inhibition of carcinogenesis has been well documented. A number of known inhibitors inducing an increase of GST, especially in liver and digestive tract, have been found to inhibit chemical carcinogenesis in experimental animal model systems (Newmark, 1987; Lam and Hasegawa, 1989). Additionally, a preponderance of epidemiological studies indicate that diets rich in cruciferous vegetables (which induce GST in animals and humans) prevent human cancer (Morse and Stoner, 1995).

The nonprotein thiols, particularly GSH, owing to their $-SH$ groups, are known to inhibit neoplastic growth by reacting with electrophilic carbon atoms of several carcinogens and thus forming excretable products containing the exceptionally stable thioether linkage (Jakoby, 1978). The anticarcinogenic activities of a number of xenobiotics, antioxidants, and dietary constituents have been observed because of their ability to enhance endogenous GSH pool (Sparnins et al., 1982; Nijhoff et al., 1993).

These considerations led us to focus our attention on the ability of this trace element to raise the level of GSH and the activity of GST in rat liver and extrahepatic tissues, namely forestomach, small and large intestine mucosa, kidney, and lung, and thereby to screen the anticarcinogenic activity of micronutrient vanadium. Since reports on the hepato- and nephrotoxicity of vanadium in rodents were available in the literature (Donaldson et al., 1985; al-Bayati et al., 1989), the toxicological evaluation of the concentration and form of vanadium studied in this experiment were adequately well taken care of.

Supplementation of drinking water with vanadium as ammonium monovanadate (NH_4VO_3, +5 oxidation state) at the concentrations of 0.2 or 0.5 ppm (w/v) vanadium for 4, 8, or 12 weeks was found to increase the GSH level with a concomitant elevation in GST activity in the liver followed by small intestine mucosa and kidney (Bishayee and Chatterjee, 1995a). The results were almost dose-dependent and most pronounced with 0.5 ppm vanadium after 12 weeks of continuous supplementation. Neither the GSH level nor GST activity was significantly altered in forestomach and lung following vanadium supplementation throughout the study. The levels of vanadium that were found to increase the content of GSH and activity of GST in the liver, intestine, and kidney did not exert any toxic manifestation as evidenced by water and food consumption as well as the growth responses of the experimental animals (Bishayee and Chatterjee, 1995a). Moreover, these concentrations of vanadium did not impair either hepatic or renal functions, as they did not alter the serum activities of glutamic oxaloacetic transaminase (EC 2.6.1.1), glutamic pyruvic transaminase (EC 2.6.1.2), sorbitol dehydrogenase (EC 1.1.1.14), or serum urea and creatinine level (Bishayee and Chatterjee, 1993; 1995a). All these results clearly indicate that vanadium, in the concentrations employed in our previous study, has a significant inducing role on GSH content with a concurrent elevation in GST activity in the liver and specific extrahepatic tissues without any apparent sign of cytotoxicity. The dose-dependent biological effect of vanadium thus suggests that this trace element may possess an important role in the detoxification of a number of xenobiotics, including electrophilic chemical carcinogens. Thus, there is a necessity to establish the correlation between GSH- or GST-inducing effect and anticarcinogenic response of vanadium and, of course, the possible evaluation of its ability to inhibit chemically induced hepatocarcinogenesis in experimental animals.

3. VANADIUM AND CHEMICAL RAT HEPATOCARCINOGENESIS

3.1. Evaluation of Morphological and Morphometric Responses to Vanadium in Preneoplastic Liver Lesions

Rat liver carcinogenesis implies important modifications of the biochemical phenotype concomitant with histological changes in the liver that are considered to be precursors of malignant tumor formation (Pitot and Sirica, 1980; Farber, 1984a). Experimental liver tumors have become a favorite model for investigating carcinogenesis. In recent years, several models of experimentally induced hepatocarcinogenesis have been developed for examination of the abilities of various compounds to modulate the carcinogenic process (Tanaka et al., 1993; Tsuda et al., 1994; Williams, 1994). In many, if not all, of these models, focal areas of altered hepatocytes become grossly visible nodules and precede the appearance of malignant tumors. The initiated hepatocytes, that is, enzyme-altered foci and large hepatocyte nodules, are rapidly induced by various hepatocarcinogens in a synchronized fashion that facilitates more detailed studies. The preneoplastic lesions can be easily identified and counted, and their number and size can be determined by quantitative morphometry as a measure of multistage hepatic carcinogenesis and its alteration by chemopreventive agents (Moreno et al., 1991; Hida et al., 1994; Rao and Fernandes, 1996).

In light of the above, the effect of vanadium on rat liver carcinogenesis induced by diethylnitrosamine (DENA) was investigated by morphological and morphometrical analysis of preneoplastic hepatic lesions in male Sprague-Dawley rats (110–120 g) during the early stages of neoplastic transformation in order to delineate the potential of vanadium as a cancer chemopreventive agent. Initiation of hepatocarcinogenesis was performed by a single intraperitoneal (ip) injection of DENA (200 mg/kg body weight). Supplementary vanadium at the level of 0.2 and 0.5 ppm (as ammonium monovanadate) added to the drinking water was given ad libitum 4 weeks before DENA administration and continued for 8 or 16 weeks after the carcinogenic insult. Rats were sacrificed at the 12th or 20th week after the experiment had begun. After sacrifice, the livers were promptly excised, weighed, and examined at the surface for subcapsular macroscopic liver lesions (hepatocyte nodules). Representative liver slices were taken from the different lobes, which were immersed and stored in 10% neutral-buffered formalin for histological and morphometrical evaluation of focal lesions. Specific hepatocellular lesions were recognized by light microscopy according to established criteria (Stewart et al., 1980).

As to the results of this study, no significant differences were observed in food and water intakes or in growth rates among the different experimental groups (Bishayee and Chatterjee, 1995b). The addition of 0.2 or 0.5 ppm vanadium in drinking water resulted in the development in fewer rats of visible hepatocyte nodules and a smaller average number of nodules per nodule-bearing liver than in the group to which DENA was administered alone (Table 1). Another important observation of this study was vanadium-mediated inhibition of the appearance of nodules more than 3 mm in size with a reduction of nodular volume and of nodular volume as a percentage of liver volume (Table 2). The morphometrical analysis of preneoplastic focal lesions observed in hematoxylin- and eosin-stained (H and E-stained) liver sections revealed the occurrence of a decreased number of altered liver cell foci per 1 cm^2 and decreased average focal area, coupled with a decrement in the percentage area of liver parenchyma occupied by foci in vanadium-supplemented groups compared with DENA controls (Table 3).

A large body of evidence indicates that hepatic preneoplastic lesions are the putative precursors of hepatocellular carcinoma in rats and perhaps in humans as well (Farber and Cameron, 1980; Pitot and Sirica, 1980). The literature data strongly suggest that liver-altered foci represent neoplasia development and can also provide information regarding the anticarcinogenic potential of an agent in an experimental model system because of the consistent manner in which foci appear during the postinitiation events of hepatocarcinogenesis (Dragan et al., 1995). Because of the strong correlation between foci and hepatocarcinogenesis, the results we observed suggest that supplementation with vanadium at the 0.5 ppm level can greatly affect the postinitiation stages of hepatocarcinogenesis. This may alter the efficiency with which DENA can initiate the appearance of foci. The ability of vanadium to reduce the number of foci per 1 cm^2 of liver area also indicates that the anticarcinogenic potential of vanadium could be mediated through an enhanced "repair" or remodeling of preneoplastic lesions. All these results were found to be more pronounced at the 20th week than at the 12th week of the study. However, the results with 0.5 ppm vanadium at the 20th week were mostly at a statistically significant level. The findings of this preliminary study suggest that vanadium at a concentration of 0.5 ppm may have a potential anticarcinogenic property against chemically induced hepatic neoplasia without any apparent sign of toxicity.

Table 1 Effect of Vanadium Supplementations on the Development of Macroscopic Hepatocyte Nodules Induced by DENA in Rats

Group	No. of Rats with Nodules/Total Rats	Nodule Incidence (%)	Total No. of Nodules	Average No. of Nodules/Nodule-Bearing Liver (Nodule Multiplicity)
First killing (12th week)				
2. DENA control	6/7	85.7	32	5.3 ± 0.3^a
3. DENA + 0.2 ppm vanadium	4/8	50	21	5.2 ± 0.3
4. DENA + 0.5 ppm vanadium	3/8	37.5	14	4.6 ± 0.1
Second killing (20th week)				
2. DENA control	10/10	100	331	33.1 ± 13.2^d
3. DENA + 0.2 ppm vanadium	6/11	54.5^b	147	$24.5 \pm 2.2^{d,e}$
4. DENA + 0.5 ppm vanadium	4/12	33.3^c	81	$20.2 \pm 1.9^{d,f}$

[a] Mean $\pm$ SE.

[b] $P < 0.02$ and [c] $P < 0.01$ compared to DENA control at the 20th week by Fisher's exact probability test.

[d] $P < 0.001$ compared to the same treatment group at the 12th versus 20th week by Student's t test.

[e] $P < 0.05$ and [f] $P < 0.01$ compared to DENA control at the 20th week by Student's t test.

Source: Bishayee and Chatterjee (1995b).

Table 2 Effect of Vanadium Supplementations on the Size Distribution and Growth of Neoplastic Nodules Induced by DENA in Rat

Group	No. of Rats	Nodules Relative to Size (% of Total No.)				Mean Nodular Volume[a,b] (cm^3)	Nodular Volume/ Liver Volume[a,c] (%)
		≥3 mm	<3–>1 mm	≤1 mm			
First killing (12th week)							
2. DENA control	6	25	34.3	40.6		0.82 ± 0.09	21.3 ± 2.1
3. DENA + 0.2 ppm vanadium	4	23.8	28.5	47.6		0.68 ± 0.06	17.3 ± 1.7
4. DENA + 0.5 ppm vanadium	3	21.4	35.7	42.8		0.73 ± 0.05	19.7 ± 1.3
Second killing (20th week)							
2. DENA control	10	36.2	39.5	24.1		1.37 ± 0.21^d	65.3 ± 7.3^e
3. DENA + 0.2 ppm vanadium	6	21.7	47.6	30.6		1.09 ± 0.13^d	52.1 ± 4.9^e
4. DENA + 0.5 ppm vanadium	4	22.2	43.2	34.5		0.70 ± 0.08^f	$43.7 \pm 5.1^{e,g}$

[a] Mean ± SE.

[b] Individual nodule volumes were calculated from two perpendicular diameters measured on each nodule.

[c] For this calculation 1 g liver was assumed to occupy 1 cm^3.

[d] $P < 0.05$ and [e] $P < 0.001$ compared to the same treatment group at the 12th versus 20th week by Student's t test.

[f] $P < 0.02$ and [g] $P < 0.01$ compared to DENA control at the 20th week by Student's t test.

Source: Bishayee and Chatterjee (1995b).

Table 3   Effect of Vanadium Supplementations on the Development of Altered Liver Cell Foci Induced by DENA in Rat

Group	No. of Rats with Foci/Total Rats	Foci Incidence (%)	No. of Foci/cm²	Mean Focal Area (mm²)	% Area of Liver Parenchyma Occupied by Foci
First killing (12th week)					
2. DENA control	7/7	100	8.35 ± 0.93[a]	0.20 ± 0.02	1.35 ± 0.09
3. DENA + 0.2 ppm vanadium	7/8	87.5	10.12 ± 1.87	0.16 ± 0.03	1.46 ± 0.08
4. DENA + 0.5 ppm vanadium	7/8	87.5	9.82 ± 1.01	0.17 ± 0.06	1.24 ± 0.06
Second killing (20th week)					
2. DENA control	10/10	100	18.81 ± 2.35[c]	0.30 ± 0.04[b]	2.80 ± 0.17[c]
3. DENA + 0.2 ppm vanadium	10/11	90.9	14.11 ± 1.53	0.21 ± 0.06	2.22 ± 0.15[c,d]
4. DENA + 0.5 ppm vanadium	10/12	83.3	12.01 ± 1.56[d]	0.15 ± 0.01[e]	1.11 ± 0.07[f]

[a] Mean ± SE.
[b] $P < 0.05$ and [c] $P < 0.001$ compared to the same treatment group at the 12th versus 20th week by Student's t test.
[d] $P < 0.05$, [e] $P < 0.01$, and [f] $P < 0.001$ compared to DENA control at the 20th week by Student's t test.

Source: Bishayee and Chatterjee (1995b).

Treatment of rats with drinking water supplemented with vanadium for 20 weeks not only decreased the number of preneoplastic foci but also caused a reduction in the focal area along with a simultaneous decrease in focal area as a percentage of liver area. This strongly suggests the potential of vanadium in inhibiting or slowing the growth of altered liver foci. The observed effect of vanadium on focal growth may represent a selective toxicity to proliferating cells by virtue of the fact that they are proliferating compared to a relative nonproliferating background, thus eventually suppressing the occurrence of hepatocarcinogenesis. In this regard, continued long-term supplementation with vanadium is necessary for obtaining the tumor inhibitory effect of this micronutrient because complete removal or repair of preneoplastic lesions was not achieved with the experimental levels of vanadium.

3.2. Biochemical Significance of Vanadium in the Morphological and Morphometric Analysis of Hepatocellular Lesions in Two-Stage Hepatocarcinogenesis

The identification of several stages in the carcinogenic process has permitted a new approach to the prevention of cancer development. Interference with one or more of the recognizable steps could disrupt the entire process (Moreno et al., 1991). The introduction of new methods based on the two-stage concept of hepatocarcinogenesis, which consists of a relatively brief exposure of rats to an initiator carcinogen followed by prolonged administration of a promoter such as phenobarbital (PB), offers the opportunity to determine whether the alleged preventive effect of any chemopreventive agent is exerted during initiation, during promotion, or during both.

The key observation of our previous study was the fact that vanadium at 0.5 ppm in drinking water was very effective in arresting the development of DENA-induced hepatocarcinogenesis in rats without any toxic manifestations (Bishayee and Chatterjee, 1995b). However, in that study, vanadium supplementation was done during the entire course of our experiment and it was not possible to ascertain at which time point this trace element was most effective. In order to explore this area, we initiated a new series of experiments in which the anticarcinogenic potential of vanadium was critically examined before the initiation as well as during the early promotion phase of experimentally induced hepatocarcinogenesis. Our study was an attempt to gain more quantitative information regarding the morphometric analysis of hepatic foci and nodules positive for γ-glutamyl transpeptidase (GGT, EC 2.3.2.2), together with remodeling and the altered enzyme activities of GGT in the presence of absence of vanadium during DENA-induced hepatocarcinogenesis in rats. The rationale behind the selection of these parameters lies in the fact that hepatic foci that demonstrate altered enzyme phenotypes including GGT expression are generally accepted to be putative preneoplastic lesions for hepatocellular carcinoma, and their quantitative analysis is a useful tool for evaluation of modulation of hepatocarcinogenesis in rats (Williams, 1989).

To investigate the chemopreventive efficacy of vanadium and to identify the stage(s) at which it could be effective against chemical hepatocarcinogenesis, the rats were divided randomly into eight experimental groups as depicted in Figure 1 according to the experimental regimen previously designed from our laboratory (Sarkar et al., 1994, 1995a,b). Animals in groups A, C, E, and G were submitted to a slightly modified two-stage hepatocarcinogenesis model. Initiation was performed by a single ip injection of DENA at a dose of 200 mg/kg body weight in 0.9% NaCl solution. Following a 2-week recovery period PB, the promoter, was incorporated into the basal diets of the above four groups at the level of 0.05% for 14 successive weeks. Group A animals were the carcinogen (DENA) controls, while group B animals served as untreated normal controls. Vanadium, as ammonium monovanadate, was added to the double-distilled, demineralized drinking water at a concentration of 0.5 ppm and given ad libitum to the rats of all groups except groups A and B. In groups C, E, and G, supplementary vanadium was provided throughout the experiment (20 weeks), before the initiation (4 weeks) or during the promotion period (15 weeks), respectively (Fig. 1). The animals from groups D, F, and H served as vanadium controls for groups C, E, and G respectively.

Table 4 shows the final body weight, liver weight, and relative liver weight of different groups of rats that were killed after 20 weeks of the study. Vanadium

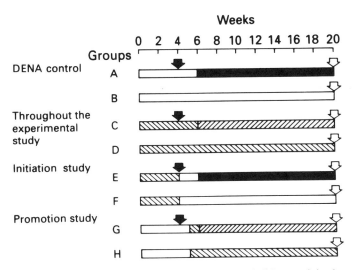

Figure 1. Schematic representation of the experimental regimen. Solid arrow, injection of DENA (200 mg/kg, ip); open column, basal diet and normal drinking water; solid column, basal diet with PB (0.05%) and normal drinking water; hatched column, descending, basal diet and vanadium supplementation (0.5 ppm) in drinking water; hatched column, ascending, basal diet with PB (0.05%) and vanadium supplementation (0.5 ppm) in drinking water; open arrow, time of sacrifice. Reproduced from Bishayee and Chatterjee (1995c), with permission.

Table 4   Body and Liver Weights of Different Groups of Rats at the End of the Study (After 20 Weeks)

Group	Effective No. of Rats	Final Body Weight (g)[a]	Liver Weight (g)[a]	Relative Liver Weight (g Liver/100 g Body)[a]
A	10	279.0 ± 30.1	13.10 ± 3.21	4.68 ± 0.41[b]
B	7	308.5 ± 22.3	10.11 ± 1.95	3.27 ± 0.25
C	12	299.8 ± 25.7	10.81 ± 2.15	3.60 ± 0.31[c]
D	8	313.3 ± 23.8	10.53 ± 1.72	3.36 ± 0.26
E	12	303.7 ± 28.5	11.93 ± 2.81	3.92 ± 0.35
F	8	299.4 ± 26.1	9.85 ± 1.89	3.28 ± 0.24
G	11	283.7 ± 32.5	12.15 ± 3.12	4.28 ± 0.37
H	7	305.2 ± 29.3	11.02 ± 2.11	3.61 ± 0.33

[a] Each value represents the mean ± *SE*.
[b] $P < 0.02$ compared with Group B.
[c] $P < 0.05$ compared with Group A.
Source: Bishayee and Chatterjee (1995c).

supplementation throughout the experiment reduced the incidence, total number, and multiplicity (Table 5), and altered the size distribution (Table 6) of visible persistent nodules (PNs), as compared with DENA control animals. Mean nodular volume and nodular volume as a percentage of liver volume were also attenuated following long-term vanadium treatment (Table 6). It also caused a large decrease in the number and surface area of GGT-positive hepatic foci and in the labeling index of focal cells, together with increased remodeling (Table 7). The activity of GGT, measured quantitatively, was found to be significantly less in the PNs and nonnodular surrounding parenchyma (NNSP) of vanadium-supplemented rats (Table 8). Similar results were

Table 5   Effect of Vanadium Supplementation (0.5 ppm) on the Development of Persistent Nodules in the Livers of Rats Initiated with DENA and Promoted by PB

Group	No. of Rats with Nodules/ Total No. of Rats	Nodule Incidence (%)	Total No. of Nodules	Average No. of Nodules/ Nodule-Bearing Liver (Nodule Multiplicity)[a]
A	10/10	100	383	38.3 ± 5.8
C	5/12	41.6[b]	52	10.4 ± 2.7[c]
E	7/12	58.3	136	19.4 ± 3.8[d]
G	8/11	72.7	241	30.1 ± 4.7

[a] Each value represents the mean ± *SE*.
[b] $P < 0.01$ compared with Group A by Fisher's exact probability test.
[c] $P < 0.001$ and [d] $P < 0.02$ compared with Group A.
Source: Bishayee and Chatterjee (1995c).

Table 6 Effect of Vanadium Supplementation (0.5 ppm) on the Size Distribution and Growth of Persistent Nodules in the Livers of Rats Initiated with DENA and Promoted by PB

| Group | No. of Rats | Nodules Relative to Size (% of Total No.) | | | Mean Nodular Volume[a,b] (cm^3) | Nodular Volume/Liver Volume (%)[a,c] |
		$\geqslant$3 mm	<3–>1 mm	$\leqslant$1 mm		
A	10	40.2	30.8	28.9	1.42 ± 0.27	68.4 ± 6.3
C	5	26.9	34.6	38.4	0.74 ± 0.09[d]	40.2 ± 4.8[e]
E	7	34.5	33.8	31.6	0.93 ± 0.15	53.2 ± 5.2
G	8	37.3	29.8	32.7	1.19 ± 0.21	60.1 ± 5.8

[a] Each value represents the mean ± SE.
[b] Individual nodule volumes were calculated from two perpendicular diameters on each nodule.
[c] For this calculation 1 g liver was assumed to occupy 1 cm^3.
[d] $P < 0.05$ and [e] $P < 0.01$ compared with Group A.

Source: Bishayee and Chatterjee (1995c).

Table 7 Influence of Vanadium Supplementation (0.5 ppm) on the Induction of GGT-Positive Liver Cell Foci in Rats Initiated with DENA Followed by Promotion with PB[a]

Group	No. of Rats	No. of Foci/cm^2	Focal Area (mm^2)	% Area of Liver Parenchyma Occupied by Foci (%)	Nonuniform Foci	Labeling Index
A	10	26.7 ± 3.5	0.48 ± 0.06	8.25 ± 0.52	11.31 ± 2.11	2.31 ± 0.08
C	12	7.3 ± 0.6[b]	0.29 ± 0.01[c]	6.12 ± 0.32[c]	32.70 ± 5.50[c]	1.62 ± 0.06[b]
E	12	16.7 ± 2.5[d]	0.34 ± 0.07	7.37 ± 0.46	18.78 ± 3.10	2.13 ± 0.05
G	11	20.1 ± 3.0	0.41 ± 0.05	7.92 ± 0.56	16.51 ± 3.22	2.21 ± 0.07

[a] Each value represents the mean ± SE.
[b] $P < 0.001$, [c]$P < 0.01$, and [d]$P < 0.05$ compared with Group A.

Source: Bishayee and Chatterjee (1995c).

Table 8 Alterations in the Enzymatic Activity of Cytosolic GGT in the Livers of Rats of Different Experimental Groups

| Group | GGT Activity[a] (nmol Product Formed/min/mg Protein) | | |
	PNs (n = 5)	NNSP (n = 5)	Control (n = 4)
A	56.90 ± 7.21^{b}	20.38 ± 3.70^{c}	0.78 ± 0.10
C	$12.15 \pm 2.90^{c,d}$	$5.12 \pm 1.11^{c,e}$	0.83 ± 0.12
E	$23.25 \pm 4.12^{b,e}$	$9.31 \pm 2.45^{c,f}$	0.72 ± 0.09
G	37.81 ± 6.41^{b}	14.17 ± 3.92^{g}	0.87 ± 0.13

[a] Each value represents the mean $\pm$ *SE.*
[b] $P < 0.001$, [c] $P < 0.01$, and [g] $P < 0.02$ compared with the corresponding controls, i.e., Groups B, D, F, and H for Groups A, C, E, and G, respectively.
[d] $P < 0.001$, [e] $P < 0.01$, and [f] $P < 0.05$ compared with Group A.
Source: Bishayee and Chatterjee (1995c).

observed when vanadium was given only before the initiation. However, supplementation of vanadium during the promotional period did not result in significant alterations of these parameters.

In this experiment the supplementation of 0.5 ppm vanadium in drinking water, especially during the entire period of the study, resulted in the development in fewer rats of visible PNs and a smaller number of nodules per nodule-bearing rat liver than those observed in DENA control animals. Another important observation of the study was the vanadium-mediated inhibition of the appearance of PNs larger than 3 mm with a concurrent attenuation of nodular volume as well as nodular volume as a percentage of liver volume. Although it is evident that not all the hepatocyte nodules become cancerous during the lifespan of the animals, numerous observations support the concept that the nodules are the precursors of hepatic cancer (Farber, 1980; Williams, 1980). Moreover, a large body of observational experience in experimental and human disease correlates the number and size of nodular hyperplasia and hepatocarcinoma (Farber and Cameron, 1980; Farber, 1990). In view of this, inhibition of nodule growth and enhancement of their regression by vanadium as observed in our study may be important for cancer prevention, especially if one considers that the PNs are easily recognizable and have a low tendency to regress spontaneously. Again, the food and fluid intakes and changes in body weights among different experimental groups were found to be statistically similar (Bishayee and Chatterjee, 1995c). This feature is of paramount importance because nutritional deprivation causing body weight loss may parallel a decrease in tumor volume (Rogers et al., 1993). Thus, the observed inhibitory effect of vanadium on nodule appearance and its growth is unlikely to be mediated through the impairment of nutritional status in the experimental animals.

Our results also showed an inhibitory role of vanadium on the number of GGT-positive preneoplastic focal lesions per 1 cm^2 of the livers of rats initiated with DENA. As GGT-positive foci represent a transient step to malignancy (Tatematsu et al., 1988), the ability of vanadium to reduce the development of GGT-positive foci suggests that this trace element can greatly affect the initiation stage of hepatocarcinogenesis by preventing the initiated cells from growing into preneoplastic foci. The potential role of vanadium in reducing the number of foci per 1 cm^2 of liver area was also reflected through a relatively high remodeling and low labeling index. This strongly indicates that a progressive loss of growth capacity by putative preneoplastic cells and their differentiation into normal-appearing hepatocytes proceed to a greater extent in the presence of vanadium.

According to the well-accepted hypothesis of Pitot et al. (1989), the number and size of altered liver cell foci indicate initiating and promoting activities, respectively. In our study, vanadium supplementation not only decreased the number of GGT-positive preneoplastic foci but also caused a decrement in the focal area, with a concomitant reduction in focal areas as a percentage of liver area, though the results were statistically more significant with respect to numbers of foci. This observation strongly supports that the anticarcinogenic potential of vanadium is primarily exerted on the initiation stage and secondarily on promotional events.

The enzymatic activity of GGT has been identified as a possible positive marker for preneoplastic hepatocytes (Hanigan and Pitot, 1985). The exponential increase in the activity of GGT in PNs and NNSP following DENA injection observed in our studies resembled a growth process that originated as a response to toxic cellular injury. As there is evidence of a close connection between GGT activation and carcinogenesis (Fiala and Fiala, 1973), a large increase in this enzyme activity could be correlated with a high nodule incidence, a high total number, and a large spread of nodules and foci in hepatic tissue. Vanadium-mediated inhibition of GGT-positive hepatic foci and PNs during rat liver carcinogenesis initiated with DENA and promoted by PB was well reflected in the relatively low level of this enzymatic activity, which was best observed in the group in which treatment with vanadium continued throughout the study. This might indicate a change in the plasma membrane of cells that could be related to the ultimate development of neoplasia, since membrane changes are most easily related theoretically to neoplastic behavior (Nicolson, 1976).

Our data thus reveal the unique protective role of vanadium against chemically induced liver tumorigenesis in rats and corroborate our previous findings (Bishayee and Chatterjee, 1995b). This time the anticarcinogenic potential of vanadium is primarily observed in the initiation phase and only secondarily in the promotion stage. In this regard, it is interesting to note the continuous long-term exposure to a low dose of vanadium would elicit a greater protection in terms of the magnitude of preneoplasia than exposure at either the initiation or promotion phase alone.

3.3. Hematological, Histological, Histochemical, and Biochemical Basis for the Anticarcinogenic Potential of Vanadium

Our knowledge of cellular and subcellular changes during liver carcinogenesis is based mainly on investigations of experimental models, especially of the rat liver intoxicated with various chemical carcinogens by histopathological techniques (Bannasch, 1978). Histochemical analysis also allows a simultaneous study of several metabolic pathways and biochemical events on frozen and fixed, but virtually unaltered, tissues. It proves helpful and simultaneously feasible in that histomorphological analysis can be done on the same tissue sample by the use of serial section examination or counterstaining of the histological slides (Kalengayi and Desmet, 1978).

In order to confirm and better understand the antineoplastic role of vanadium, our further attempt was correctly aimed at elucidating the role of this interesting element at the specified concentration (0.5 ppm) on the hematological status and hematopoietic system of rats during chemically induced hepatocarcinogenesis. Attempts have also been made to investigate further whether the effect of vanadium on the above parameters may have any consequences for histopathological and histochemical changes in rat liver during DENA-induced two-stage hepatocarcinogenesis and the eventual survival of the tumor-bearing animals.

For this study, the rats were treated according to the experimental regimen described in detail in Section 3.2. (Fig. 1). The results of this investigation clearly demonstrated that in a DENA-initiated and PB-promoted two-stage hepatocarcinogenic regimen in male Sprague-Dawley rats, supplementation of drinking water with vanadium at 0.5 ppm during the entire course of the study, before initiation and during promotion, afforded a significant prolongation of survival time (Fig. 2). Supplementary vanadium also resulted in the development in fewer rats of both neoplastic nodules and hepatocellular carcinomas and in fewer numbers of tumors per rat than in the DENA control group (Table 9). The protective role of vanadium against liver cancer could thus be corroborated with the slower growth rate of hepatic nodules, to a considerable extent delaying the appearance of neoplasia because of the increased latency period. This eventually may result in a significant delay in hepatoma-related mortality in rats treated with a DENA-PB regimen as observed in our study (Fig. 2).

A single ip injection of DENA resulted in an anemia probably due to severe red cell reduction, which was also reflected in the significantly lower hematocrit (Hct) value compared with normal animals (Table 10). The lower Hct value observed here is in perfect agreement with the reduced body weight of DENA-treated animals, as noted previously (Section 3.2.; Table 4), since there is a direct relationship between depression in Hct value and weight loss in rodents (Dible et al., 1987). Now, the effect of vanadium in reducing the severity of anemia through improvement of red cell count and Hct value might have maintained the normal body weight of DENA-injected animals. The

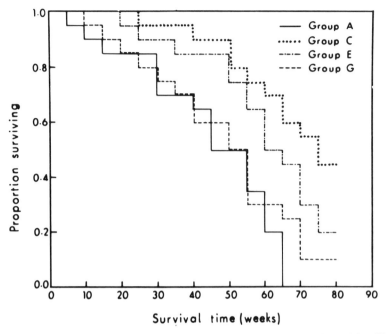

Figure 2. Kaplan-Meier survival curves for different experimental groups of rats. After 20 weeks of usual treatment as per Figure 1, all animals were maintained on normal drinking water and basal diet. Statistical significance of Group C versus group A (by Willcoxin rank test): $P < 0.01$. Reproduced from Bishayee et al. (1997a), with permission.

Table 9   Effect of Supplementary Vanadium on Liver Tumors Induced by DENA in Rats[a]

| Group | No. of Rats with Tumors/ Total Rats | | No. of Tumors/Rat | |
	Neoplastic Nodules	Hepatocellular Carcinomas	Neoplastic Nodules	Hepatocellular Carcinomas
A	6/6 (100)	6/6 (100)	47.3 ± 4.1	4.15 ± 0.51
C	5/9 (55.5)	4/9 (44.4[b])	11.5 ± 1.9[c]	0.98 ± 0.11[c]
E	7/11 (63.6)	7/11 (63.6)	30.9 ± 4.6[d]	2.98 ± 0.46
G	8/10 (80)	7/10 (70)	40.7 ± 3.6	3.92 ± 0.41

[a] Animals were sacrificed 56 weeks after DENA injection, i.e., 60 weeks after the start of the study. Numbers in parentheses indicate tumor incidence in percent.
[b] Significance of difference versus Group A by Fisher's exact probability test at $P < 0.05$.
[c,d] Significance of difference versus group A by Student's t test at [c]$P < 0.001$ and [d]$P < 0.02$.
Source: Bishayee et al. (1997a).

Table 10   Effect of Vanadium Supplementation on Some Hematological Indices during DENA-Induced Hepatocarcinogenesis in Rats[a]

Group	RBC (10^6/mm^3)	Hb (g/dL)	Hct (%)
B	7.3 ± 0.5	12.4 ± 0.7	33.1 ± 4.2
A	4.3 ± 0.1[b]	7.0 ± 0.2[b]	17.5 ± 1.9[c]
C	6.1 ± 0.4[d]	10.5 ± 0.6[e]	28.7 ± 2.7[d]
E	5.3 ± 0.3[f]	8.3 ± 0.4[f]	21.3 ± 1.7
G	4.6 ± 0.1	7.7 ± 0.3	22.0 ± 2.1

[a] Animals were sacrificed 16 weeks after DENA injection for hematological studies. Each value represents the mean ± *SE* of six animals.
[b,c] Significance of difference versus Group B at [b]$P < 0.001$ and [c]$P < 0.01$.
[d,e,f] Significance of difference versus Group A at [d]$P < 0.01$, [e]$P < 0.001$, and [f]$P < 0.02$.
Source: Bishayee et al. (1997a).

beneficial effect of vanadium was also apparent in elevating the hemoglobin (Hb) level (Table 10) with an improvement of the derived parameters, namely, mean corpuscular volume (MCV), mean corpuscular hemoglobin (MCH) and mean corpuscular hemoglobin concentration (MCHC), which were otherwise depressed in DENA control animals (Table 11). The exact mechanism of vanadium-mediated increasing hemoglobin level during chemical hepatocarci-

Table 11   Effect of Vanadium Supplementation on Red Cell Indices during DENA-Induced Hepatocarcinogensis in Rats[a]

Group	MCV (μm^3)	MCH (pg)	MCHC (%)
B	40.2 ± 3.9	15.2 ± 0.8	32.3 ± 1.7
A	41.1 ± 2.9	12.3 ± 0.5[b]	26.1 ± 1.0[b]
C	39.7 ± 3.0	15.5 ± 0.6[c]	29.9 ± 1.2[d]
E	41.8 ± 3.1	14.0 ± 0.7	27.1 ± 2.1
G	42.7 ± 3.7	12.9 ± 0.4	28.8 ± 1.9

[a] Animals were sacrificed 16 weeks after DENA injection for hematological studies. Each value represents the mean ± *SE* of six animals.
[b] Significance of difference versus Group B at [b]$P < 0.02$.
[c,d] Significance of difference versus Group A at [c]$P < 0.01$ and [d]$P < 0.05$.
Source: Bishayee et al. (1997a).

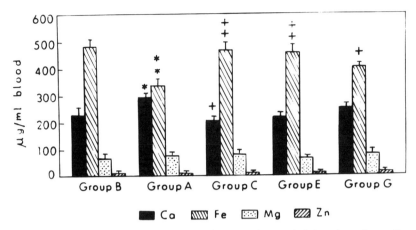

Figure 3. Redistribution of blood level of various metals on administration of vanadium in DENA-induced hepatocarcinogenesis in rats. For treatment of different groups, see Figure 1. Each bar represents the mean $\pm$ *SE* for six animals. *$P < 0.05$ and **$P < 0.01$ compared with Group B. [+]$P < 0.05$ and [++]$P < 0.01$ compared with Group A. Although Mg and Zn data are included in this figure, they are excluded from discussion in the text.

nogenesis is not clear at the present time. As shown in Figure 3, the level of Fe was found to be reduced in the blood of rats following DENA injection (Bishayee et al., unpublished results). Supplementary vanadium at different phases was able to elevate the blood iron content in DENA-treated rats (Fig. 3), which may be related to the improved hemoglobin status during DENA-induced hepatocarcinogenesis (Table 10). However, as depicted in Figure 4, the level of Fe was found to be increased in livers of DENA-treated rats (Bishayee et al., unpublished results), supporting its positive role in the growth of the neoplastic cell population. Now, the antitumor role of vanadium could be conceptualized as a limiting of the uptake of Fe by the hepatocytes, resulting in a reduced magnitude of neoplastic transformation. Whatever may be the mechanism, vanadium-mediated increase in hemoglobin as observed here could have a broad implication with respect to the antitumor efficacy of this dietary trace element, particularly in the light of the fact that high hemoglobin level has been found to possess an inhibitory influence on tumor growth (Bush et al., 1978).

On the other hand, both blood Ca levels and hepatic concentrations of Ca were found to be significantly elevated in DENA the control group (Figs. 3 and 4; Bishayee et al., unpublished results). This was in agreement with the earlier report that neoplastic cells had elevated levels of cytosolic Ca (Tsuruo et al., 1984). Now, vanadium-mediated redistribution of subcellular Ca concentration towards a decrement (Figs. 3 and 4) may contribute to the antineoplastic potential of this trace element.

Figure 5 showed a significant increase in the total leukocyte count 16 weeks following DENA injection, as compared with normal animals. Although no

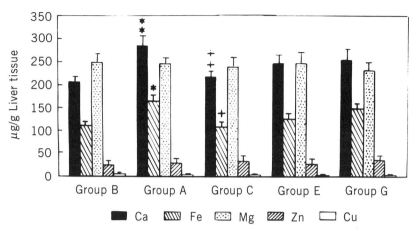

Figure 4. Redistribution of hepatic level of various metals on administration of vanadium in DENA-induced hepatocarcinogenesis in rats. For treatment of different groups, see Figure 1. Each bar represents the mean $\pm$ SE for six animals. *$P < 0.01$ and **$P < 0.01$ compared with Group B. +$P < 0.05$ and ++$P < 0.01$ compared with Group A. Although Mg, Zn, and Cu data are included in this figure, they are excluded from discussion in the text.

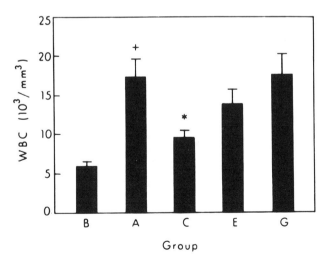

Figure 5. Alterations in total count of leukocytes during DENA-induced hepatocarcinogenesis in rats in the presence or absence of vanadium supplementation. For treatment of different groups, see Figure 1. Each bar indicates the mean $\pm$ SE for six animals. +$P < 0.001$ compared with Group B. *$P < 0.01$ compared with Group A. Reproduced from Bishayee et al. (1997a), with permission.

appreciable change in monocyte and eosinophil counts was found to have resulted from DENA treatment, a decrease in lymphocyte count with a concurrent increase in neutrophil count (Fig. 6) resulted in an overall rise in total leukocyte count. The role of vanadium supplementation in inhibiting the total count of leukocytes, especially in the group that received vanadium for 20 successive weeks (Fig. 5), could have been effected through a reversal of the lymphoid-myeloid ratio (Fig. 6). Thus, the stabilization of leukocyte count with a parallel retaining of normal hemoglobin level and Hct value following vanadium supplementation in DENA-treated animals supports the idea that vanadium may have a paramount importance in modulating these factors, possibly by an unknown immunohematopoietic mechanism. These findings

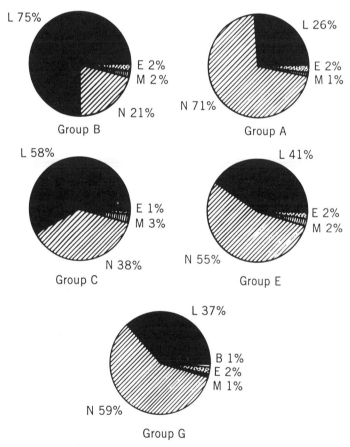

Figure 6. Changes in differential count (N, neutrophils; L, lymphocytes; E, eosinophils; M, monocytes; B, basophils) of leukocytes during DENA-induced hepatocarcinogenesis in rats in the presence or absence of vanadium supplementation. For treatment of different groups, see Figure 1. Reproduced from Bishayee et al. (1997a), with permission.

are in accordance with an earlier report that indicated vanadium's effect on the resistance of the organism through regulation of its leukocyte system (Barr-David et al., 1992). Furthermore, supplementary vanadium for 20 weeks registered a slight increment in monocyte count in DENA-treated animals (Fig. 6). This finding may have a greater significance in regard to the anticarcinogenic potential of vanadium in view of a recent report in which such an effect (increase of monocyte count) was considered to be a potential mechanism of action in the cancer chemopreventive role of a naturally occurring dietary constituent (Brevard, 1994).

In our experimental conditions, the concentration of vanadium given to untreated normal rats in drinking water over a period of 4–20 weeks did not induce any significant alteration in the hematological parameters studied (Bishayee et al., 1996a). Our finding, thus, is in keeping with the data of Russanov et al. (1994) but is at odds with the results of Zaporowska and Wasilewski (1989, 1992). This discrepancy in the vanadium effect might be due to a large difference between the vanadium concentrations used in our study and that studied elsewhere with respect to chronic vanadium intoxication (Zaporowska and Wasilewski, 1989, 1992). Based on the present study as well as the previous results on general toxicity studies and organ function tests (Bishayee and Chatterjee, 1993; 1995a), it can be concluded that our experimental level of vanadium is safe, nontoxic, and suitable even for chronic treatment.

A single ip injection of DENA also induced a significant alteration in two plasma proteins, namely albumin and globulin, though no noticeable change was found in the level of total protein (Table 12). The decrease in albumin level following DENA treatment may be taken as the consequence of reduced albumin synthesis due to DENA intoxication. In view of this, an increase in albumin level upon vanadium supplementation (Table 12) indicates a repair

Table 12   Effect of Vanadium Supplementation on Plasma Protein Levels During DENA-Induced Hepatocarcinogenesis in Rats[a]

Group	Total Plasma Protein (g/dL)	Albumin (g/dL)	Globulin (g/dL)	Albumin : Globulin
B	8.3 ± 0.4	5.4 ± 0.3	2.1 ± 0.1	2.57 ± 0.21
A	7.3 ± 0.3	2.6 ± 0.1[b]	4.0 ± 0.3[b]	0.65 ± 0.05[b]
C	8.0 ± 0.4	4.7 ± 0.3[c]	2.3 ± 0.1[c]	2.04 ± 0.11[c]
E	7.8 ± 0.4	3.2 ± 0.2[d]	3.1 ± 0.2[e]	1.03 ± 0.09[d]
G	7.5 ± 0.3	2.9 ± 0.1	3.4 ± 0.2	0.85 ± 0.08

[a] Animals were sacrificed 16 weeks after DENA injection for biochemical estimations. Each value represents the mean ± SE of six animals.
[b] Significance of difference versus group B at $P < 0.001$.
[c,d,e] Significance of difference versus group A at [c]$P < 0.001$, [d]$P < 0.01$, and [e]$P < 0.05$.
Source: Bishayee et al. (1997a).

of impaired protein synthesis by DENA, and a decrease in globulin level suggests an improvement in the functional status of hepatic Kupffer cells. In this context, the anticarcinogenic effect of vanadium may be implicated in the repair of altered protein synthesis in microsomes, with a concurrent improvement of the phagocytic activity of the Kupffer cells of the reticuloendothelial system of rat liver. However, further delineation of these responses of vanadium is necessary and warrants further study.

Histological examination of liver sections following H and E staining revealed a grossly altered hepatocellular architecture with oval hepatocytes 16 weeks after DENA injection (Fig. 7B) compared to normal counterpart cells (Fig. 7A). The altered liver cells of foci and nodules were considerably enlarged, largely vesiculated, and mostly binucleated (Fig. 7B). There was an excessive vacuolation in cytoplasm that contained masses of acidophilic material. Supplementation of drinking water with vanadium for the entire study elicited a maximum protection against DENA-induced hepatocarcinogenesis, which was reflected in the almost normal hepatocellular structure (Fig. 7C). Hepatocytes from this group were found to contain compact cytoplasmic material having fewer binucleated cells. A moderate improvement in the hepatic histological picture was observed in the group to which vanadium was given only before initiation (Fig. 7D). On the other hand, supplementary vanadium during the promotional event only marginally improved the hepatocellular phenotype vis-à-vis DENA control (data not shown).

The retention of glycogen on fasting has been well established as one of the most important positive markers for preneoplastic hepatocytes (Enzmann

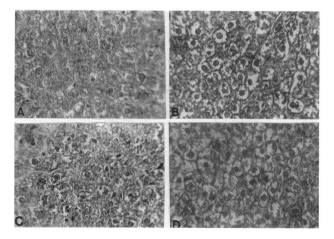

Figure 7. Histological picture of liver tissue of rats subjected to the following conditions: A. Normal untreated controls (Group B). B. DENA controls (Group A). C. DENA with long-term supplementary vanadium (Group C). D. DENA with vanadium before initiation (Group E). H and E, ×50. For treatment of different groups, see Figure 1.

et al., 1995; Mayer et al., 1996). It is well known that both the clear and the eosinophilic cells form multiple foci during the preneoplastic phase. These lesions correspond to glycogen storage foci, which are characterized by an excessive storage of glycogen (hepatocellular glycogenesis), which are observable under light microscope by periodic acid–Schiff reaction. A considerable accumulation of glycogen in cells of neoplastic nodules has also been demonstrated, suggesting that the nodules originate from the foci storing glycogen in excess (Bannasch, 1978). However, the results of our study showed a large number of glycogen storage cells in foci and nodules 16 weeks following a single dose of DENA (Bishayee, 1995). The observed DENA-induced hepatocellular glycogenesis indicates a considerable extent of preneoplasia, which is also evident from our histological evaluation (Fig. 7B). Vanadium-mediated reduction of glycogen storage in altered hepatocytes, especially following its continuous long-term treatment (Bishayee, 1995), adds to the evidence regarding the ability of vanadium to limit the size of hepatic preneoplasia (as observed in Fig. 7c) and thereby ultimately the incidence of liver cancer.

All these results thus establish that continuous long-term supplementation of drinking water with vanadium under very low concentration has a unique antitumorigenic potency that has subsequently been confirmed by histopathological, histochemical, and biochemical observations of the hepatic tissue of DENA-treated rats. The critical involvement of vanadium in modulating several factors associated with erythropoiesis under the carcinogenic challenge may thus have a possible impact on an eventual increased survival of the host.

4. PROBABLE MECHANISM OF THE ANTITUMOR EFFECT OF VANADIUM

4.1. Role of Hepatic Biotransformation Enzymes in Vanadium-Mediated Inhibition of Hepatocarcinogenesis

Drug-metabolizing enzymes have been recognized as important markers in characterizing the metabolic patterns of preneoplastic cells and in the development of strategies for prevention of carcinogenesis and early treatment of cancer. Attention has been concentrated on altered phenotypes during the early phases of neoplastic development, and new markers, especially enzymes or isoenzymes, have proved useful for detection of preneoplastic lesions in varous organs including liver (Sato, 1990). The cytochrome P-450 (Cyt P-450)-dependent enzymes, known as mixed-function oxidases, carry out myriad biotransformations, including C-, N-, and S-hydroxylations and dehalogenations as well as dealkylations and reductions. Some of the Cyt P-450 isoforms are fairly specific in their choice of substrates, but many (and particularly those in the hepatic endoplasmic reticulum) catalyze a surprisingly large number of chemical reactions with an unlimited number of biologically occurring and xenobiotic compounds (Wrighton and Stevens, 1992). Cyt P-450 and cytochrome b_5 (Cyt b_5) species are markedly decreased in hepatic hyperplastic nodules induced by DENA (Farber, 1984b). Aryl hydrocarbon hydroxylase (AHH; EC 1.14.14.2), yet another phase I drug-metabolizing enzyme, was also decreased in the nodules during chemical hepatocarcinogenesis (Sarkar et al., 1994). Again, alterations in the pattern of UDP-glucuronyl transferase (UDPGT; EC 2.4.1.17), a phase II drug-metabolizing enzyme, have been investigated in detail as a possible preneoplastic and neoplastic marker (Sato, 1990). Inducers of flavoenzyme quinone reductase (QR; EC 1.6.99.2) have recently been known to exert a cancer chemoprotective potential (Tawfiq et al., 1994). GSTs have also been reported as reliable markers for preneoplastic lesions and neoplastic tissues in the liver as well as in the other organs of the rat and other species including humans (Sato, 1990; Dragan et al., 1994). Thus, the study of these biotransforming enzymes has been found very useful and particularly relevant in assessing the chemopreventive efficacy of a drug against a tumorigenic assault. Inhibition of chemically induced tumor initiation may involve inhibition and/or induction of enzymes involved in both metabolic activation and detoxication in addition to other potential mechanisms (DiGiovanni, 1990). Compounds may induce both enzyme systems (bifunctional inducers) or only the phase II system (monofunctional inducers), and the induction of phase II enzymes is the predominant mechanism by which these compounds act as chemoprotective agents (Wattenberg, 1990). We therefore sought to characterize the effect of vanadium on phase I and phase II enzymes in order to better understand the role of the detoxification enzyme system in the antitumor effects of vanadium.

Data collected from this study demonstrated that there was a significant decrease in hepatic Cyt P-450 and Cyt b_5 contents in the DENA control group compared with the normal group (Table 13). In vanadium-supplemented

Table 13   Changes in the Levels and Activities of Several Hepatic Microsomal Biotransformation Enzymes in Different Groups of Rat[a]

Group	Cyt P-450 (nmol/mg Protein)	Cyt b_5 (nmol/mg Protein)	AHH (pmol 3-OH-B[a]P/ min/mg protein)	UDPGT (nmol Glucuronide/ min/mg Protein
B	0.67 ± 0.07	0.25 ± 0.03	320.1 ± 18.3	25.5 ± 5.0
A	0.30 ± 0.03^c	0.12 ± 0.01^b	201.7 ± 12.1^c	8.3 ± 1.1^b
C	0.44 ± 0.06	0.19 ± 0.04	182.3 ± 10.3	30.4 ± 5.4^d
E	0.37 ± 0.06	0.15 ± 0.03	210.4 ± 15.7	15.6 ± 3.3
G	0.35 ± 0.04	0.10 ± 0.02	195.7 ± 11.5	16.8 ± 3.7

[a] Each value represents the mean $\pm$ *SE* of six animals.
[b] $P < 0.01$ and [c] $P < 0.001$ compared to Group B.
[d] $P < 0.01$ compared to Group A.

Source: Bishayee et al. (1994).

groups, the unaltered levels of these two components in comparison with DENA controls (Table 13) indicates that the observed tumor-inhibiting effect of vanadium may have very little influence on Cyt P-450-mediated biotransformation of the particular carcinogen employed in this study. Although hepatic AHH activity was found to be significantly less in the DENA-treated group compared with the normal group, no concomitant alteration in this enzyme activity was found following vanadium supplementations (Table 13). This indicated that the mechanism by which vanadium induced the phase II enzymes of this study (Table 14) could be an electrophilic signal that was independent of Ah locus (Prochaska and Talalay, 1988). The markedly decreased activity of UDPGT in DENA controls compared with the normal group (Table 13) could be considered a failure of the detoxification process. On the other hand, vanadium-mediated maintenance of UDPGT activity, especially in the group to which vanadium was given throughout the study (Table 13), reflected the important role of this trace element in the acceleration of glucuronidation reactions. This could be effected through induction of UDP-glucose dehydrogenase (UDPGDH; EC 1.1.1.22) activity (Table 14), thereby providing more UDP-glucuronic acid to be utilized by the enhanced UDPGT. The elevated activity of NADPH-dependent QR following vanadium treatment (Table 14) may account for oxidation of NADH to NAD, which can be used as a cofactor for UDPGDH. As shown in Table 14, vanadium elevated the GST activity substantially in the long-term supplemented group and moderately in the group to which it was given only before initiation, which was otherwise severely decreased in comparison with the DENA control group. This may have had a major impact on the increased detoxificational status of the animals that had received the single combination of vanadium-DENA. Vanadium-dependent increased hepatic GST activity, thus, may increase the ability of the organism

Table 14   Changes in the Activities of Several Hepatic Cytosolic Xenobiotic Metabolizing Enzymes in Different Groups of Rat[a]

Group	UDPGDH (nmol NADH/min/ mg Protein)	QR (nmol Dichlorophenol Reduced/min/mg Protein)	GST (μmol CDNB Conjugate/ min/mg Protein)
B	58.1 ± 7.3	450.5 ± 21.3	1.15 ± 0.23
A	20.1 ± 2.9[c]	350.6 ± 12.5[b]	0.40 ± 0.05[b]
C	49.1 ± 5.0[f]	401.3 ± 17.8[d]	1.01 ± 0.15[e]
E	32.7 ± 3.8[d]	389.1 ± 15.3	0.75 ± 0.12[d]
G	28.9 ± 4.1	371.5 ± 13.7	0.64 ± 0.10

[a] Each value represents the mean ± SE of six animals.
[b] $P < 0.01$ and [c]$P < 0.001$ compared to Group A.
[d] $P < 0.05$, [e]$P < 0.01$, and [f]$P < 0.001$ compared to Group A.

Source: Bishayee et al. (1994).

to inactivate DENA-derived carcinogenic metabolites and thereby reduce the risk of tumor induction.

On the basis of the results illustrated here, we conclude that the mechanism of the anticarcinogenic effect of vanadium may involve (a) inhibition of metabolic activation of the procarcinogen, leading to reduced generation and/or binding of the ultimate carcinogen to DNA; and/or (b) elevated detoxification of the precarcinogen and/or its reactive metabolites through specific induction of activities of some of the xenobiotic biotransforming enzymes.

4.2. Cytogenetic, Biochemical, and Molecular Aspects of the Anticancer Effect of Vanadium

Chromosomal aberrations (CAs) are known to be important somatic mutations and are clearly involved in the origin as well as progression and diversification of some cancers (Land et al., 1983), but it is difficult to characterize the aberrations that are crucial for the initiation and the early stages of tumor development. It is well known that most tumors are characterized by predetermined patterns of chromosomal deviation, often accidentally concealed by chromosomal disturbances that are associated with all malignant developments (Ising and Levan, 1957). DENA-evoked chromosome damage in rat liver cells has been implicated as one of the earliest host reactions to this particular hepatocarcinogen (Grover and Fisher, 1971).

In view of the acknowledged importance of chromosomal alterations as the stage-of-progression markers in multistage carcinogenesis, we have tried to characterize the basic mechanism of vanadium-mediated suppression of DENA-induced hepatocarcinogenesis and thereby to monitor the effect of vanadium on CAs in rat liver cells induced by DENA. As we have previously observed that supplementary vanadium induced an increase in hepatic GSH content and GST activity in normal rats (Section 2), the present work endeavored to probe into a possible relationship between hepatic GSH or GST status and CAs in the host during DENA-induced neoplastic transformation in liver in the presence or absence of vanadium. All these studies may entail the underlying mechanism of the anticarcinogenic effect of vanadium from the biochemical as well as the cytogenetic point of view.

Figure 8 indicates that rats that had received supplementary vanadium at 0.5 ppm for 30 days (Group C) significantly suppressed DENA-induced CAs in their hepatic cells as early as 96 h after a single ip injection of DENA (200 mg/kg) in comparison with DENA controls (Group B). Control rats that received only the vanadium supplementation for 30 days (Group D) did not show any significant increase in CAs in hepatocytes in comparison with the normal control (group A) (Fig. 8). Supplementary vanadium that started 30 days prior to DENA injection and continued for 15, 30, or 45 days following DENA administration displayed a considerably suppressed incidence of DENA-evoked CAs in rat hepatocytes (Table 15). A maximum beneficial effect of vanadium against DENA-induced CAs was achieved when it contin-

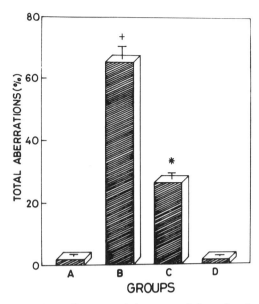

Figure 8. Inhibitory effect of vanadium on total chromosomal aberrations in rat liver cells induced by DENA. Chromosome aberrations (50 metaphase plates per rat) were prepared 72 h following partial hepatectomy, which was performed 24 h after DENA treatment. Each bar indicates the mean ± *SE* for five rats. $^+P < 0.001$ compared with Group A. $*P < 0.001$ compared with Group B.

ued till 45 days after the carcinogenic treatment, that is, for a total period of 75 days. Vanadium-mediated protection of CAs was predominantly reflected in its ability to reduce the structural aberrations followed quantitatively and qualitatively, by physiological changes, irrespective of the duration of vanadium treatment (Table 15). Rats that had received only vanadium supplementation for different time periods showed no significant increase in CAs in their hepatic cells compared with that of normal control groups (Bishayee et al., 1997b). However, subchronic supplementation with a low level of vanadium that was found to afford a unique protection against DENA-induced chromosomal abnormalities may partly explain its previously observed antitumor efficacy (Sections 3.2 and 3.3) through modulation of cytogenetic changes. Since CAs are generally assumed to be lethal to cells, the anticlastogenic effect of vanadium as noted here indicates that pretreatment with this trace element may inhibit the fatal toxicity of DENA to cells. Now, the positive correlations among CAs, mutagenicity, and carcinogenicity, as well as between DNA repair and CAs following exposure to carcinogens, have been well established. In view of this, the fact can be taken into consideration that the number of viable cells with subtle genetic changes might be attenuated by vanadium in proportion to the decrease in the aberrant cells induced by DENA.

Table 15 Influence of Vanadium on Frequency Distribution of CAs in Liver Cells of Rats Treated with DENA

Time (days)	Treatment[a]	Structural Aberrations				Numerical Aberrations		Physiological Aberrations		Total Aberrations		Protection (%)
		Individual Type		Exchange Type								
		No.	%	No.	%	No.	%	No.	%	No.	%[b]	
15	DENA	50	20	50	20	43	17.2	80	32	223	89.2 ± 2.27	—
	Vanadium + DENA	33	13.2	36	14.4	30	12	71	28.4	170	68.0 ± 3.28^{c}	21.2
30	DENA	60	24	48	19.2	42	16.8	50	20	200	80.0 ± 0.23	—
	Vanadium + DENA	32	12.8	27	10.8	21	8.4	31	12.4	111	44.4 ± 3.81^{c}	35.6
45	DENA	65	26	50	20	45	18	45	18	205	82.0 ± 2.22	—
	Vanadium + DENA	31	12.4	25	10	22	8.8	27	10.8	105	42.0 ± 2.12^{c}	40.0

[a] DENA (200 mg/kg) was injected ip and chromosome specimens (50 metaphase plates/rat, i.e., 250 plates/group) were prepared 15, 30, or 45 days after DENA injection. Vanadium supplementation (0.5 ppm) was started 30 days before DENA treatment and continued for 15, 30, or 45 days after treatment.
[b] Values represent means $\pm$ SE for five rats.
[c] $P < 0.001$ compared to corresponding DENA control.

Source: Bishayee et al. (1997b).

Another striking observation of this study was vanadium-mediated suppression of structural as well as numerical aberrations. It has been well established that in the majority of malignant tumors, the neoplastic cells have undergone chromosomal alterations that are viable in extent but are often highly complex, usually indicates structural and numerical changes (Atkin, 1991). Again, a high rate of chromosome breakage that amounts to structural aberration has been etiologically associated with the initiation of a carcinogenic process. Now, if the breakage lesions are found to be nonrandom and if the breakage loci happen to be those that are virtually linked to tumorigenesis, then the probability of tumorigenesis would increase drastically in the target tissue (Dave et al., 1994). In the light of the above, the role of vanadium in suppressing the structural and numerical aberrations observed here may reflect the ability of vanadium to counteract the initiation of rat liver tumorigenesis. This finding corroborates our earlier observation that the anticarcinogenic potential of vanadium is maximally observed in the initiation phase during DENA-induced hepatocarcinogenesis in rats (Section 3.2).

Furthermore, supplementary vanadium at 0.5 ppm alone for 45–75 days did not induce CAs in rat liver cells (Bishayee et al., 1997b), which is in line with the findings of Giri et al. (1979), who obtained a similar result with 4 mg/kg of oral vanadium pentoxide for 21 days. Based upon the present study and those reported recently from other laboratories (Roldán and Altamirano, 1990; Migliore et al., 1993; Léonard and Gerber, 1994), it has been confirmed that vanadium at low concentration is not clastogenic in experimental animals. On the other hand, the results of Wagner and Plewa (1994) show the antimutagenic potential of ammonium metavanadate, which supports our present finding concerning the anticlastogenic effect of ammonium monovanadate.

The chromosomal breakage that characterizes each of the chromosome breakage diseases can be measured rapidly by monitoring the formation of micronuclei (Countryman and Heddle, 1976). The micronucleus (MN) assay has also been extensively used in routine mutagen/carcinogen screening programs to detect agents that exert chromosomal breakage and spindle dysfunction in the bone marrow of rodents, especially in mice and rats (MacGregor et al., 1987). In particular, the MN assay registers as positives the majority of human carcinogens and antineoplastic agents. As noted here, vanadium treatment was found to be effective in reducing the toxicity of DENA as reflected in its ability to normalize the ratio of number of polychromatic erythrocytes (PCE) to number of normochromatic erythrocytes (NCE) in rat liver cells, which had been severely decreased following DENA injection. Our results also showed a significant induction of micronucleated polychromatic erythrocytes (MNPCE) and micronucleated normochromatic erythrocytes (MNNCE), due to a single ip injection of DENA, that was measured 15, 30, or 45 days after the initiation of hepatocarcinogenesis (Table 16). This DENA-induced induction of micronucleated erythocytes in rat hepatocytes is in agreement with the findings of a recent epidemiological study in which such an effect of mutagens present in drinking water coincided with the incidence of

Table 16 Influence of Vanadium on MN Assay and Frequency Distribution of Micronucleated Erythrocytes in Liver Cells of Rats Treated with DENA

Time (days)	Treatment[a]	Scored Cells[b]		PCE:NCE	MNPCE/1,000 PCE[c]	MNNCE/1,000 NCE[c]
		PCE	NCE			
15	Normal	2,416	2,584	0.93	0.98 ± 0.13	0.22 ± 0.02
	DENA	1,850	3,150	0.59	8.11 ± 0.17[d]	15.87 ± 0.43[d]
	Vanadium + DENA	2,399	2,601	0.92	3.83 ± 0.08[e]	11.32 ± 0.09[e]
30	Normal	2,502	2,579	0.97	1.12 ± 0.25	0.38 ± 0.04
	DENA	1,872	3,125	0.60	16.03 ± 0.52[d]	23.02 ± 0.98[d]
	Vanadium + DENA	2,650	2,380	1.11	7.97 ± 0.48[e]	13.59 ± 0.67[e]
45	Normal	2,540	2,470	1.02	2.03 ± 0.30	0.32 ± 0.03
	DENA	2,000	3,000	0.67	34.00 ± 1.15[d]	17.67 ± 0.99[d]
	Vanadium + DENA	2,699	2,350	1.14	14.48 ± 1.00[e]	13.62 ± 0.56[f]

[a] Treatment schedules are the same as described in Table 15.
[b] Approximately 5,000 bone marrow cells were examined, i.e., about 1,000 cells each from five rats.
[c] Values represent means ± SE for five rats.
[d] $P < 0.001$ compared to corresponding normal group.
[e] $P < 0.001$ and [f] $P < 0.01$ compared to corresponding DENA control.

Source: Bishayee et al. (1997b).

liver cancer (Ruan and Chen, 1994). In view of this, vanadium-mediated reduction in the frequency of induced MNPCE and MNNCE as observed in the present study is consistant with an antitumor potential of this trace element. Further, such an effect of vanadium not only reflects its anticlastogenic and antimutagenic effects in vivo but also emphasizes vanadium-mediated suppression of structural and numerical CAs as observed here, since the MN assay provides an indirect measure of the induction of these two types of chromosomal abnormalities especially (Mavournin et al., 1990).

Carcinogen-induced DNA damage and DNA repair have been established as significant events that occur during the initiation stage of carcinogenesis (Popescu et al., 1984). Cellular DNA is generally considered to be the most critical cellular target when considering the lethal carcinogenic and mutagenic effects of drugs, radiation, and environmental chemicals. These agents may damage DNA by altering bases or disrupting the sugar-phosphate backbone. Although base damage may have serious consequences for a cell, low levels of base damage are difficult to measure by physical or chemical means. In contrast, DNA strand breaks can be detected with great sensitivity by methods that utilize observation of the role of the unwinding of the two DNA strands. By applying our recently developed modified fluorimetric analysis of DNA unwinding (FADU) technique directly after isolation and purification of hepatic DNA (Sarkar et al., 1997), we have evaluated the response of vanadium in DNA chain break during the early preneoplastic steps of DENA-induced hepatocarcinogenesis in rats. Our results showed that 24 h after a single ip injection of DENA (200 mg/kg) (Group B), the native double-stranded DNA were found to be significantly less than in the normal controls (Group A) (Fig. 9). On the other hand, the aberrant single-stranded regions were drastically increased in Group B animals over those of Group A (Fig. 9). Supplementary vanadium for 30 days prior to DENA injection (Group C) elicited a statistically significant increment in the native double-stranded DNA with a concurrent decrement in the total single-stranded DNA generation compared with that of Group B animals (Fig. 9). However, rats that received only vanadium supplementation for 30 days (Group D) showed practically no adverse effect on the DNA strand breaks (Fig. 9). A notable feature of the analysis (Fig. 10) is the number of single strand breaks per DNA 24 h after DENA injection in the presence or absence of vanadium supplementation. The supplementary vanadium group (Group C) registered a 56% inhibition in the number of single strand breaks per DNA in comparison with numbers for the DENA controls, which registered a drastic increment (15-fold) of this parameter compared with numbers for the normal counterparts (Fig. 10). Supplementation of vanadium alone for 30 days did not show any DNA damaging efficacy as demonstrated by the insignificant difference in the generation of single strand breaks per DNA over the normal control (Fig. 10).

It is well known that DNA strand breaks responsible for chromosomal alterations are generated from DNA-base lesions induced by most of the

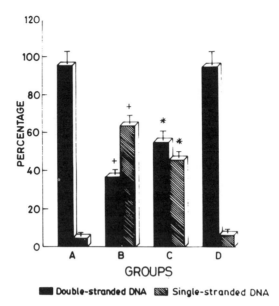

Figure 9. Effect of vanadium on the generation of DNA chain breaks in the presence or absence of DENA treatment. Each bar indicates the mean $\pm$ SE for five rats. $^{+}P < 0.001$ compared with Group A. $^{*}P < 0.02$ compared with Group B.

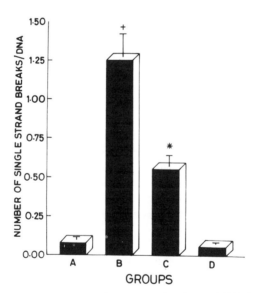

Figure 10. Inhibition of the number of single-strand breaks per DNA in rat hepatocytes by vanadium 24 h after DENA treatment. Each bar indicates the mean $\pm$ SE for five rats. $^{+}P < 0.001$ compared with Group A. $^{*}P < 0.01$ compared with Group B.

chemical mutagens. These DNA-base lesions are generally repaired by the excision-repair system (Friedberg et al., 1979). So as one hypothesis, it may be assumed that the anticlastogenic effect of vanadium observed in vivo may be due to the promotion of excision-repair activity. Again, it has also been established that DNA double-strand breaks (DDBs) are generated from mutagen-induced DNA lesions in the S phase of the cell cycle. It is suggested that DDBs are repaired by postreplicational repair in the G_2 phase and that unrepaired DDBs result in breakage-type CAs (Kihlman et al., 1982). In this context, the suppression of breakage-type aberrations by vanadium may be due to a modification of the capability of the postreplicational repair of DDBs.

Recent observations from our laboratory showed a significant depletion of GSH level in liver (Fig. 11) with a concurrent decrease in hepatic GST activities towards different substrates, for example 1-chloro-2,4-dinitrobenzene (CDNB) (Fig. 12) and 1,2-dichloro-4-nitrobenzene (DCNB) (Fig. 13) following DENA injection. This may enhance the covalent binding of DENA-derived alkylating agents to cellular DNA and the degree of cell damage, leading to neoplastic growth. Vanadium-dependent maintenance of GSH content towards near-normalization (Fig. 11) together with induction of a steady high level of GST activities (Figs. 12 and 13) during the entire length of our study may have led to an enhanced carcinogen elimination and reduction of carcinogen-DNA adduct formation, as well as subsequent expression of preneoplastic lesions and ultimately neoplasia. This may be regarded as one

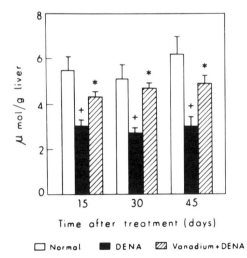

Figure 11. Hepatic cytosolic GSH content in different groups of rats. Treatment schedules are the same as mentioned in Table 15. GSH estimations were performed 15, 30, or 45 days following DENA injection. Each bar represents the mean $\pm$ SE for five rats. $^+P < 0.001$ compared with corresponding normal group. $^*P < 0.001$ compared with corresponding DENA control. Reproduced from Bishayee et al. (1997b), with permission.

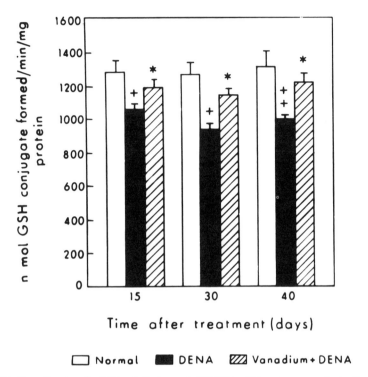

Figure 12. Hepatic cytosolic GST activity towards CDNB in different groups of rats. Treatment schedules are the same as mentioned in Table 15. GST activities were measured 15, 30, and 45 days following DENA injection. Each bar represents the mean $\pm$ SE for five rats. $^{+}P < 0.01$ and $^{++}P < 0.001$ compared with corresponding normal group. $^{*}P < 0.001$ compared with corresponding DENA control. Reproduced from Bishayee et al. (1997b), with permission.

of the underlying biochemical mechanisms of the antihepatocarcinogenic effect of vanadium.

Vanadium-dependent induction of hepatic GSH levels and GST activities in normal rats (Section 2) and in DENA-treated rats as observed here is not fully understood at the present time. However, it is interesting that the reduction of vanadium (+5) to vanadium (+4) species within the cells has been implicated in the physiological action of this trace element (Sabbioni et al., 1991) and one-electron reduction of vanadium (+5) to vanadium (+4) has recently been reported to be GSH-dependent (Sabbioni et al., 1993). Further studies are needed on this aspect to explore the involvement of GSH in the observed response of vanadium.

The results of the present study provide substantial evidence that vanadium triggers a unique protective effect against the induction of DNA strand breaks and CAs by a potent hepatocarcinogen (DENA) that is presumably related to the induction of a GSH-mediated GST-catalyzed detoxificational capacity of the host.

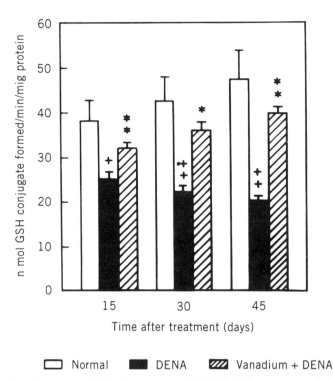

Figure 13. Hepatic cytosolic GST activity towards DCNB in different groups of rats. Treatment schedules are the same as mentioned in Table 15. GST activities were measured 15, 30, and 45 days following DENA injection. Each bar represents the mean ± *SE* for five rats. $^{+}P < 0.05$ and $^{++}P < 0.001$ compared with corresponding normal group. $^{*}P < 0.01$ and $^{**}P < 0.001$ compared with corresponding DENA controls. Reproduced from Bishayee et al. (1997b), with permission.

5. CONCLUSION

This study provides promising leads and good reasons to suspect that micronutrient vanadium may well be considered a potential cancer chemopreventive agent. Experimental mammals, such as rats, are very much like ourselves. The differences are embarrassingly small at the most fundamental level. Because of this, experimental mammals have become very powerful research tools, and obviously they provide us with the ability to explore human biology in ways that would be otherwise impossible. In everyday life, with constant low levels of exposure to environmental and endogenous carcinogens, protection by micronutrients against damage could be of considerable importance. The observed anticarcinogenic biological response of vanadium against chemical rat hepatocarcinogenesis may have the broadest implication if one considers its availability and easy acceptability in the form of possible supplementation in drinking water. The apparent nontoxicity of vanadium at the specified level may add to its future potential. With the success of this strategy developed in animal model, if truly functional, we would have the beginning of a new chemotherapeutic program that could be of utmost importance to the health and well-being of our society.

ACKNOWLEDGMENTS

We would like to gratefully acknowledge the financial support for the original work considered in this chapter from the Council of Scientific and Industrial Research of the Government of India [Grant Nos. 9/96(177)/91-EMR-I and 9/96(262)/95-EMR-I]. We are also indebted to Dr. A. Sarkar, Dr. R. Karmakar, and Dr. A. Mandal of our laboratory for their excellent assistance during the execution of this study.

REFERENCES

al-Bayati, M. A., Giri, S. N., Raabe, O. G., Rosenblatt, L. S., and Shifrine, M. (1989). Time and dose-response study of the effects of vanadate on rats: Morphological and biochemical changes in organs. *J. Environ. Pathol. Toxicol. Oncol.* **9,** 435–455.

Atkin, W. B. (1991). Nonrandom chromosomal changes in human neoplasia. In R. C. Socete and G. Obe (Eds.), *Eukaryotic Chromosome. Structural and Functional Aspects.* Narosa Publishing House, New Delhi, pp. 102–109.

Bannasch, P. (1978). Cellular and subcellular pathology of liver carcinogenesis. In H. Remmer, H. M. Bolt, P. Bannasch, and H. Popper (Eds.), *Primary Liver Tumors.* MTP Press, Lancaster, pp. 87–111.

Barr-David, G., Hambley, T. W., Irwin, J. A., Judd, R. J., Lay, P. A., Martin, B. D., Brambey, R., Dixon, N. E., and Hendry, P. (1992). Suppression by vanadium(IV) of chromium mediated DNA cleavage and chromium(VI/V) induced mutagenesis. *Inorg. Chem.* **31,** 4906–4908.

Bishayee, A. (1995). *Biological and Biochemical Role of Vanadium in the Chemoprevention of Neoplastic Transformation against Chemically-Induced Hepatocarcinogenesis in Rats.* Ph.D. thesis, Jadavpur University, Calcutta, India.

Bishayee, A., and Chatterjee, M. (1993). Selective enhancement of glutathione S-transferase activity in liver and extrahepatic tissues of rat following oral administration of vanadate. *Acta Physiol. Pharmacol. Bulg.* **19,** 83–89.

Bishayee, A., and Chatterjee, M. (1994a). Increased lipid peroxidation in tissues of the catfish *Clarias batrachus* following vanadium treatment: In vivo and in vitro evaluation. *J. Inorg. Biochem.* **54,** 277–284.

Bishayee, A., and Chatterjee, M. (1994b). Dose-related enhancement of cytosolic glutathione S-transferase activity and glutathione content in liver and extrahepatic tissues in mice with *Mikania cordata* root extract. *Aust. J. Med. Herbalism* **6,** 9–13.

Bishayee, A., and Chatterjee, M. (1994c). Anticarcinogenic biological response of *Mikania cordata:* Reflections on hepatic biotransformation systems. *Cancer Lett.* **81,** 193–200.

Bishayee, A., and Chatterjee, M. (1995a). Time course effects of vanadium supplement on cytosolic reduced glutathione level and glutathione S-transferase activity. *Biol. Trace Elem. Res.* **48,** 275–285.

Bishayee, A., and Chatterjee, M. (1995b). Inhibition of altered liver cell foci and persistent nodule growth by vanadium during diethylnitrosamine-induced hepatocarcinogenesis in rats. *Anticancer Res.* **15,** 455–462.

Bishayee, A., and Chatterjee, M. (1995c). Inhibitory effect of vanadium on rat liver carcinogenesis initiated with diethylnitrosamine and promoted by phenobarbital. *Br. J. Cancer* **71**, 1214–1220.

Bishayee, A., Mandal, A., Chatterjee, M., and Choudhuri, P. (1994). Antitumour potential of vanadium against chemically induced hepatocarcinogenesis: Reflection on hepatic drug detoxification. In R. S. Rao, M. G. Deo, and L. D. Sanghvi (Eds.), *Proceedings of the International Cancer Congress.* New Delhi, India, October 30–November 5. Monduzzi Editore, Bologna, pp. 3071–3076.

Bishayee, A., Karmakar, R., Mandal, A., Kundu, S. N., and Chatterjee, M. (1997a). Vanadium-mediated chemoprotection against chemical rat hepatocarcinogenesis: Haematological and histological characteristics. *Eur. J. Cancer Prev.* **6**, 58–70.

Bishayee, A., Banik, S., Mandal, A., Marimuthu, P., and Chatterjee, M. (1997b). Vanadium-mediated suppression of diethylnitrosamine-induced chromosomal aberrations in rat hepatocytes and its correlation with induction of hepatic glutathione and glutathione S-transferase. *Int. J. Oncol.* **10**, 413–423.

Brevard, P. B. (1994). Beta-carotene increases monocyte numbers in peripheral rat blood. *Int. J. Vitam. Nutr. Res.* **64**, 21–25.

Bush, R. S., Jenkin, R. D. T., Allt, W. E. C., Beale, F. A., Bean, H., Dembo, A. J., and Pringle, J. F. (1978). Definitive evidence for hypoxic cell influencing cure in cancer therapy. *Br. J. Cancer* **37**, 302–306.

Chakraborty, A., and Chatterjee, M. (1994). Enhanced erythropoietin and suppression of γ-glutamyl transpeptidase (GGT) activity in murine lymphoma following administration of vanadium. *Neoplasma* **41**, 291–296.

Chakraborty, A., Bhattacharjee, S., and Chatterjee, M. (1994). Effects of vanadium salts on lipid peroxidation, GSH levels and catalase activity in Indian catfish, *Clarias batrachus* (L.). *Philippine J. Sci* **123**, 251–266.

Chakraborty, A, Bhattacharjee, S., and Chatterjee, M. (1995a). Alterations in enzymes in an Indian catfish, *Clarias batrachus* (Linn.) exposed to vanadium. *Bull. Environ. Contam. Toxicol.* **54**, 281–288.

Chakraborty, A., Ghosh, R., Roy, K., Ghosh, S., Chowdhury, P., and Chatterjee, M. (1995b). Vanadium: A modifier of drug-metabolizing enzyme patterns and its critical role in cellular proliferation in transplantable murine lymphoma. *Oncology* **52**, 310–314.

Chasseaud, L. F. (1979). The role of glutathione and glutathione S-transferase in the metabolism of chemical carcinogenesis and other electrophilic agents. *Adv. Cancer Res.* **29**, 175–274.

Countryman, P. I., and Heddle, J. A. (1976). The production of micronuclei from chromosome aberrations in irradiated cultures of human lymphocytes. *Mutat. Res.* **41**, 321–322.

Dave, B. J., Hsu, T. C., Hong, W. K., and Pathak, S. (1994). Nonrandom distribution of mutagen-induced chromosome breaks in lymphocytes of patients with different malignancies. *Int. J. Oncol.* **5**, 733–740.

Dible, S., Siddik, Z. H., Boxall, F. E., and Harrap, K. R. (1987). The effect of diethyldithiocarbamate on the haematological toxicity and antitumour activity of carboplatin. *Eur. J. Cancer Clin. Oncol.* **23**, 813–818.

DiGiovanni, J. (1990). Inhibition of chemical carcinogenesis. In C. S. Cooper and P. L. Grover (Eds.), *Handbook of Experimental Pharmacology.* Springer-Verlag, New York, Vol. 94/II, pp. 159–223.

Djordjevic, C. (1995). Antitumor activity of vanadium compounds. *Met. Ions Biol. Syst.* **31**, 595–616.

Djordjevic, C., and Wampler, G. L. (1985). Antitumor activity and toxicity of peroxo heteroligand vanadates(V) in relation to biochemistry of vanadium. *J. Inorg. Biochem.* **25**, 51–55.

Donaldson, J., Hemming, R., and Labella, F. (1985). Vanadium exposure enhances lipid peroxidation in the kidney of rats and mice. *Can. J. Physiol. Pharmacol.* **63,** 196–199.

Dragan, Y. P., Campbell, H. A., Baker, K., Vaughan, J., Mass, M., and Pitot, H. C. (1994). Focal and non-focal expression of placental glutathione S-transferase in carcinogen-treated rats. *Carcinogenesis* **15,** 2587–2591.

Dragan, Y., Teeguarden, J., Campbell, H., Hsia, S., and Pitot, H. (1995). The quantitation of altered hepatic foci during multistage hepatocarcinogenesis in the rat: Transforming growth factor expression as a marker for the stage of progression. *Cancer Res.* **93,** 73–83.

Enzmann, H., Zerban, H., Kopp-Schneider, A., Loser, E., and Bannasch, P. (1995). Effect of low doses of N-nitrosomorpholine on the development of early stages of hepatocarcinogenesis. *Carcinogenesis* **16,** 1513–1518.

Farber, E. (1980). The sequential analysis of liver cancer induction. *Biochim. Biophys. Acta* **605,** 149–166.

Farber, E. (1984a). Cellular biochemistry of the stepwise development of cancer with chemicals. *Cancer Res.* **44,** 5463–5474.

Farber, E. (1984b). The biochemistry of preneoplastic liver: A common metabolic pattern in hepatocyte nodules. *Can. J. Biochem. Cell Biol.* **62,** 486–494.

Farber, E. (1990). Clonal adaptation during carcinogenesis. *Biochem. Pharmacol.* **39,** 1837–1846.

Farber, E., and Cameron, R. (1980). The sequential analysis of cancer development. *Adv. Cancer Res.* **31,** 125–226.

Fiala, S., and Fiala, E. S. (1973). Activation by chemical carcinogens of gamma-glutamyltranspeptidase in rat and mouse liver. *J. Natl. Cancer Inst.* **51,** 151–158.

Friedberg, E. C., Ehmann, U. K., and Williams, J. I. (1979). Human diseases associated with defective DNA repair. *Adv. Radiat. Biol.* **8,** 85–174.

Giri, A. K., Sanyal, R., Sharma, A., and Talukder, G. (1979). Cytological and cytochemical changes induced through certain heavy metals in mammalian systems. *Natl. Acad. Sci. Lett.* **2,** 391–394.

Grover, S., and Fisher, P. (1971). Cytogenetic studies in Sprague-Dawley rats during the administration of a carcinogenic nitroso compound—diethylnitrosamine. *Eur. J. Cancer* **7,** 77–82.

Hanauske, U., Hanauske, A.-R., Marshall, M. H., Muggia, V. A., and Von Hoff, D. D. (1987). Biphasic effect of vanadium salts on in vitro tumor colony growth. *Int. J. Cell Clon.* **5,** 170–178.

Hanigan, M. H., and Pitot, H. C. (1985) Gamma-glutamyltranspeptidase—Its role in hepatocarcinogenesis. *Carcinogenesis* **6,** 165–172.

Hida, T., Ohtaki, Y., Sudo, K., Aizawa, T., Aburada, M., and Miyamoto, K. I. (1994). Gomisin A, a ligandin component of Schizandora fruits, inhibits developments of preneoplastic lesions in rat liver by 3′-methyl-4-dimethylaminoazobenzene. *Cancer Lett.* **76,** 11–18.

Ising, U., and Levan, A. (1957). The chromosomes of two highly malignant human tumors. *Acta Pathol. Microbiol. Scand.* **40,** 13–24.

Iwai, K., Kimura, S., Ido, T., and Iwata, R. (1990). Tumor uptake of [^{48}V] vanadyl-chlorine e6Na as tumor-imaging agent in tumor-bearing mice. *Int. J. Rad. Appl. Instrum.* **17,** 775–780.

Jakoby, W. B. (1978). The glutathione S-transferase: A group of multifunctional detoxification proteins. *Adv. Enzymol.* **46,** 383–414.

Jakoby, W. B. (1980). *Enzymatic Basis of Detoxification.* Academic Press, New York, Vol. 2.

Jones, T. R., and Reid, T. W. (1984). Sodium orthovanadate stimulation of DNA synthesis in Nakano mouse lens epithelial cells in serum free medium. *J. Cell. Physiol.* **124,** 199–205.

Kachinskas, D. J., Phillips, M. A., Qin, Q., Stokes, J. D., and Rice, R. H. (1994). Arsenate perturbation of human keratinocyte differentiation. *Cell Growth Differ.* **5,** 1235–1241.

Kalengayi, M. M. R., and Desmet, V. J. (1978). Liver cell populations and histochemical patterns of adult and fetal type proteins during aflatoxin B$_1$ hepatocarcinogenesis. In H. Remmer,

H. M. Bolt, P. Bannasch, and H. Popper (Eds.), *Primary Liver Tumors*. MTP Press, Lancaster, pp. 467–483.

Kihlman, B. A., Hansson, K., and Andersson, H. C. (1982). The effects of post-treatments with caffeine during S and G_2 on the frequencies of chromosomal aberrations by thiotepa in root tips of *Vicia fava. Environ. Exp. Bot.* **20,** 271–286.

Kingsnorth, A. N., LaMuraglia, G. M., Ross, J. S., and Malt, R. A. (1986). Vanadate supplements and 1,2-dimethylhydrazine induced colon cancer in mice. *Br. J. Cancer* **53,** 683–686.

Köpf-Maier, P. (1987). Cytostatic non-platinum metal complexes: New perspective for the treatment of cancer? *Naturwissenschaften* **74,** 374–382.

Köpf-Maier, P. (1994). Complexes of metals other platinum as antitumour agents. *Eur. J. Clin. Pharmacol.* **47,** 1–16.

Köpf-Maier, P., and Köpf, H. (1979). Vanadocene dichloride: Another antitumor agent from the metallocene series. *Z. Naturforsch.* **34b,** 805–807.

Köpf-Maier, P., and Köpf H. (1988). Transition and main-group metal cyclopentadienyl complexes: Preclinical studies on a series of antitumor agents of different structural type. *Struct. Bond.* **70,** 103–185.

Lam, L. K. T., and Hasegawa, S. (1989). Inhibition of benzo(a)pyrene-induced forestomach neoplasia by citrus limonoids in mice. *Nutr. Cancer* **12,** 43–47.

Land, H., Parada, L. F., and Weinberg, R. A. (1983). Cellular oncogenes and multistep carcinogenesis. *Science* **222,** 771–778.

Léonard, A., and Gerber, G. B. (1994). Mutagenicity, carcinogenicity and teratogenicity of vanadium compounds. *Mutat. Res.* **317,** 81–88.

MacGregor, J., Heddle, J., Hite, M., Margolin, B., Ramel, C., Salamone, M., Tice, R., and Wild, D. (1987). Guidelines for the conduct of micronucleus assay in mammalian bone marrow erythrocytes. *Mutat. Res.* **189,** 103–112.

Mavournin, K. H., Blakey, D. H., Climino, M. C., Salamone, M. F., and Heddle, J. A. (1990). The in vivo micronucleus assay in mammalian bone marrow and periferal blood. A report of the U.S. Environmental Protection Agency gene-tox program. *Mutat. Res.* **239,** 29–30.

Mayer, D., Reuter, S., Hoffmann, H., Bocker, T., and Bannasch, P. (1996). Dehydroepiandrosterone reduces expression of glycolytic and gluconeogenesis enzymes in the liver of male and female rats. *Int. J. Oncol.* **8,** 1069–1078.

Migliore, L., Bocciardi, R., Macri, C., and Lo Jacono, F. (1993). Cytogenetic damage induced in human lymphocytes by four vanadium compounds and micronucleus analysis by fluoroscence in situ hybridization with a centrometric probe. *Mutat. Res.* **319,** 205–213.

Moreno, F. S., Rizzi, M. B. S. L., Dagli, M. L. Z., and Penteado, M. V. C. (1991). Inhibitory effects of β-carotene on preneoplastic lesions induced in Wistar rats by the resistant hepatocyte model. *Carcinogenesis* **12,** 1817–1822.

Morse, M. A., and Stoner, G. D. (1995). A response to: On cancer chemopreventive complications and limitations of some proposed strategies. *Carcinogenesis* **16,** 973.

Murthy, M. S., Rao, L. N., Kuo, L. Y., Toney, J. H., and Marks, T. J. (1988). Antitumor and toxicologic properties of the organometallic anticancer agent vanadocene dichloride. *Inorg. Chim. Acta* **152,** 117–124.

Newmark, H. L. (1987). Plant phenolics as inhibitors of mutational and precarcinogenic events. *Can. J. Physiol. Pharmacol.* **65,** 461–466.

Nicolson, G. (1976). Trans-membrane control of the receptors on normal and malignancy. *Biochim. Biophys. Acta* **458,** 1–71.

Nijhoff, W. A., Groen, G. M., and Peters, W. H. M. (1993). Induction of rat hepatic and intestinal glutathione S-transferases and glutathione by dietary naturally occurring anticarcinogens. *Int. J. Oncol.* **3,** 1131–1139.

Pitot, H. C., Campbell, H. A., and Maronpot, R. (1989). Critical parameters in the quantitation of the stages of initiation, promotion and progression in one model of hepatocarcinogenesis. *Toxicol. Pathol.* **17,** 594–605.

Pitot, H. C., and Sirica, A. E. (1980). The stages of initiation and promotion in hepatocarcinogenesis. *Biochim. Biophys. Acta* **605,** 191–215.

Popescu, N. C., Ansbaugh, S. C., and Dipaolo, J. A. (1984). Correlation of morphological transformation to sister chromatid exchanges induced by split doses of chemical or physical carcinogens in cultured Syrian hamster cells. *Cancer Res.* **44,** 1933–1938.

Popper, H. (1978). Introduction: The increasing importance of liver tumors. In H. Remmer, H. M. Blot, P. Bannasch, and H. Popper (Eds.), *Primary Liver Tumors.* MTP Press, Lancaster, pp. 3–10.

Prochaska, H. J., and Talalay, P. (1988). Regulatory mechanisms of monofunctional and bifunctional anticarcinogenic enzyme inducers in murine liver. *Cancer Res.* **48,** 4776–4782.

Rao, K. V. K., and Fernandes, C. L. (1996). Progressive effects of malachite green at varying concentrations on the development of N-nitrosodiethylamine induced hepatic preneoplastic lesions in rats. *Tumori* **82,** 280–286.

Rogers, A. E., Zeisel, S. H., and Groopman, J. (1993). Diet and carcinogenesis. *Carcinogenesis* **14,** 2205–2217.

Roldán, R. E., and Altamirano, L. M. A. (1990). Chromosomal aberrations, sister chromatid exchange, cell-cycle kinetics and satellite associations in human lymphocyte cultures exposed to vanadium pentoxide. *Mutat. Res.* **245,** 61–65.

Ruan, C. C., and Chen, Y. H. (1994). Relationship between mutagenic effects of drinking water and incidence of liver cancer in Fusui County—A prevalent area. *Chung. Hua. Yu. Fng. I. Hsueh. Tsa. Chih.* **28,** 78–80.

Russanov, E., Zaporowska, H., Ivancheva, E., Kirkova, M., and Konstantinova, S. (1994). Lipid peroxiodation and antioxidant enzymes in vanadate-treated rats. *Comp. Biochem. Physiol.* **107C,** 415–421.

Sabbioni, E., Pozzi, G., Devos, S., Pinter, A., Casella, L., and Fischbach, M. (1993). The intensity of vanadium(V)-induced cytotoxicity and morphological transformation in BALB/3T3 cells is dependent on glutathione-dependent bioreduction to vanadium(IV). *Carcinogenesis* **14,** 2565–2568.

Sabbioni, E., Pozzi, G., Pinter, A., Casella, L., and Garattini, S. (1991). Cellular retention, cytotoxicity and morphological transformation by vanadium(IV) and vanadium(V) in BALB/3T3 cell lines. *Carcinogenesis* **12,** 47–52.

Sakurai, H., Tamura, H., and Okatani, K. (1995). Mechanism for a new antitumor vanadium complex: Hydroxyl radical dependent DNA cleavage by 1,10-phenanthroline-vanadyl complex in the presence of hydrogen peroxide. *Biochem. Biophys. Res. Commun.* **206,** 113–137.

Sardar, S., and Chatterjee, M. (1993). Vanadium: A possible role in the protection of host cells bearing transplantable murine lymphoma. *Tumor Res.* **28,** 51–61.

Sardar, S., Ghosh, R., Mondal, A., and Chatterjee, M. (1993). Protective role of vanadium in the survival of hosts during the growth of a transplantable murine lymphoma and its profound effects on the rates and patterns of biotransformation. *Neoplasma* **40,** 27–30.

Sarkar, A., Basak, R., Bishayee, A., Basak, J., and Chatterjee, M. (1997). β-Carotene inhibits rat liver chromosomal aberrations and DNA chain break after a single injection of diethylnitrosamine. *Br. J. Cancer* **76,** 855–861.

Sarkar, A., Mukherjee, B., and Chatterjee, M. (1994). Inhibitory effect of β-carotene on chronic 2-acetylaminofluorene induced hepatocarcinogenesis in rat: Reflections on hepatic drug metabolism. *Carcinogenesis* **15,** 1055–1060.

Sarkar, A., Mukherjee, B., and Chatterjee, M. (1995a). Inhibition of 3'-methyl-4-dimethylamino-azobenzene-induced hepatocarcinogenesis in rat by dietary β-carotene: Changes in hepatic anti-oxidant defence enzyme levels. *Int. J. Cancer* **61,** 799–805.

Sarkar, A., Bishayee, A., and Chatterjee, M. (1995b). Beta-carotene prevents lipid peroxidation and red blood cell membrane protein damage in experimental hepatocarcinogenesis. *Cancer Biochem. Biophys.* **15,** 111–125.

Sato, K. (1990). Glutathione transferases as markers of preneoplasia and neoplasia. *Adv. Cancer Res.* **52,** 205–255.

Smith, J. B. (1983). Vanadium ions stimulate DNA synthesis in Swiss mouse 3T3 and 3T6 cells. *Proc. Natl. Acad. Sci. USA* **80,** 6162–6166.

Sparnins, V. L., Venegas, P. L., and Wattenberg, L. W. (1982). Glutathione S-transferase activity enhancement by compounds inhibiting chemical carcinogenesis and by dietary constituents. *J. Natl. Cancer Inst.* **68,** 493–496.

Stewart, H. L., Williams, G. M., Keysser, C. H., Lombard, L. S., and Montali R. J. (1980). Histological typing of liver tumors of the rat. *J. Natl. Cancer Inst.* **64,** 177–207.

Tanaka, T., Kojima, T., Kawamori, T., Yoshimi, N., and Mori, H. (1993). Chemoprevention of diethylnitrosamine-induced hepatocarcinogenesis by a simple phenolic acid protocatechuic acid in rats. *Cancer Res.* **53,** 2775–2779.

Tatematsu, M., Mera, Y., Inoue, T., Satoh, K., Sato, K., and Ito, N. (1988). Stable phenotypic expression of glutathione-S-transferase placental type and unstable phenotypic expression of γ-glutamyltranspeptidase in rat liver preneoplastic and neoplastic lesions. *Carcinogenesis* **9,** 215–220.

Tawfig, N., Wanigatunga, S., Heaney, R. K., Musk, S. R. R., Williamson, G., and Fenwick, G. R. (1994). Induction of the anti-carcinogenic enzyme quinone reductase by food extracts using murine hepatoma cells. *Eur. J. Cancer Prev.* **3,** 285–292.

Thompson, H. J., Chasteen, N. D., and Meeker, L. D. (1984). Dietary vanadyl(IV) sulfate inhibits chemically-induced mammary carcinogenesis. *Carcinogenesis* **5,** 849–851.

Toney, J. H., Murthy, M. S., and Marks, T. J. (1985). Biodistribution and pharmacokinetics of vanadium following intraperitoneal administration of vanadocene dichloride to mice. *Chem.-Biol. Interact.* **56,** 45–54.

Tsuda, H., Uehara, N., Iwahori, Y., Asamoto, M., Iigo, M., Nagao, M., Matsumoto, K., Ito, M., and Hirono, I. (1994). Chemopreventive effects of beta-carotene, alpha-tocopherol and five naturally occurring antioxidants on initiation of hepatocarcinogenesis by 2-amino-3-methylimi-dazo [4,5-f]quinoline in the rat. *Jpn. J. Cancer* **85,** 1214–1219.

Tsuruo, T., Iida, H., Kawabata, H., Tsukagoshi, S., and Sakurai, Y. (1984). High calcium content of pleiotropic drug-resistant P388 and K562 leukemia and Chinese hamster ovary cells. *Cancer Res.* **44,** 5095–5099.

Wagner, E. D., and Plewa, M. J. (1994). Induction of somatic mutations in Tradescantia clone 4430 by three phenylenediamine isomers and the antimutagenic mechanisms of diethyldithio-carbamate and ammonium meta-vanadate. *Mutat. Res.* **306,** 165–172.

Wattenberg, L. W. (1990). Inhibition of carcinogenesis by naturally occurring and synthetic compounds. In Y. Kuroda, D. M. Shankel, and M. D. Waters (Eds.), *Antimutagenesis and Anticarcinogenesis, Mechanisms, II.* Plenum, New York, pp. 155–166.

Wei, L., and Yung, B. Y. M. (1995). Effect of okadylic acid and vanadate on TPA-induced monocyte differentiation in human promyelocytic leukemia cell line HL-60. *Cancer Lett.* **90,** 199–206.

Williams, G. M. (1980). The pathogenesis of rat liver cancer caused by chemical carcinogenesis. *Biochim. Biophys. Acta* **605,** 167–189.

Williams, G. M. (1989). The significance of chemically-induced hepatocellular altered foci in rat liver and application to carcinogen detection. *Toxicol. Pathol.* **17,** 663–680.

Williams, G. M. (1994). Review: Interventive prophylaxis of liver cancer. *Eur. J. Cancer Prev.* **3,** 89–99.

Wrighton, S. A., and Stevens, J. C. (1992). The human hepatic cytochromes P450 involved in drug metabolism. *Crit. Rev. Toxicol.* **22,** 1–21.

Zaporowska, H., and Wasilewski, W. (1989). Some selected peripheral blood and haemopoietic system indices in Wistar rats with chronic vanadium intoxication. *Comp. Biochem. Physiol.* **93C,** 175–180.

Zaporowska, H., and Wasilewski, W. (1992). Haematological results of vanadium intoxication in Wistar rats. *Comp. Biochem. Physiol.* **101C,** 57–61.

Zheng, G.-Q., Kenney, P. M., and Lam, L. K. T. (1992). Myristicin: A potential cancer chemopreventive agent from parsley leaf oil. *J. Agric. Food Chem.* **40,** 107–110.

INDEX

DATE DUE

DEMCO 13829810